# Stochastic Dominance and Applications to Finance, Risk and Economics

# Stochastic Dominance and Applications to Finance, Risk and Economics

**Songsak Sriboonchitta**
Chiang Mai University
Chiang Mai, Thailand

**Wing-Keung Wong**
Hong Kong Baptist University
Hong Kong, People's Republic of China

**Sompong Dhompongsa**
Chiang Mai University
Chiang Mai, Thailand

**Hung T. Nguyen**
New Mexico State University
Las Cruces, New Mexico, U.S.A.

**CRC Press**
Taylor & Francis Group
Boca Raton London New York

CRC Press is an imprint of the
Taylor & Francis Group an **informa** business
A CHAPMAN & HALL BOOK

Chapman & Hall/CRC
Taylor & Francis Group
6000 Broken Sound Parkway NW, Suite 300
Boca Raton, FL 33487-2742

First issued in paperback 2017

© 2010 by Taylor and Francis Group, LLC
Chapman & Hall/CRC is an imprint of Taylor & Francis Group, an Informa business

No claim to original U.S. Government works

ISBN 13: 978-1-138-11799-0 (pbk)
ISBN 13: 978-1-4200-8266-1 (hbk)

---

### Library of Congress Cataloging-in-Publication Data

---

Stochastic dominance and applications to finance, risk and economics / Songsak
    Sriboonchitta.
       p. cm.
    Includes bibliographical references and index.
    ISBN 978-1-4200-8266-1 (hardcover : alk. paper)
    1. Finance--Mathematical models. 2. Stochastic processes. 3. Risk--Mathematical
models. 4. Statistical decision. I. Songsak Sriboonchitta. II. Title.

HG106.S753 2010
330.01'519232--dc22                                            2009031871

---

Visit the Taylor & Francis Web site at
http://www.taylorandfrancis.com

and the CRC Press Web site at
http://www.crcpress.com

# Contents

# Preface

This book is essentially a text for a course on stochastic dominance for beginners as well as a solid reference book for researchers. The material in this book has been chosen to provide basic background on the topic of stochastic dominance for various areas of applications. The material presented is drawn from many sources in the literature. For further reading, there is a bibliography. The text is designed for a one-semester course at the advanced undergraduate or beginning graduate level. The minimum prerequisites are some calculus and some probability and statistics. However, we start from the ground up, and background material will be reviewed at appropriate places. Clearly the course is designed for students from fields such as economics, decision theory, statistics and engineering.

A great portion of this text (Chapters 1–7) is devoted to a systematic exposition of the topic of stochastic dominance, emphasizing rigor and generality. The other portion of the text (Chapters 8–10) reports some new applications of stochastic dominance in finance, risk and economics.

The exercises at the end of each chapter will deepen the students' understanding of the concepts and test their ability to make the necessary computations. After completing the course, the students should be able to read more specialized and advanced journal papers or books on the subject.

This is an introduction to the topic of stochastic dominance in investment decision science. Several points about this text are as follows.

This text is written for students, especially for students in economics. Since the main applications are in financial economics, we have the students in mind in the sense that they should enjoy the text and be able to read through the text without tears. This includes self study. Even students in economics are aware of the fact that the underlying language of economics is largely mathematical, and thus we proceed to explain clearly at the beginning of any economic concepts why necessary mathematics enters the picture.

Our pedagogy is this: We only bring in mathematics when needed and carefully review or present the mathematical background at the places where it is immediately needed. We should not give the impression that economics is reserved for mathematicians! Our point of view is this: Students should think about the material as useful concepts and techniques for economics, rather than a burden of heavy mathematics.

We want students to be able to read the text, line by line, with enthusiasm and interest, so we wrote the text in its most elementary (and yet general and rigorous) and simple manner. For example, if a result can be proved by several different methods, we choose the simplest one, since students need a nice, simple proof that they can understand. To assist them in their reading, we will not hesitate to provide elementary arguments, as well as to supply review material as needed. Also, we trust that in reading the proofs of results, the students will learn ideas and proof techniques which are essential for their future studies and research.

In summary, this text is not a celebration of how great the mathematical contributions to economics are, but simply a friendly guide to help students build a useful repertoire of mathematical tools in decision-making under uncertainty, especially in investment science.

W.K. Wong would like to thank Z. Bai, H. Liu, U. Broll, J.E. Wahl, R. Chan, T.C. Chiang, Y.F. Chow, H.E. Thompson, S.X. Wei, M. Egozcue, W.M. Fong, D. Gasbarro, J.K. Zumwalt, P.L. Leung, D. Lien, H.H. Lean, C.K. Li, M. McAleer, K.F. Phoon, C. Ma, R. Smyth and B. Xia for their contributions to Chapters 8, 9 and 10 and Appendix A.

All the authors would like to express thanks to their families for their love and support during the writing of this book. We thank graduate student Hien Tran at New Mexico State University for his contributions to many proof details in this text as well as for his patience in checking our writing. We are grateful to Professor Carol Walker of New Mexico State University for assisting us in producing the manuscript for this book. We especially thank Bob Stern, our editor, and his staff for their encouragement and help in preparing this text.

Songsak Sriboonchitta, Wing-Keung Wong,
Sompong Dhompongsa and Hung T. Nguyen
Chiang Mai, Hong Kong,
Chiang Mai and Las Cruces

# Chapter 1

# Utility in Decision Theory

*This opening chapter sets out to review the well-known rational decision theory from which stochastic dominance rules are derived. An extensive review of probability theory is presented, not only for the problem of choice under uncertainty in this chapter, but also for reading the rest of the text at a rigorous level.*

## 1.1 Choice under certainty

This text is about how to make choices among alternatives when uncertainty is present. Such activities are frequent in human societies, especially in economic contexts. A simpler situation is choice under *certainty* such as in the social choice or consumer choice that you have encountered in a course in microeconomics. The problem is this. You wish to choose an element in a given set $\mathcal{Z}$ of things. Whatever element in $\mathcal{Z}$ you decide to choose, you will get that element for sure. That is the meaning of "choice under certainty."

In the following, we present the essentials of the theory of choice under certainty and uncertainty. For more details, an excellent text is Kreps [81].

In order to perform your choice, you possess your own *preferences* and use them to select, say, an "optimal" element in $\mathcal{Z}$. Now your preferences are expressed as a *preference relation*, denoted as $\succ$, where for $x, y \in \mathcal{Z}$, $x \succ y$ means "you strictly prefer $x$ to $y$." Of course, your preference relation should reflect your interest of some sort, for example, your valuations of things. Such valuations "represent" your preference relation in a quantitative way. This intuitive observation is made rigorous as follows.

First your preference relation $\succ$ is a binary relation on $\mathcal{Z}$, i.e., a subset $A$ of the Cartesian product $\mathcal{Z} \times \mathcal{Z}$, where $(x, y) \in A$ when $x \succ y$. Since it is a "preference" relation, it has to have some appropriate properties. Clearly, the relation $\succ$ should have at least the following properties:

(i) *Asymmetry*: $x \succ y$ implies $y \nsucc x$ (read "not $y \succ x$" or "$y$ is not strictly preferred to $x$")

(ii) *Negative transitivity*: $x \nsucc y$ and $y \nsucc z$ imply $x \nsucc z$

From a strict preference relation, we define two related ones. We write $x \approx y$ when $x \nsucc y$, and $y \nsucc x$, and write $x \succeq y$ when $y \nsucc x$ (i.e., either $x \succ y$ or $x \approx y$). The binary relation $\succeq$ is referred to as a *weak preference relation* (*complete* and *transitive*) and $\approx$ is called an *indifference relation* or an *equivalence relation* (reflexive, symmetric and transitive). Note that, for any $x, y$, exactly one of $x \succ y$, $y \succ x$, or $x \approx y$ holds. These facts can be verified as follows.

First, the transitivity of $\succ$ follows from the negative transitivity and the fact that $z \succ x$ and $x \approx y$ imply $z \succ y$, $z \succ x$ and $z \approx y$ imply $y \succ x$.

Next, $\approx$ is an equivalence relation: by asymmetry, $x \nsucc x$, which in turn implies that $x \approx x$, for any $x \in \mathcal{Z}$, so that $\approx$ is reflexive. If $x \approx y$ then $x \nsucc y$ and $y \nsucc x$, so that $y \approx x$, i.e., $\approx$ is symmetric. For transitivity, if $x \approx y$ and $y \approx z$, we have $x \nsucc y$, $y \nsucc x$, $y \nsucc z$, $z \nsucc y$ implying, by negative transitivity, that $x \nsucc z$ and $z \nsucc x$, so that $x \approx z$.

The weak preference relation $\succeq$ is complete: for any $x, y$ in $\mathcal{Z}$, we have $x \succ y$, $y \succ x$, or $x \approx y$. If $x \succ y$ or $x \approx y$, then $x \succeq y$, if $y \succ x$, then $y \succeq x$. Also, $\succeq$ is transitive: Given $x \succeq y$ and $y \succeq z$. Suppose $x \nsucc z$ and $x \not\approx z$ (not equivalent to $z$), i.e., $x$ is not $\succeq z$. Now, $x \not\approx z$ implies $x \succ z$ or $z \succ x$ which in turn implies $z \succ x$ since $x \nsucc z$. From $z \succ x$ and $x \succeq y$, we have $z \succ y$ by transitivity of $\succ$, contradicting the hypothesis $y \succeq z$.

We could start out with a weak preference relation $\succeq$ and define $\succ$ and $\approx$ from it too. We write $x \succ y$ when $y \nsucceq x$ and $x \approx y$ when $x \succeq y$ and $y \succeq x$. Thus, it's just a matter of taste to start out with a strict or a weak preference relation.

We seek a numerical representation for $\succ$.

**Theorem 1.1** *If $\mathcal{Z}$ is finite or infinitely countable then there is a function, called a utility function, $u : \mathcal{Z} \to \mathbb{R}$ such that*

$$x \succ y \iff u(x) > u(y). \tag{1.1}$$

**Proof.** Let $\mathcal{Z} = \{a_1, a_2, ...\}$. The case where $\mathcal{Z}$ is finite can be similarly proved. Without loss of generality, we assume $a_i \sim a_j$ for all $i \neq j$. Set $u(a_1) = 0$. Suppose $u(a_i)$ has been defined for each $a_i$, $i \leq m-1$ for some $m$, and Equation 1.1 holds for $a_1, a_2, ..., a_{m-1}$. We are going to define $u(a_m)$. Observe that exactly one of the following holds:

1. $a_i \prec a_m$ for all $i < m$; if so, set $u(a_m) = m$

2. $a_i \succ a_m$ for all $i < m$; if so, set $u(a_m) = -m$

3. $a_i \prec a_m \prec a_j$ for some $i, j < m$ and $a_i \not\prec a_h$, $a_h \not\prec a_j$ for $h < m$; if so, set $u(a_m)$ be any number in the interval $(u(a_i), u(a_j))$

Note that $u(a_i) \in [-i, i]$ for all $i \leq m$. To see Equation 1.1 holds for $a_i$, $i = 1, 2, ..., m$, we consider three cases:

1. $a_i \prec a_m$ for all $i < m$: clearly, $u(a_i) \leq i < m = u(a_m)$

2. $a_i \succ a_m$ for all $i < m$: as above, $u(a_i) \geq -i > -m = u(a_m)$

3. $a_i \prec a_m \prec a_j$ for some $i, j < m$ and no $h < m$ such that $a_i \prec a_h \prec a_j$: clearly, $u(a_i) < u(a_m) < u(a_j)$. If $a_h \prec a_i$ then $u(a_h) < u(a_i)$ by inductive assumption, and if $a_j \prec a_h$, then $u(a_j) < u(a_h)$ as well

Thus, for $i, j \leq m$, $a_i \succ a_j \iff u(a_i) < u(a_j)$. Now the function $u(.)$ can be defined for all elements of $\mathcal{Z}$ following the mathematical induction.

For uncountable $\mathcal{Z}$, $\succ$ should possess some additional properties to be represented by a utility function. Specifically, a necessary and sufficient condition for a preference relation $\succ$ to be represented numerically is that $\mathcal{Z}$ has a countable $\succ$-order dense subset of $\mathcal{Z}$. Recall that a subset $A$ of $\mathcal{Z}$ is said to be *preference-order dense* ($\succ$-order dense) if for any $x, y \in \mathcal{Z}$ such that $x \succ y$, there exists some $z \in \mathcal{Z}$ such that $x \succeq z \succeq y$.

Note that the utility function $u$ is unique up to a strictly increasing transformation: If $u$ represents $\succ$, then $f \circ u$ also represents $\succ$ for any $f : \mathbb{R} \to \mathbb{R}$ strictly increasing.

First, we show that $\succ$ is a preference relation. Assume the existence of such a function $u(.)$.

1. $x \succ y \iff u(x) > u(y) \iff u(y) \not> u(x) \iff y \not\succ x$

2. $x \not\succ y$ and $y \not\succ z \implies u(x) \not> u(y)$ and $u(y) \not> u(z) \implies u(x) \leq u(y) \leq u(z) \implies u(x) \not> u(z) \longrightarrow x \not\succ z$

Secondly, we show that $\mathcal{Z}$ contains a countable $\succ$-order dense subset. Let $\mathcal{T}$ be the set of all closed intervals with rational endpoints. Clearly, $\mathcal{T}$ is countable. For each $I \in \mathcal{T}$, if $I \cap u(\mathcal{Z}) \neq \varnothing$, choose $a_I \in \mathcal{Z}$ such that $u(a_I) \in I$. Let $A = \{a_I : I \cap u(\mathcal{Z}) \neq \varnothing\}$. Then $A$ is countable as $\mathcal{T}$ is. Let $\mathcal{B}$ be the family of closed intervals of the form $[u(x), u(y)]$ for which $x \prec y$ and there is no $a \in A$ such that $x \prec a \prec y$. Suppose for some different intervals $[u(x), u(y)], [u(x'), u(y')]$ in $\mathcal{B}$ such that $[u(x), u(y)] \cap [u(x'), u(y')] \neq \varnothing$, there are some points $p, q, r$ in $\mathcal{Z}$ that are not in $A$ such that $p \prec q \prec r$ and there is no point $a \in A$ such that $p \prec a \prec r$. Choose an interval $[r_1, r_2]$ with rationals $r_1, r_2$ such that $u(p) < r_1 < u(q) < r_2 < u(r)$. As $I \cap u(\mathcal{Z}) \neq \varnothing$, we have $p \prec a_I \prec r$ which is a contradiction since $a_I \in A$. This shows that distinct intervals in $\mathcal{B}$ have disjoint interiors. As a consequence, $\mathcal{B}$ is countable. Now if we let $B$ be the set of endpoints of intervals in $\mathcal{B}$, then $B$ is countable. Finally, let $C = A \cup B$. Clearly, $C$ is a countable $\succ$-order dense subset of $\mathcal{Z}$.

Conversely, suppose there is a countable dense subset $A = \{a_1, a_2, ...\}$ of $\mathcal{Z}$. For each $a_n \in A$, let $r(a_n) = 1/2^n$. Note that, if $B \subseteq A$, then $\sum_{z \in B} r(z) < \infty$. For each $z \in \mathcal{Z}$, let $A^*(z) = \{a \in A : a \succ z\}$, and $A_*(z) = \{a \in A : a \prec z\}$. Clearly, if $z \succeq z'$, then $A^*(z) \subseteq A^*(z')$. Moreover, if $z \succ z'$, then for some $a \in A$, $z \succ a \succeq z'$ or $z \succeq a \succ z'$. Consequently,

$$\sum_{a \in A_*(z')} r(a) > \sum_{a \in A_*(z)} r(a)$$

or

$$\sum_{a \in A^*(z')} r(a) < \sum_{a \in A^*(z)} r(a).$$

Note also that if $z \approx z'$, then $A_*(z) = A_*(z')$ and $A^*(z) = A^*(z')$. This implies $u(z) = u(z')$. To see, for example, that $A_*(z) = A_*(z')$, we observe that if for some $a \in A$, $a \prec z$, then we must have $a \prec z'$ (otherwise, $a \approx z'$ which contradicts $a \approx z$ and $a \prec z$).

Let

$$u(z) = \sum_{a \in A_*(z')} r(a) - \sum_{a \in A^*(z')} r(a).$$

Then we have that $z \succ z' \implies u(z) > u(z')$. On the other hand, if $z \not\succ z'$, then $z \approx z'$ or $z \prec z'$. In the later case, $u(z) < u(z')$ as above. In the former case, it follows from the previous observation that $u(z) = u(z')$. We conclude that

$$z \succ z' \iff u(z) > u(z').$$

∎

The above utility function $u(.)$ is unique up to a strictly increasing transformation. Indeed, if $u$ and $v$ represent $\succ$, then there exists an increasing function $f : \mathbb{R} \to \mathbb{R}$ such that

(a) $f$ is strictly increasing on the range $u(\mathcal{Z})$ of $u$, and

(b) $v = f \circ u$.

Moreover, for any strictly increasing function $g : \mathbb{R} \to \mathbb{R}$, $g \circ u$ represents $\succ$.

A proof of the above assertion is this. Assume that for $a \neq b$ in $\mathcal{Z}$, $a \not\succ b$. Define $f(u(a)) = v(a)$, for $a \in \mathcal{Z}$. Clearly $f$ is strictly increasing on $u(\mathcal{Z})$. We extend $f$ to $\mathbb{R}$ by putting

$$f(x) = \begin{cases} \sup\{v(a) : u(a) \le x\} & \text{if } \{v(a) : u(a) \le x\} \neq \varnothing \\ \inf\{v(a) : u(a) > x\} & \text{otherwise} \end{cases}$$

To see why $f$ is increasing, let $x < y$ and consider $f(x)$, $f(y)$.

**Case 1:** When $u(a) > x$ for all $a \in \mathcal{Z}$, we have $f(x) \le v(a)$ for all $a \in \mathcal{Z}$. If $u(a) \le y$ for some $a$, then $a \preceq y$ and thus $f(x) \le v(a) \le f(y)$. If $u(a) > y$ for all $a \in \mathcal{Z}$, then $f(x) = \inf_{a \in \mathcal{Z}} v(a) = f(y)$.

**Case 2:** When $u(a) \le x$ for some $a \in \mathcal{Z}$, we have

$$f(x) = \sup_{u(a) \le x} v(a) \le \sup_{u(a) \le y} v(a) = f(y).$$

## 1.2   Basic probability background

The problem of choice under uncertainty is the problem of choosing among uncertain quantities, such as future returns in various investment portfolios. We need first to review the *mathematical modeling of uncertain quantities*.

The following extensive review of probability background is necessary for reading this whole text at a rigorous level.

### 1.2.1   Probability measures and distributions

The mathematical model of a random phenomenon is a probability space $(\Omega, \mathcal{A}, P)$ where $\Omega$ is a set, $\mathcal{A}$ is a $\sigma$-field of subsets of $\Omega$, and $P$ is a probability measure on $\mathcal{A}$. Specifically, $\mathcal{A}$ satisfies the following axioms,

(i) $\Omega \in \mathcal{A}$

(ii) if $A \in \mathcal{A}$ then its set complement $A^c \in \mathcal{A}$

(iii) if $A_n, n \geq 1$, is a countable collection of elements of $\mathcal{A}$, then $\cup_{n \geq 1} A_n \in \mathcal{A}$ ($\mathcal{A}$ is a *field* if (iii) is only satisfied for finite unions)

and $P$ is a mapping from $\mathcal{A}$ to $[0, 1]$ such that

(i) $P(\Omega) = 1$

(ii) ($\sigma$-additivity) if $A_n$, $n \geq 1$, is a sequence of finite or infinitely countable pairwise disjoint elements of $\mathcal{A}$, then

$$P(\cup_{n \geq 1} A_n) = \sum_{n \geq 1} P(A_n).$$

Uncertain quantities are modeled as *random variables*. Specifically, a random variable $X$ is a (measurable) map from $\Omega$ to its range, say, the real line $\mathbb{R}$ (a real-valued random variable) such that $X^{-1}(A) = \{\omega \in \Omega : X(\omega) \in A\} \in \mathcal{A}$, for any Borel set $A$, i.e., for $A \in \mathcal{B}(\mathbb{R})$ which is the smallest $\sigma$-field containing the intervals $(a, b]$, $-\infty < a < b < \infty$. We also say that $X$ is $\mathcal{A}$-$\mathcal{B}(\mathbb{R})$-measurable.

The probability law of $X$ is the probability measure $P_X$ on $\mathcal{B}(\mathbb{R})$ where

$$P_X(A) = P(X^{-1}(A)) = P(\{\omega : X(\omega) \in A\}).$$

If $P_X$ is given, then the function $F_X : \mathbb{R} \to [0, 1]$ defined by

$$F_X(x) = P_X((-\infty, x])$$

is called the *distribution function* of $X$. In view of the properties of the probability measure $P_X$, $F_X$ satisfies the following characteristic properties:

1. $F_X$ is monotone non-decreasing, i.e., $x \leq y$ implies $F(x) \leq F(y)$

2. $\lim_{x \to -\infty} F_X(x) = 0$ and $\lim_{x \to \infty} F_X(x) = 1$

3. $F_X$ is right continuous on $\mathbb{R}$

The upshot is this. Each function $F : \mathbb{R} \to [0, 1]$ satisfying the above properties (1)–(3) determines uniquely a probability measure $P$ on $\mathcal{B}(\mathbb{R})$ via $F(x) = P((-\infty, x])$, meaning that the random evolution of a random variable is completely determined by its distribution function. The unique probability measure $P$ so associated is denoted as $dF(x)$ and is called the

*Stieltjes probability measure associated with* $F$. In other words, $dF(x)$ is the probability law of the random variable whose distribution function is $F$. This result is referred to as the Lebesgue-Stieltjes characterization theorem. We include it here for completeness.

Axiomatically, let's call any function $F$ satisfying the above properties (1)–(3) a *distribution function* on $\mathbb{R}$.

**Theorem 1.2 (Lebesgue-Stieltjes characterization)** *There exists a bijection between distribution functions on* $\mathbb{R}$ *and probability measures on* $\mathcal{B}(\mathbb{R})$.

**Proof.** It suffices to show that if $F$ is a distribution function on $\mathbb{R}$, then there is a unique probability measure $P$ on $\mathcal{B}(\mathbb{R})$ such that $F(x) = P((-\infty, x])$, for any $x \in \mathbb{R}$.

Given a distribution function $F$, we first define a set-function $P$ on subsets of $\mathbb{R}$ of the form $(-\infty, x]$, $x \in \mathbb{R}$ by $P((-\infty, x]) = F(x)$. Next, we extend $P$ to the class $\mathcal{C} = \{(a, b] : -\infty \le a \le b \le \infty\}$ (where by $(a, \infty]$ we mean $(a, \infty)$) by $P((a, b]) = F(b) - F(a)$. Note that both $\varnothing$ and $\mathbb{R}$ are in $\mathcal{C}$, with $P(\mathbb{R}) = 1$ (we set $F(-\infty) = 0, F(\infty) = 1$). Extend $P$ further to the field $\mathcal{B}_0$ consisting of finite disjoint unions of intervals in $\mathcal{C}$ by

$$P\left( \bigcup_{i=1}^{n}(a_i, b_i] \right) = \sum_{i=1}^{n} F(b_i) - F(a_i).$$

That $P$ so defined is additive on $\mathcal{B}_0$ is obvious from its construction. Now since $\mathcal{B}(\mathbb{R})$ is the $\sigma$-field generated by the field $\mathcal{B}_0$, $P$ can be extended uniquely to a probability measure on $\mathcal{B}(\mathbb{R})$ if $P$ is $\sigma$-additive on $\mathcal{B}_0$ (Carathéodory theorem), i.e., if $A_n \in \mathcal{A}$, $n \ge 1$, pairwise disjoint and $\bigcup_{i=1}^{\infty} A_n \in \mathcal{A}$, then

$$P\left( \bigcup_{n \ge 1} A_n \right) = \sum_{n \ge 1} P(A_n).$$

Just like in the case of $\sigma$-fields, to establish $\sigma$-additivity, it suffices to show that $P$ is continuous at $\varnothing$. Let

$$A_n = \bigcup_{i=1}^{k_n}(a_{i,n}, b_{i,n}] \searrow \varnothing$$

with $a_{i,n} < b_{i,n} < a_{i+1,n} < b_{i+1,n}$ for $i = 1, \ldots, k_n - 1$. Here $A_n \searrow A$ means that, for every $n \in \mathbb{N}$,

$$A_1 \supseteq A_1 \supseteq \cdots \supseteq A_n \quad \text{and} \quad \bigcap_{n=1}^{\infty} A_n = A.$$

Also $A_n \nearrow A$ means that, for every $n \in \mathbb{N}$,

$$A_1 \subseteq A_1 \subseteq \cdots \subseteq A_n \quad \text{and} \quad \bigcup_{n=1}^{\infty} A_n = A.$$

Now let $\varepsilon > 0$. Since $\lim_{x \to -\infty} F(x) = 1 - \lim_{x \to \infty} F(x) = 0$, there exists $z = z(\varepsilon) > 0$ such that $F(-z) \leqslant \varepsilon/3$ and $1 - F(z) \leqslant \varepsilon/3$. Since $\lim_{x \to a_{i,n}^+} F(x) = F(a_{i,n})$, there exists $x_{i,n} \in (a_{i,n}, b_{i,n}]$ such that

$$F(x_{i,n}) - F(a_{i,n}) \leqslant \frac{\varepsilon}{3} \frac{1}{2^{i+n}}.$$

Now let

$$B_n = \left( \bigcup_{i=1}^{k_n} (x_{i,n}, b_{i,n}] \right) \cap (-z, z] \quad \text{and} \quad C_n = \bigcap_{j=1}^{n} B_j.$$

Then $B_n, C_n \in \mathcal{A}$ and $C_n \subseteq A_n$. Also

$$A_n \setminus C_n \subseteq \bigcup_{j=1}^{n} (A_j \setminus B_j).$$

Since $P$ is additive (and hence sub-additive) on $\mathcal{A}$, we have, for each $n$,

$$P(A_n \setminus B_n) = P(A_n) - P(C_n) \leqslant P\left( \bigcup_{j=1}^{n} (A_j \setminus B_j) \right) \cap \mathbb{R}$$

$$= P\left( \bigcup_{j=1}^{n} (A_j \setminus B_j) \cap [(-z, z] \cup (-\infty, -z] \cup (z, \infty)) \right]$$

$$\leqslant P((-\infty, -z]) + P((z, \infty)) + \sum_{j=1}^{n} P((A_j \setminus B_j) \cap (-z, z])$$

$$= F(-z) + 1 - F(z) + \sum_{j=1}^{n} P((A_j \setminus B_j) \cap (-z, z])$$

$$= F(-z) + 1 - F(z) + \sum_{j=1}^{n} \sum_{i=1}^{k_n} [F(x_{i,n}) - F(a_{i,n})] \leqslant \varepsilon.$$

Note that not only $C_n \subseteq A_n$ but $\overline{C_n} \subseteq A_n$ so that $\bigcap_{n \geqslant 1} \overline{C_n} = \varnothing$ since $\bigcap_{n \geqslant 1} A_n = \varnothing$ by hypothesis. But each $\overline{C_n}$ is compact (closed and bounded, $\overline{C_n} \subseteq [-z, z]$) so that there exists an integer $N$ such that $\bigcap_{n \leqslant N} \overline{C_n} = \varnothing$. But since $C_n$ is decreasing and hence $\overline{C_n}$ is decreasing, we have $\overline{C_N} = \bigcap_{n \leqslant N} \overline{C_n} = \varnothing$ so that for any $n \geqslant N$, $\overline{C_n} \subseteq \overline{C_N} = \varnothing$, i.e., $\overline{C_n} = \varnothing$ for any $n \geqslant N$. Thus for any $n \geqslant N$, $C_n = \varnothing$ as $C_n \subseteq \overline{C_n}$, and hence $P(C_n) = 0$ and

$$P(A_n) - P(C_n) = P(A_n) \leqslant \varepsilon,$$

which is equivalent to

$$\lim_{n \to \infty} P(A_n) = 0.$$

∎

**Remark 1.3** *The above proof is based on the* finite intersection property *of compact sets, which is stated as follows:*
   *In a compact space, a system of closed sets* $\{F_\alpha, \ \alpha \in A\}$ *has*

$$\bigcap_{k=1}^{n} F_\alpha \neq \varnothing$$

*if and only if every finite intersection satisfies*

$$\bigcap_{k=1}^{n} F_k \neq \varnothing.$$

*Thus if* $\bigcap_{\alpha \in A} F_\alpha = \varnothing$, *then there exists* $N$ *such that* $\bigcap_{k=1}^{N} F_k = \varnothing$.

   Note however that a distribution function $F$ need not be left continuous. Clearly $F$ is continuous at a point $x_0$ (or $x_0$ is a point of continuity of $F$) if and only if $F(x_0^-) = F(x) = F(x_0^+)$. A point $x$ is a discontinuity point of $F$ if $F(x^+) - F(x^-) > 0$. Discontinuity points of $F$ are all of the first kind (called jump points), i.e., each point $x$ of discontinuity of $F$ is such that $F(x^-)$ and $F(x^+)$ exist but are unequal. The jump size is the difference $F(x^+) - F(x^-)$. Moreover, the set of discontinuity points of a distribution function $F$ is at most countable. Indeed, for each positive integer $k$, if we denote by $J_k$ the set of jump points whose jump sizes are greater than $1/k$, then each $J_k$ is finite, since the sum of any $n$ jump sizes is bounded by 1, the variation $F(\infty) - F(-\infty)$. But the set of all possible jump points on $\mathbb{R}$ of $F$ is $\bigcup_{k \geqslant 1} J_k$ and hence at most countable.

Having characterized probability measures on $(\mathbb{R}, \mathcal{B}(\mathbb{R}))$ by distribution functions, we turn now to the classification of distribution functions. A point $x$ is a *point of increase* of $F$ if $F(x + \varepsilon) - F(x - \varepsilon) > 0$ for any $\varepsilon > 0$. Note that discontinuity points are points of increase, but a point of increase can be a point of continuity. A distribution function $F$ is said to be *discrete* (or of *discrete type*) if all its points of increase are discontinuity points. A discrete distribution function $F$ is written as

$$F(x) = \sum_{y \in D, \, y \leqslant x} F(y) - F(y^-),$$

where $D$ is the (finite or countable) set of discontinuity points of $F$. If we let

$$f : D \to [0, 1] \quad \text{with} \quad f(y) = F(y) - F(y^-),$$

then $f \geqslant 0$ and $\sum_{y \in D} f(y) = 1$, i.e., $f$ is a *probability density function*.

If all points of increase of a distribution function $F$ are continuity points then $F$ is a *continuous distribution function* (or of *continuous type*). In this case,

$$f(x) = P(\{x\}) = F(x^+) - F(x^-) = 0, \text{ for } x \in \mathbb{R},$$

where $P$ is the probability measure on $(\mathbb{R}, \mathcal{B}(\mathbb{R}))$ associated with $F$. A probability measure $P$ having this property (i.e., does not charge any singleton set) is said to be *non-atomic*.

A distribution function can be a mixture of the two above types. Here is an example of a distribution function having a discrete part and a continuous part:

$$F(x) = \begin{cases} 0 & \text{for} \quad x < -2 \\ 0.2 & \text{for} \quad -2 \leqslant x < 1 \\ 0.6x - 0.4 & \text{for} \quad 1 \leqslant x < 2 \\ 0.8 & \text{for} \quad 2 \leqslant x < 3 \\ 1.0 & \text{for} \quad x \geqslant 3. \end{cases}$$

Being monotone, any distribution function $F$ is always *differentiable almost everywhere* (a.e.). Let $f = dF/dx$ be the derivative a.e. of $F$. The question is: can we recover $F$ from $f$? In other words, is it true that, for any $x \in \mathbb{R}$,

$$F(x) = \int_{-\infty}^{x} f(t) dt? \tag{1.2}$$

If such a situation holds, then we have a counterpart of the discrete case where the derivative $f$ of $F$ will play the role of a density function in the case of infinitely uncountable ranges. Note however that, in this case, $f(x) \neq P(X = x)$ since the latter is always zero!

It turns out that the above question has a positive answer only for a certain class of distribution functions $F$. This is simply the *Fundamental Theorem of Calculus*, namely, Equation (1.2) is true if and only if $F$ is *absolutely continuous*. Recall that the absolute continuity is a stronger condition than continuity. A function $F$ is said to be *absolutely continuous* if for each $\varepsilon > 0$, there exists $\delta > 0$, such that

$$\sum_{j=1}^{n} |F(b_j) - F(a_j)| < \varepsilon$$

for any non-overlapping open intervals $(a_j, b_j)$, $j = 1, 2, \ldots, n$, $n \geqslant 1$ with $\sum_{j=1}^{n}(b_j - a_j) < \delta$. It is clear that, for $n = 1$, we get the (ordinary) continuity of $F$.

**Remark 1.4** *For an absolutely continuous distribution $F$, $f \geqslant 0$ since $F$ is non-decreasing, and*

$$F(\infty) = \int_{-\infty}^{\infty} f(x)dx = 1.$$

*Any function $f$ with these two properties is called a probability density function. The class of absolutely continuous distribution functions (or of absolutely continuous type) is a subclass of continuous distribution functions.*

**Definition 1.5** *A distribution function $F$ is said to be* singular *(or of singular type) if it is continuous but has all of its points of increase on a set of zero Lebesgue measure.*

In summary, probability measures on $(\mathbb{R}, \mathcal{B}(\mathbb{R}))$ are characterized by distribution functions. Each special type of distribution gives rise to a special type of laws for random experiments (discrete, absolutely continuous and singular). Mixtures of these types are possible since, in general, a distribution function $F$ can be decomposed uniquely as a convex combination of the above types of distributions.

**Example 1.6 (Singular distribution function)** *Consider the random experiment of tossing a fair coin an infinite number of times independently. Let $X_n$ denote the outcome on the nth toss. The $X_n$'s are independent and identically distributed random variables. We write 0 and 2* *for Tails and Heads, respectively. Formally, let $(\Omega, \mathcal{A}, P)$ be a probability space on which all the $X_n$'s are defined. You can take*

$$\Omega = \{0,2\}^{\mathbb{N}} = \{\omega = (\omega_1, \omega_2, \ldots), \ \omega_j = 0, 2, \ j \geqslant 1\},$$

*and, $X_n(\omega) = \omega_n$, the nth projection. Viewing $\{0,2\}$ as a probability space with mass $1/2$ as each point (i.e., $P(X_n = 0) = P(X_n = 2) = 1/2$), $(\Omega, \mathcal{A}, P)$ is the infinite product of this probability space. Consider*

$$Y : \Omega \to [0,1], \qquad Y(\omega) = \sum_{n \geqslant 1} \frac{X_n(\omega)}{3^n}.$$

*Note that this random series is convergent everywhere, i.e., in view of the ternary expansions of numbers in $[0,1]$, or simply by comparison with the geometric series:*

$$0 \leqslant \sum_{n \geqslant 1} \frac{X_n(\omega)}{3^n} \leqslant 2 \sum_{n \geqslant 1} \frac{1}{3^n} = 1.$$

*So its sum $Y$ is a random variable.*

We are going to show that the distribution function $F$ of $Y$ is continuous, but $F$ is not the integral of its (a.e.) derivative, i.e., $F$ is not absolutely continuous. There are several ways to show that $F$ is continuous. We choose to prove it by looking at the geometry of the range $\mathcal{R}(Y)$ of $Y$ to bring out the *fractal nature* of it.

In view of the ternary expansions of numbers in $[0,1]$, $\mathcal{R}(Y)$ is precisely the *(middle third) Cantor set*. Geometrically, the Cantor set $\mathcal{C}$ is

$$\mathcal{C} = [0,1] \setminus \bigcup_{n \geqslant 1} \bigcup_{k=1}^{3^{n-1}} \left( \frac{3k-2}{3^n}, \frac{3k-1}{3^n} \right) = \bigcap_{n \geqslant 1} E_n,$$

where

$$(E_n)^c = \bigcup_{k=1}^{3^{n-1}} \left( \frac{3k-2}{3^n}, \frac{3k-1}{3^n} \right).$$

Each compact set $E_n$ is the union of $2^n$ closed intervals, each has length $1/3^n$. Since $E_{n+1} \subseteq E_n$, the length or *Lebesgue measure* $\lambda$ of $\mathcal{C}$ is zero:

$$\lambda(\mathcal{C}) = \lim_{n \to \infty} \lambda(E_n) = \lim_{n \to \infty} \left( \frac{2}{3} \right)^n = 0.$$

Note that $\mathcal{C}$ is an *uncountable subset* of $[0,1]$. Since $\mathcal{C} = \mathcal{R}(Y)$, $P_Y(\mathcal{C}) = 1$.

Let $F$ be the distribution function of $Y$. Since $Y = \sum_{n \geqslant 1} X_n/3^n$, $F$ is referred to as the *infinite convolution* of the distributions of the independent random variables $X_n/3^n$, $n \geqslant 1$. From the geometric construction of the middle third Cantor set, we see that $\mathcal{C}$ is *self-similar* in the sense that can be written as a union of linear transforms of itself. Specifically,

$$\mathcal{C} = \phi_1(\mathcal{C}) \cup \phi_2(\mathcal{C}),$$

where $\phi_1, \phi_2 : \mathbb{R} \to \mathbb{R}$ with $\phi_1(x) = x/3$ and $\phi_2(x) = x/3 + 2/3$. Now

$$Y = \sum_{n \geqslant 1} \frac{X_n}{3^n} = \frac{X_1}{3} + \frac{Z}{3}, \quad Z = \sum_{n \geqslant 1} \frac{X_{n+1}}{3^n}.$$

Note that $Y$ and $Z$ have the same distribution, and $X_1$ is independent of $Z$.

For any Borel set $A$ in $[0, 1]$, we have:

$$
\begin{aligned}
P_Y(A) &= P(Y \in A) = P\left(\frac{X_1}{3} + \frac{Z}{3} \in A\right) \\
&= P(X_1 = 0)P\left(\frac{Z}{3} \in A\right) + P(X_1 = 2)P\left(\frac{2}{3} + \frac{Z}{3} \in A\right) \\
&= \frac{1}{2}\left(P(Y \in 3A) + P(Y \in 3A - 2)\right) \\
&= \frac{1}{2}\left[P_Y\left(\phi_1^{-1}(A)\right) + P_Y\left(\phi_2^{-1}(A)\right)\right].
\end{aligned}
\tag{1.3}
$$

Here $\alpha A + \beta = \{\alpha x + \beta : x \in A\}$.

We express Equation (1.3) by saying that the probability $P_Y$ is *self-similar*. It is precisely (1.3) that will be used to show that $P_Y\{x\} = 0$, for any $x \in \mathcal{C}$, i.e., $F$ is continuous. For this purpose, suppose there exists $x_0 \in \mathcal{C}$ such that $P_Y(\{x_0\}) > 0$. Under this assumption, we have the following results.

**Claim 1.7** *The* $\sup_{x \in \mathcal{C}} P_Y(\{x\})$ *is attained.*

**Proof.** It is clear that

$$\sup_{x \in \mathcal{C}} P_Y(\{x\}) = \sup_{x \in B} P_Y(\{x\}) \leqslant 1,$$

where $B = \{x \in \mathcal{C} : P_Y(\{x\}) \geqslant P_Y(\{x_0\})\}$. For any *finite* subset $\{x_1, x_2, \ldots, x_n\}$ of $B$, we have

$$n P_Y(\{x_0\}) \leqslant \sum_{j=1}^{n} P_Y(\{x_j\}) \leqslant 1.$$

Thus, $n \leqslant 1/[P_Y(\{x_0\})]$, and hence $B$ is finite. Therefore, there exists $a \in B$ such that

$$P_Y(\{a\}) = \sup_{x \in \mathcal{C}} P_Y(\{x\}) > 0.$$

■

**Claim 1.8** *The assumption $P_Y(\{x_0\}) > 0$ leads to a contradiction.*

**Proof.** Applying the self-similarity of $P_Y$ in (1.3) to $A = \{a\}$, we obtain

$$P_Y(\{a\}) = 1/2P_Y(\{3a\}) + 1/2P_Y(\{3a - 2\}).$$

If $a \in [0, 1/3]$, then $3a - 2 \leqslant -1$ and hence $P_Y(\{3a - 2\}) = 0$, so that

$$P_Y(\{3a\}) = 2P_Y(\{a\}) > P_Y(\{a\}) = \sup_{x \in \mathcal{C}} P_Y(\{x\}),$$

which is impossible. Similarly, if $a \in [2/3, 1]$, then $3a \geqslant 2$, and hence $P_Y(\{3a\}) = 0$, so that

$$P_Y(\{3a - 2\}) = 2P_Y(\{a\}) > P_Y(\{a\}),$$

also impossible. Thus $F$ is continuous. Now, if $F$ is absolutely continuous, then for *any* Borel set $A$ of $[0, 1]$, $P_Y(A) = \int_A f(x)dx$, where $f$ is the (a.e.) derivative of $F$ (shortly, we will indicate how to translate the Fundamental Theorem of Calculus to this formula involving measures, in which $F$ is replaced by its associated measure $dF = P_Y$, and $dx$ stands for the Lebesgue measure $\lambda$). But this is impossible since for $A = \mathcal{C}$ (a compact set of $[0, 1]$, and hence a Borel set), $P_Y(\mathcal{C}) = 1$, while $\int_{\mathcal{C}} f(x)dx = 0$ since $\lambda(\mathcal{C}) = 0$. ■

In summary, the law $P_Y$ of $Y$ on $[0, 1]$ is *purely singular* (with respect to the Lebesgue measure $\lambda$), i.e., $P_Y$ is *non-atomic* and there exists a set $A$ (here $A = \mathcal{C}$), such that $P_Y(A) = 1$ while $\lambda(A) = 0$. We write "purely" to emphasize the non-atomicity property. If we drop this requirement, then $P_Y$ is singular (with respect to $\lambda$). Of course, *discrete probability measures* are *singular* but not purely singular.

In the above example, the distribution function $F$ of $Y$, called the *Cantor function*, can be explicitly determined. In fact, for $x \in [0, 1]$,

$$F(x) = \sum_{k=1}^{N(x)-1} \frac{x_k}{2^{k+1}} + \frac{1}{2^{N(x)}},$$

where $x = \sum_{n \geq 1} x_n / 3^n$, $x_n \in \{0, 1, 2\}$, its ternary expansion, and

$$N(x) = \begin{cases} \infty & \text{if } x \in C \\ \min\{k : x_k = 1\} & \text{if } x \notin C. \end{cases}$$

This can be seen as follows. If we exclude ternary expansions of numbers $x \in [0, 1]$ terminating either with $1, 0, 0, 0, \ldots$ or $1, 2, 2, 2, \ldots$, then $[0, 1]$ can be identified with $\{0, 1\}^{\mathbb{N}}$, i.e., we identify $x$ with the sequence $\{x_1, x_2, \ldots\}$ where $x_j \in \{0, 1, 2\}$. We specify the random variable $Y : \Omega \to \mathbb{R}$ concretely as follows. Take $(\Omega, \mathcal{A}, P)$ to be the infinite product of probability spaces $\{0, 2\}$, with mass $1/2$ on each point. By identification, $Y(\omega) = \omega$.

The function $F$ is continuous (this can be also seen by observing that $P$ does not change any points in $\Omega$), and as such,

$$F(x) = P(\omega \leq x) = P(\omega < x), \quad \omega = (\omega_1, \omega_2, \ldots).$$

For $x \in C$, we consider the two following cases first.

(a) The set $\{k : x_k = 2\}$ is finite. Let $M = \max\{k : x_k = 2\}$, then, for $M = 1$, i.e., $x = (2, 0, 0, 0, \ldots)$,

$$F(x) = P(\omega < x) = P(\omega : \omega_1 - 0) = 1/2 = \sum_{k \geq 1} \frac{x_k}{2^{k+1}}.$$

Suppose this is true for $M > 1$. Consider $x_1, x_2, \ldots, x_{M+1} = 2$. Let $y = (x_1, x_2, \ldots, x_M, 0, 0, \ldots)$, then according to the induction hypothesis $F(y) = \sum_{k=1}^{M} x_k / 2^{k+1}$. Now,

$$
\begin{aligned}
F(x) &= P(\omega < x) = P(\omega < y) + P(y \leq \omega < x) \\
&= P(\omega \leq y) + P(\omega : \omega_1 = x_1, \ldots, \omega_M = x_M, \omega_{M+1} = 0) \\
&= \sum_{k=1}^{M} \frac{x_k}{2^{k+1}} + \frac{1}{2^{M+1}} = \sum_{k=1}^{M+1} \frac{x_k}{2^{k+1}} = \sum_{k=1}^{\infty} \frac{x_k}{2^{k+1}},
\end{aligned}
$$

as $x_k = 0$ for $k > M + 1$ and $1/2^{M+1} = 2/2^{M+2} = x_{M+1}/2^{M+2}$.

(b) The set $\{k : x_k = 0\}$ is finite. Again, we claim that

$$F(x) = \sum_{k=1}^{\infty} \frac{x_k}{2^{k+1}}.$$

Indeed, suppose that $x = (x_1, x_2, \ldots, x_M, 2, 2, 2, \ldots)$. We consider $y = (x_1, x_2, \ldots, x_M, 0, 0, 0, \ldots)$. Then from (a), $F(y) = \sum_{k=1}^{M} x_k/2^{k+1}$. But

$$
\begin{aligned}
F(x) &= P(\omega < x) = P(\omega < y) + P(y \leqslant \omega < x) \\
&= F(y) + P(\omega_1 = x_1, \ldots, \omega_M = x_M) \\
&= \sum_{k=1}^{M} \frac{x_k}{2^{k+1}} + \frac{1}{2^M} \\
&= \sum_{k=1}^{M} \frac{x_k}{2^{k+1}} + \sum_{k=M+1}^{\infty} \frac{2}{2^{k+1}} = \sum_{k=1}^{\infty} \frac{x_k}{2^{k+1}}.
\end{aligned}
$$

Now for arbitrary $x \in \mathcal{C}$, $x = (x_1, x_2, \ldots)$ with $x_j \in \{0, 2\}$, we let

$$a = (x_1, x_2, \ldots, x_M, 0, 0, 0, \ldots) \quad \text{and} \quad b = (x_1, x_2, \ldots, x_M, 2, 2, 2, \ldots).$$

Then since $a \leqslant x \leqslant b$, we have

$$F(a) = \sum_{k=1}^{M} \frac{x_k}{2^{k+1}} \leqslant F(x) \leqslant F(b) = \sum_{k=1}^{M} \frac{x_k}{2^{k+1}} + \frac{1}{2^M}.$$

Letting $M \to \infty$, we have proved our claim.

Finally, let $x \notin \mathcal{C}$, $x_n = 1$ for some $n$, i.e.,

$$x = (x_1, x_2, \ldots, x_{N-1}, 1, x_{N+1}, \ldots), \quad N = N(x) = \min\{k : x_k = 1\}.$$

If we let $y = (x_1, x_2, \ldots, x_{N-1}, 2, 0, 0, \ldots)$, then we see that, for $\omega \in \Omega$, $\omega \leqslant x$ if and only if $\omega < y$. Thus, from (a),

$$F(x) = P(\omega \leqslant x) = P(\omega < y) = F(y) = \sum_{k=1}^{N-1} \frac{x_k}{2^{k+1}} + \frac{1}{2^N}.$$

From the above, we see that $F$ takes constant values on $[0, 1] \setminus \mathcal{C}$. For example, for $x \in [1/3, 2/3]$, $F(x) = 1/2$; for $x \in [1/9, 2/9]$, $F(x) = 1/4$; for $x \in [7/9, 8/9]$, $F(x) = 3/4$, $\ldots$, and more generally, there are $2^n - 1$ intervals on which $F$ is constant, with values $(2k-1)/2^n$, $k = 1, 2, \ldots, 2^n - 1$. Thus, the (a.e.) derivative $F'$ of $F$ is identically zero. As such, there are numbers $u, v$ such that $0 \leqslant u < v \leqslant 1$ and

$$\int_u^v F'(x)dx = 0 < F(v) - F(u).$$

This explains the fact that $F$ increases only on a set of Lebesgue measure zero.

Now, we turn to *distributions of random vectors*. Let $\mathbf{X} = (X_1, \ldots, X_d)^{\mathrm{T}}$ be a random vector. Its probability law $P_\mathbf{X}$ is a probability measure on $\mathcal{B}(\mathbb{R}^d)$ and hence is characterized by its joint (or multivariate) distribution function $F : \mathbb{R}^d \to [0, 1]$ where

$$
\begin{aligned}
F(x_1, x_2, \ldots, x_d) &= P(X_1 \leqslant x_1, X_2 \leqslant x_2, \ldots, X_d \leqslant x_d) \\
&= P_\mathbf{X} \left( \prod_{i=1}^{d} (-\infty, x_i] \right).
\end{aligned}
$$

The distribution function of each $X_i$, say, $F_i$, is called a *marginal distribution function* and is derived from $F$ as

$$
F_i(x_i) = F(\infty, \ldots, \infty, x_i, \infty, \ldots, \infty).
$$

If $F$ is differentiable $d$ times, then $\mathbf{X}$ has a *joint probability density function* given by

$$
f(x_1, x_2, \ldots, x_d) = \frac{\partial^d F(x_1, x_2, \ldots, x_d)}{\partial x_1 \partial x_2 \cdots \partial x_d}.
$$

In this case, each $X_i$ also has a *marginal density function* given by

$$
f_i(x_i) = \int_{\mathbb{R}^d} f(x_1, x_2, \ldots, x_d) dx_1 dx_2 \ldots dx_{i-1} dx_{i+1} \ldots dx_d.
$$

## 1.2.2  Integration

We also need some convergence results in measure theory for deriving stochastic dominance rules from approximations of utility functions in Chapter 2. Finally, we need integration theory for studying *risk measures* for risk assessment and management in Chapters 4 and 5. Of course, basic probability results are used to embark on *statistical tests for stochastic dominance* in Chapter 6.

As such, let's handle all that right here! Let $(\Omega, \mathcal{A}, P)$ be a probability space on which real-valued random variables are defined. The following is valid even when we relax the condition $P(\Omega) = 1$, i.e., valid for any nonnegative measures.

For a simple random variable $X(\omega) = \sum_{i=1}^{k} x_i I_{A_i}(\omega)$, with $x_i \in \mathbb{R}$ and $A_i \in \mathcal{B}(\mathbb{R})$, $I_A$ being the indicator function of the set $A$, its expectation (mean) is defined to be

$$EX = \sum_{i=1}^{k} x_i P(A_i).$$

Note that this definition does not depend on this particular choice of canonical representation of $X$.

For $X \geq 0$, we can approximate $X(\omega)$ by an increasing sequence of simple functions, namely

$$X_n(\omega) = \sum_{i=1}^{n2^n}[(i-1)/2^n]I_{A_{i,n}}(\omega) + nI_{B_n}(\omega)$$

where $B_n = \{\omega : X(\omega) \geq n\}$ and

$$A_{i,n} = \{\omega : (i-1)/2^n \leq X(\omega) < 1/2^n\}.$$

More specifically, $X_n \leq X_{n+1}$ and

$$X(\omega) = \lim_{n \to \infty} X_n(\omega).$$

It can be shown that, for *any* sequence of nondecreasing, simple and nonnegative random variables $(X_n, n \geq 1)$ converging pointwise to $X$, we have

$$\lim_{n \to \infty} X_n(\omega) = \sup\{EY : Y \text{ is simple and } 0 \leq Y \leq X\},$$

so that $X(\omega) = \lim_{n \to \infty} X_n(\omega)$ does not depend on any particular choice of the approximating sequence.

Thus, for $X \geq 0$, its expectation is defined as

$$EX = \lim_{n \to \infty} E(X_n),$$

where $(X_n, n \geq 1)$ is *any* sequence of nondecreasing, simple and nonnegative random variables converging pointwise to $X$.

For real-valued $X$, we write $X = X^+ - X^-$, where $X^+(\omega) = \max\{X(\omega), 0\}$ (the positive part of $X$), and $X^-(\omega) = -\min\{X(\omega), 0\}$ (the negative part of $X$). Clearly $X^+, X^-$ are both nonnegative. We define multiplication and addition for the extended reals as follows:

1. $0(\infty) = (\infty)0 = (-\infty)0 = 0(-\infty) = 0$

2. $x(\infty) = (\infty)x = \infty$ if $x > 0$

3. $x(-\infty) = (-\infty)x = \infty$ if $x < 0$

4. $x + (\infty) = (\infty) + x = \infty$ for all $x \in \mathbb{R}$

5. $\infty + (-\infty)$ is undefined

We start out with simple random variables. A *simple random variable* is a random variable $X$ which has a finite set of values. If $x_1, x_2, \ldots, x_k$ are distinct values of $X$ in $\mathbb{R}$, the $X$ can be written in the canonical form

$$X = \sum_{i=1}^{k} x_i I_{A_i},$$

where $A_i = \{\omega : X(\omega) = x_i\}$, noting again that $\{A_1, A_2, \ldots, A_k\}$ forms an $\mathcal{A}$-measurable partition of $\Omega$.

**Definition 1.9** *The expected value of a simple random variable $X$, denoted as $E(X)$, is defined as*

$$E(X) = \sum_{i=1}^{k} x_i P(A_i).$$

This definition does not depend on this particular choice of canonical representation of $X$. This can be seen as follows. As a simple function, $X$ can be represented in many different forms, such as $X = \sum_{j=1}^{m} y_j I_{B_j}$, where $\{B_1, B_2, \ldots, B_m\}$ form another $\mathcal{A}$-measurable partition of $\Omega$, in which the $y_j$'s can be repeated. With

$$\Omega = \bigcup_{i=1}^{k} A_i = \bigcup_{j=1}^{m} B_j,$$

we have

$$A_i = A_i \cap \Omega = A_i \cap \left( \bigcup_{j=1}^{m} B_j \right) = \bigcup_{j=1}^{m} (A_i \cap B_j),$$

and

$$B_j = B_j \cap \Omega = B_j \cap \left( \bigcup_{i-1}^{k} A_i \right) = \bigcup_{i=1}^{k} (A_i \cap B_j).$$

Also, $X(\omega) = x_i = y_j$ for $\omega \in A_i \cap B_j \neq \varnothing$, since every function is, by definition, single-valued. Thus,

$$
\sum_{i=1}^{k} x_i P(A_i) \;=\; \sum_{i=1}^{k} x_i P\left(\bigcup_{j=1}^{m}(A_i \cap B_j)\right) = \sum_{i=1}^{k} x_i \sum_{j=1}^{m} P(A_i \cap B_j)
$$

$$
=\; \sum_{i=1}^{k}\sum_{j=1}^{m} y_j P(A_i \cap B_j) = \sum_{j=1}^{m} y_j \sum_{i=1}^{k} P(A_i \cap B_j)
$$

$$
=\; \sum_{j=1}^{m} y_j P\left(\bigcup_{i=1}^{k}(A_i \cap B_j)\right) = \sum_{j=1}^{m} y_j P(B_j).
$$

To pursue further analysis, we need the following facts which are easy to derive (and hence left as exercises):

1. Simple random variables form a vector space, i.e., for $\alpha, \beta \in \mathbb{R}$ and $X, Y$ simple, $\alpha X + \beta Y$ is simple

2. The operator $E(.)$ is linear on this vector space, i.e.,

$$
E(\alpha X + \beta Y) = \alpha E(X) + \beta E(Y)
$$

3. $E(.)$ is a positive operator, i.e., if $X \geqslant 0$, then $E(X) \geqslant 0$ (as a consequence, if $X \leqslant Y$ then $E(X) \leqslant E(Y)$)

   Next, for *nonnegative random variables*, we make two basic observations:

**Lemma 1.10** *Let $X$ be a nonnegative random variable. Then there exists a sequence of nonnegative simple random variables $\{X_n, n \geqslant 1\}$, such that*

$$
X_n(\omega) \leqslant X_{n+1}(\omega), \quad \text{for} \quad \omega \in \Omega,
$$

*i.e., the sequence $\{X_n, n \geqslant 1\}$ is non-decreasing, and*

$$
X(\omega) = \lim_{n \to \infty} X_n(\omega), \quad \text{for} \quad \omega \in \Omega,
$$

*denoted by $X_n \nearrow X$ or $X_n(\omega) \nearrow X(\omega)$.*

**Proof.** Consider

$$
X_n(\omega) = \sum_{i=1}^{n2^n} \frac{i-1}{2^n} I_{A_i,n}(\omega) + n I_{B_n}(\omega),
$$

where $B_n = \{\omega : X(\omega) \geqslant n\}$ and

$$A_{i,n} = \left\{\omega : \frac{i-1}{2^n} \leqslant X(\omega) < \frac{i}{2^n}\right\}.$$

Then, by inspection, we have that $X_n \leqslant X_{n+1} \leqslant \cdots \leqslant X$ and $X_n \nearrow X$ on $\Omega$.  ∎

**Lemma 1.11** *For any sequence of non-decreasing, simple and nonnegative random variables $\{X_n, n \geqslant 1\}$ converging to a nonnegative random variable $X$, we have*

$$\lim_{n\to\infty} E(X_n) = \sup\{E(Y) : Y \text{ is simple and } 0 \leqslant Y \leqslant X\}.$$

**Proof.** First, by properties of expectation of simple random variables, the sequence of numbers $\{E(X_n), n \geqslant 1\}$ is non-decreasing and hence has a limit (which could be $\infty$). Denote this limit by $a$, and let $b = \sup\{E(Y) : Y$ simple and $0 \leqslant Y \leqslant X\}$ (which exists in $[-\infty, \infty]$). By definition of $b$, and by Lemma 1.10,

$$\lim_{n\to\infty} E(X_n) = a \leqslant b.$$

On the other hand, for any $Y$ simple and nonnegative such that $Y \leqslant X$, we have that $E(Y) \leqslant a$, so that, $b \leqslant a$, and the lemma is proved. Indeed, let $0 \leqslant \varepsilon < 1$. Consider $A_n = \{\omega : \varepsilon Y(\omega) \leqslant X_n(\omega)\}$. Then $\varepsilon Y I_{A_n}$ is simple and

$$E(\varepsilon Y I_{A_n}) \leqslant E(X_n) \leqslant a.$$

Now, let $Y$ be written in its canonical form $Y = \sum_{i=1}^{k} y_i I_{B_i}$. We have

$$E(\varepsilon Y I_{A_n}) = \varepsilon \sum_{i=1}^{k} y_i P(B_i \cap A_n).$$

Note that $A_n \nearrow \bigcup_{n\geqslant 1} A_n = \Omega$. Indeed, clearly for $\omega$ such that $X(\omega) = 0$, $\omega \in \bigcup_{n\geqslant 1} A_n$ in view of $Y \leqslant X$, and $X_n \leqslant X$ for all $n$. For $\omega$ such that $X(\omega) > 0$, we have $\varepsilon Y(\omega) < X(\omega)$. But since $X_n(\omega) \nearrow X(\omega)$, we have $X_n(\omega) \geqslant \varepsilon Y(\omega)$ for $n$ sufficiently large. As such, $B_i \cap A_n \nearrow B_i$ for each $i = 1, 2, \ldots, k$. Thus, by continuity of $P$,

$$\lim_{n\to\infty} E(\varepsilon Y I_{A_n}) = \varepsilon \sum_{i=1}^{k} y_i P(B_i) = \varepsilon E(Y) \leqslant a,$$

implying that $E(Y) \leqslant a$ since $\varepsilon$ is arbitrary in $[0, 1)$.  ∎

The implication of the above two lemmas is this. We can define the *expected value of a nonnegative* random variable $X$ either by the $a$ or $b$ given above. If $a$ is chosen, the definition is well-defined since $a$ does not depend on any particular choice of the approximate sequence, i.e., for any two sequences of nonnegative simple random variables such that $X_n \nearrow X$, and $Y_n \nearrow X$, we have

$$\lim_{n \to \infty} E(X_n) = \lim_{n \to \infty} E(Y_n)$$

since they are both equal to $b$. We choose $a$ for convenience of analysis, namely for using the "monotone convergence property."

**Definition 1.12** *The **expected value** of a nonnegative random variable $X$ is defined to be*

$$E(X) = \lim_{n \to \infty} E(X_n),$$

*where $\{X_n, n \geq 1\}$ is any sequence of non-decreasing, nonnegative, and simple random variables such that $X_n \nearrow X$. Finally, for an arbitrary random variable $X$, we write $X = X^+ - X^-$, where $X^+(\omega) = \max\{X(\omega), 0\}$ (positive part of $X$) and $X^-(\omega) = -\min\{X(\omega), 0\}$ (negative part of $X$). Note that both $X^+ \geq 0$ and $X^- \geq 0$. We then define*

$$E(X) = E(X^+) - E(X^-)$$

*and we write $E(X) = \int_\Omega X(\omega) dP(\omega)$, provided, of course, that not both $E(X^+)$ and $E(X^-)$ are $\infty$.*

**Remark 1.13** *Clearly $E(X) < \infty$ when $E(X^+) < \infty$ and $E(X^-) < \infty$. Now $|X(\omega)| = X^+(\omega) + X^-(\omega)$, we see that $E(X) < \infty$ if and only if $E|X| < \infty$, in which case, we say that $X$ has a finite expected value, or $X$ is integrable (with respect to $P$). If only one of $E(X^+)$, $E(X^-)$ is $\infty$, then $X$ has an infinite expected value, and if both $E(X^+)$ and $E(X^-)$ are $\infty$, then we say that $X$ does not have an expected value or $E(X)$ does not exist.*

To emphasize the dependence of $E(X)$ on $P$, when needed, we write $E_P(X)$. Also, for $A \in \mathcal{A}$,

$$E(XI_A) = \int_A X(\omega) dP(\omega).$$

**Theorem 1.14 (Monotone convergence)** *Let $\{X_n,\ n \geqslant 1\}$ be a non-decreasing sequence of nonnegative random variables with measurable limit $X$ almost surely (a.s.), i.e.,*

$$P\left(\omega:\ \lim_{n\to\infty} X_n(\omega) \neq X(\omega)\right) = 0,$$

*then*

$$\lim_{n\to\infty} \int_A X_n(\omega)dP(\omega) = \int_A X(\omega)dP(\omega).$$

**Proof.** By hypothesis, $\{\int_A X_n(\omega)dP(\omega),\ n \geqslant 1\}$ is a non-decreasing sequence of (extended) real numbers which has a limit, say $L \leqslant \infty$. Since $X_n \leqslant X$, we have

$$\int_A X_n(\omega)dP(\omega) \leqslant \int_A X(\omega)dP(\omega),$$

and hence $L \leqslant \int_A X(\omega)dP(\omega)$. Fix $0 < r < 1$, and let $Y$ be simple such that $0 \leqslant Y \leqslant X$. For each integer $n$, let

$$A_n = \{\omega:\ rY(\omega) \leqslant X_n(\omega)\} \cap A.$$

From

$$X_n(\omega)I_A(\omega) \geqslant X_n(\omega)I_{A_n}(\omega) \geqslant rY(\omega),$$

we obtain

$$
\begin{aligned}
\int_A X_n(\omega)dP(\omega) &\geqslant \int_{A_n} X_n(\omega)dP(\omega) \geqslant \int_{A_n} rY(\omega)dP(\omega) \\
&= r \int_{A_n} Y(\omega)dP(\omega) \\
&= r \int_{A_n} Y(\omega)I_{A_n}(\omega)dP(\omega). \quad (1.4)
\end{aligned}
$$

Now the sequence $\{A_n,\ n \geqslant 1\}$ increases to $B = \bigcup_{n\geqslant 1} A_n$, and by the convergence of $\{X_n, n \geqslant 1\}$, $P(A \setminus B) = 0$. Then $0 \leqslant YI_{A_n} \nearrow YI_B$ so that

$$
\begin{aligned}
\lim_{n\to\infty} r \int_\Omega Y(\omega)I_{A_n}(\omega)dP(\omega) &= r \int_\Omega Y(\omega)I_B(\omega)dP(\omega) \\
&= r \int_\Omega Y(\omega)I_A(\omega)dP(\omega) = r \int_A Y(\omega)dP(\omega).
\end{aligned}
$$

As $n \to \infty$ in Equation (1.4), we get $r \int_A Y(\omega)dP(\omega) \leqslant L$. Letting $r \to 1$ yields $\int_A Y(\omega)dP(\omega) \leqslant L$ for all such simple $Y$. Take a sequence of such

$Y$, say $\{Y_n, n \geqslant 1\}$, increasing to $X$ with one more limit operation, we obtain $\int_A X(\omega)dP(\omega) \leqslant L$. It follows that

$$\int_A X(\omega)dP(\omega) = L.$$

∎

**Theorem 1.15 (Dominated convergence)** *Let $\{X_n, n \geqslant 1\}$ be a sequence of random variables with $\lim_{n\to\infty} X_n = X$ a.s., and $|X_n| \leqslant Y$ for any $n \geqslant 1$, with $\int_A Y(\omega)dP(\omega) < \infty$. Then*

$$\int_A |X(\omega)|dP(\omega) < \infty$$

*and*

$$\int_A X(\omega)dP(\omega) = \lim_{n\to\infty} \int_A X_n(\omega)dP(\omega).$$

**Proof.** Clearly $|X| \leqslant Y$   a.s., so that $X$ and $X_n$'s are integrable on $A$. Apply Fatou's lemma (see Exercise 7) to

$$\liminf_{n\to\infty}(X_n + Y) = X + Y \quad \text{a.s.,}$$

yielding

$$\int_A (X(\omega) + Y(\omega))dP(\omega) \leqslant \liminf_{n\to\infty} \int_A (X_n(\omega) + Y(\omega))dP(\omega)$$

$$= \int_A Y(\omega)dP(\omega) + \liminf_{n\to\infty} \int_A X_n(\omega)dP(\omega).$$

By linearity of the integral, we obtain

$$\int_A Y(\omega)dP(\omega) \;+\; \int_A X(\omega)dP(\omega) \leqslant \int_A Y(\omega)dP(\omega)$$

$$+ \; \liminf_{n\to\infty} \int_A X_n(\omega)dP(\omega),$$

from which it follows that

$$\int_A X(\omega)dP(\omega) \leqslant \liminf_{n\to\infty} \int_A X_n(\omega)dP(\omega).$$

Similarly, from the sequence of nonnegative random variables $\{Y - X_n, n \geqslant 1\}$, we get

$$\int_A (-X(\omega))dP(\omega) \leqslant \liminf_{n\to\infty} \int_A (-X_n(\omega))dP(\omega),$$

which implies

$$\limsup_{n\to\infty} \int_A X_n(\omega)dP(\omega) \leqslant \int_A X(\omega)dP(\omega).$$

Since $\liminf \leqslant \limsup$, a combination of the above yields equality. ∎

We now consider methods for computing expected values of random variables.

**Lemma 1.16** *Let $(U,\mathcal{U})$ be a measurable space and $g : U \to \mathbb{R}$ be $\mathcal{U}$-$\mathcal{B}(\mathbb{R})$-measurable and nonnegative. Let $X : \Omega \to U$ be $\mathcal{A}$-$\mathcal{U}$-measurable. Then*

$$E[g(X)] = \int_U g(u)dP_X(u). \tag{1.5}$$

**Proof.** Note that $P_X = PX^{-1}$ is the probability measure on $\mathcal{U}$ defined by

$$P_X(A) = P(X \in A), \qquad A \in \mathcal{U}.$$

Also $g(X) = g \circ X : \Omega \to \mathbb{R}^+$ is a nonnegative random variable, so that $E(g(X))$ exists (possibly $\infty$). The following reasoning is referred to as the *standard argument of measure theory*.

First if $g$ is an indicator function, $g(u) = I_A(u)$ for some $A \in \mathcal{U}$, then

$$\int_U I_A(u)dP_X(u) = P_X(A) = P(X \in A) = E(I_A(X))$$

so that Equation (1.5) is true.

Next, if $g \geqslant 0$ and *simple*, $g(u) = \sum_{i=1}^k u_i I_{A_i}(u)$, then by linearity of the integral,

$$
\begin{aligned}
E[g(X)] &= E\left(\sum_{i=1}^k u_i I_{A_i}(X)\right) = \sum_{i=1}^k u_i P(X \in A_i) \\
&= \sum_{i=1}^k u_i P_X(A_i) = \sum_{i=1}^k \int_U u_i I_{A_i}(u)dP_X(u) \\
&= \int_U \left(\sum_{i=1}^k u_i I_{A_i}(u)\right) dP_X(u).
\end{aligned}
$$

Finally, let $g \geqslant 0$ and measurable, and $\{g_n, n \geqslant 1\}$ be a sequence of nonnegative simple functions such that $g_n \nearrow g$ on $U$. For each $n$, we have

$$E[g_n(X)] = \int_U g_n(u)dP_X(u).$$

Then, by *monotone convergence theorem*,

$$
\begin{aligned}
E[g(X)] &= \lim_{n\to\infty} E[g_n(X)] = \lim_{n\to\infty} \int_U g_n(u)dP_X(u) \\
&= \int_U \left( \lim_{n\to\infty} g_n(u) \right) dP_X(u) = \int_U g(u)dP_X(u).
\end{aligned}
$$

∎

**Remark 1.17** *For $A \in \mathcal{U}$, the above lemma takes the form*

$$
E\left(g(X)I_{X^{-1}(A)}\right) = \int_A g(u)dP_X(u).
$$

*If we let $(U,\mathcal{U}) = (\mathbb{R}, \mathcal{B}(\mathbb{R}))$, $X \geqslant 0$, $g : \mathbb{R} \to \mathbb{R}$ with $g(x) = x$, and $A = \mathbb{R}^+$, then*

$$
E(X) = \int_0^\infty x \, dF(x).
$$

*Thus the knowledge of the distribution function $F$ of $X$ is sufficient for the calculation of $E(X)$.*

**Lemma 1.18** *Let $F$ be a distribution function on $\mathbb{R}$ of absolutely continuous type with density function $f$. For any $g : \mathbb{R} \to \mathbb{R}^+$, measurable, we have*

$$
\int_{-\infty}^\infty g(x)dF(x) = \int_{-\infty}^\infty g(x)f(x)dx. \tag{1.6}
$$

**Proof.** For $g = I_A$, we have

$$
\int_{-\infty}^\infty I_A dF(x) = dF(A) = \int_A f(x)dx.
$$

For $g \geqslant 0$ and simple, Equation (1.6) follows by linearity. For $g \geqslant 0$ measurable, let $g_n \nearrow g$, $g_n \geqslant 0$ and simple. Then for each $n$,

$$
\int_{-\infty}^\infty g_n(x)dF(x) = \int_{-\infty}^\infty g_n(x)f(x)dx
$$

and the formula (1.6) follows by the monotone convergence theorem. Note that both sides of (1.6) might be $\infty$. ∎

We now establish the important justification for practical calculations of expected values of *arbitrary random variables*.

**Theorem 1.19** *Let $X$ be a random element with values in $(U, \mathcal{U})$ and let $g : U \to \mathbb{R}$. Then $g \circ X$ is $P$-integrable if and only if $g$ is $P_X$-integrable, in which case we have*

$$E\left(g(X)\right) = \int_{\Omega} g(X(\omega))dP(\omega) = \int_{U} g(u)dP_X(u). \qquad (1.7)$$

**Proof.** We write $g = g^+ - g^-$, then by Lemma 1.16, (1.7) holds for $g^+$ and for $g^-$, respectively. Hence if $g$ is $P_X$-integrable, i.e.,

$$\int_{U} g^+(u)dP_X(u) < \infty \quad \text{and} \quad \int_{U} g^-(u)dP_X(u) < \infty,$$

then

$$\int_{\Omega} (g(X))^+(\omega)dP(\omega) < \infty \quad \text{and} \quad \int_{\Omega} (g(X))^-(\omega)dP(\omega) < \infty$$

so that $g(X)$ is $P$-integrable. ∎

More generally, for $A \in \mathcal{U}$, we have

$$\int_{X^{-1}(A)} g(X(\omega))dP(\omega) = \int_{A} g(u)dP_X(u).$$

Consider $(U, \mathcal{U}) = (\mathbb{R}, \mathcal{B}(\mathbb{R}))$ and $g : \mathbb{R} \to \mathbb{R}$ with $g(x) = x$. Then Equation (1.7) becomes

$$E(X) = \int_{-\infty}^{\infty} x dF(x),$$

provided that $g(x) = x$ is $dF$-integrable. The important point is this. Use

$$E(X) = \int_{-\infty}^{\infty} x dF(x) \quad \text{only when} \quad \int_{-\infty}^{\infty} |x|dF(x) < \infty.$$

Also Lemma 1.18 is easily extended to any random variable $X$ (non-negative or not). If $F$ has a density $f$, then

$$\int_{-\infty}^{\infty} g(x)dF(x) = \int_{-\infty}^{\infty} g(x)f(x)dx,$$

whenever $g$ is $dF$-integrable, or equivalently, $g(x)f(x)$ is $dx$-integrable. Thus, for example, in the discrete case when $X$ takes values $\{x_n, n \geqslant 1\}$ with probability density function $f(x_n)$, we use

$$E(X) = \sum_{n \geqslant 1} x_n f(x_n) \quad \text{only when} \quad \sum_{n \geqslant 1} |x_n| f(x_n) < \infty,$$

(absolutely convergent series). When $E(|X|) = \infty$, $E(X)$ is either $\infty$ or does not exist.

As an application of Theorem 1.19, consider the computation of expected values of *functions of several random variables*. We illustrate this for the case of two random variables. Consider random variables $X$ and $Y$, defined on $(\Omega, \mathcal{A}, P)$, or equivalently the random vector $(X, Y) : \Omega \to \mathbb{R}^2$. The probability law of $(X, Y)$ is the probability $P_{(X,Y)} = P(X, Y)^{-1}$ on $\mathcal{B}(\mathbb{R}^2)$, where for any $A \in \mathcal{B}(\mathbb{R}^2)$,

$$(X, Y)^{-1}(A) = \{\omega : (X(\omega), Y(\omega)) \in A\},$$
$$P_{(X,Y)}(A) = P(\omega : (X(\omega), Y(\omega)) \in A).$$

Let $g : \mathbb{R}^2 \to \mathbb{R}$ be measurable. Then if $g$ is $P_{(X,Y)}$-integrable, we have

$$E[g(X, Y)] = \int_{\mathbb{R}^2} g(x, y) dP_{(X,Y)}(x, y).$$

We discuss now the computational aspects of the right-hand side of this equation. As in the case of random variables, suppose that the joint distribution function of $(X, Y)$ on $\mathcal{B}(\mathbb{R}^2)$ is of absolutely continuous type, i.e., for any $A \in \mathcal{B}(\mathbb{R}^2)$,

$$P_{(X,Y)}(A) = dF(A) = \int_A f(x, y) dx \otimes dy,$$

where $f : \mathbb{R}^2 \to \mathbb{R}^+$ is its joint probability density function, and $dx \otimes dy$ denotes the *product Lebesgue measure* on $\mathbb{R}^2$. In this case,

$$E(g(X, Y)) = \int_{\mathbb{R}^2} g(x, y) f(x, y) dx \otimes dy.$$

Let $h(x, y) = g(x, y) f(x, y)$, then we are led to consider the computation of the quantity $\int_{\mathbb{R}^2} h(x, y) dx \otimes dy$, which is a *double integral*.

Note that in previous analysis, we consider only special types of measures, namely probability measures. A (general measure) $\mu$ is a set-function just like $P$, except that $\mu(\Omega)$ need not equal one, and in fact it can be $\infty$. For example, $dx$ is the Lebesgue measure on $\mathbb{R}$ with $dx(\mathbb{R}) = \infty$. Moreover, $dx$ is special in the sense that it is $\sigma$-*finite*, i.e., there exists a sequence of measurable sets $\{A_n, n \geqslant 1\}$, elements of $\mathcal{B}(\mathbb{R})$, such that $\bigcup_{n \geqslant 1} A_n = \mathbb{R}$ and $dx(A_n) < \infty$ for each $n$. Indeed, let $A_n = (-n, n)$, then $\mathbb{R} = \bigcup_{n \geqslant 1} A_n$ and $dx(A_n) = 2n < \infty$, $n \geqslant 1$. Of course, probability measures are $\sigma$-finite. With a (positive) measure $\mu$ on any abstract measure space $(U, \mathcal{U})$, the concept of integral of numerical functions with respect to $\mu$, denoted by $\int_U h(u)d\mu(u)$, is defined similarly to $\int_\Omega X(\omega)dP(\omega)$, and $(U, \mathcal{U}, \mu)$ is called a *measure space*.

The following important theorem allows us to compute *double integrals* of the form $\int_{\mathbb{R}^2} h(x,y)dx \otimes dy$ via *iterated integrals* (or sequential single integrals). Again, to cover all possible situations, we state it in a general form. Also some technical details are necessary for the validity of the theorem. The proof of the theorem is not quite hard, but lengthy, so we refer the students to a text like Billingsley ([16], pp. 232-235). We state it for the case of two measurable spaces, but the general case is similar.

**Theorem 1.20 (Fubini)** *Let $\mu_i$ be $\sigma$-finite measures on $(\Omega_i, \mathcal{A}_i)$, $i = 1, 2$ and $h : \Omega_1 \times \Omega_2 \to \mathbb{R}$ measurable (i.e., $\mathcal{A}_1 \otimes \mathcal{A}_2$-$\mathcal{B}(\mathbb{R})$-measurable).*

*(i) If $h$ is nonnegative, then*

$$\int_{\Omega_1 \times \Omega_2} h(\omega_1, \omega_2)d(\mu_1 \otimes \mu_2)(\omega_1, \omega_2)$$

$$= \int_{\Omega_1} \left( \int_{\Omega_2} h(\omega_1, \omega_2)d\mu_2(\omega_2) \right) d\mu_1(\omega_1)$$

$$= \int_{\Omega_2} \left( \int_{\Omega_1} h(\omega_1, \omega_2)d\mu_1(\omega_1) \right) d\mu_2(\omega_2).$$

*(ii) If $h$ is arbitrary (not necessarily nonnegative), but integrable with respect to $\mu_1 \otimes \mu_2$, i.e.,*

$$\int_{\Omega_1 \times \Omega_2} |h(\omega_1, \omega_2)|d(\mu_1 \otimes \mu_2)(\omega_1, \omega_2) < \infty,$$

*then the equalities in (i) hold.*

The hypothesis of "$\sigma$-finite measures" is used to establish the existence and measurability of quantities involved in the above formula.

The practical example is $(\Omega_i, \mathcal{A}_i) = (\mathbb{R}, \mathcal{B}(\mathbb{R}))$, $i = 1, 2$ and $\mu_1 = \mu_2 =$ the Lebesgue measure $dx$ on $\mathbb{R}$. The Fubini's theorem says that, for appropriate $h$, the double integral of $h$ is computed by an iterated integral in *any order* of preference. But it should be noted that if $h$ is *not* integrable (with respect to $\mu_1 \otimes \mu_2$), it can happen that the above two iterated integrals (in different orders) are not equal. Also, these two integrals can be equal even if $h$ is not integrable. See Exercises.

To apply Fubini's theorem in computations, besides the case where $h \geqslant 0$, we should first check the *integrability* of $h$ with respect to the product measure $\mu_1 \otimes \mu_2$. This is done as follows. Since $|h| \geqslant 0$, the equalities in (i) of Fubini's theorem hold for $|h|$ in the place of $h$. Thus $h$ is $\mu_1 \otimes \mu_2$ integrable if and only if any of the iterated integral of $|h|$ is finite. Thus, simply compute, say,

$$\int_{\Omega_1} \left( \int_{\Omega_2} |h(\omega_1, \omega_2)| \, d\mu_2(\omega_2) \right) d\mu_1(\omega_2),$$

to see whether it is finite or not.

In the discrete case, this is known in the context of *absolutely convergent series* (see any calculus text), namely the following. Let $\sum_{n \geqslant 1} x_n$ be an infinite series of numbers. If the sequence of partial sums $s_n = \sum_{i=1}^{n} x_i$ converges to $s < \infty$, then we say that the series $\sum_{n \geqslant 1} x_n$ is convergent and its sum is denoted by $\sum_{n \geqslant 1} x_n = s$. Of course a series of non-negative numbers such as $\sum_{n \geqslant 1} |x_n|$ always has a sum $\leqslant \infty$. When $\sum_{n \geqslant 1} |x_n| < \infty$, we say that the series $\sum_{n \geqslant 1} x_n$ is *absolutely convergent*. (For that to happen, it is necessary and sufficient that the increasing sequence $\sum_{i=1}^{n} |x_n|$ is bounded from above.) Note that if $\sum_{n \geqslant 1} |x_n| < \infty$, then $\sum_{n \geqslant 1} x_n$ is finite and is the same for any reordering of the sequence $\{x_n, n \geqslant 1\}$. For discrete random vectors, we are using *double series*. The double series

$$\sum_{m \geqslant 1, n \geqslant 1} x_{mn}$$

is absolutely convergent when

$$\sum_{m \geqslant 1, n \geqslant 1} |x_{mn}| < \infty,$$

and in this case, we have

$$\sum_{m \geqslant 1, n \geqslant 1} x_{mn} = \sum_{n \geqslant 1} \left( \sum_{m \geqslant 1} x_{mn} \right) = \sum_{m \geqslant 1} \left( \sum_{n \geqslant 1} x_{mn} \right) < \infty,$$

i.e., the finite value of the double series is computed via its iterated series and its value is independent of the order of summations. For this to happen, it is necessary and sufficient that $\sum_{m \geqslant 1} |x_{mn}| < \infty$ for each $n$ and

$$\sum_{n \geqslant 1} \left( \sum_{m \geqslant 1} |x_{mn}| \right) < \infty.$$

Note that the limit sum of the double series given above is the limit of the convergent double sequence

$$\{S_{m,n}\}_{m,n=1}^{\infty}, \qquad S_{m,n} = \sum_{i \leqslant m, j \leqslant n} x_{ij},$$

i.e., for any $\varepsilon > 0$, there exist $N, M$ such that

$$|S_{m,n} - L| < \varepsilon \quad \text{for all} \quad m \geqslant M, \ n \geqslant N,$$

and we write

$$\lim_{m \to \infty, n \to \infty} S_{m,n} = L.$$

**Example 1.21** *Let $(X, Y)$ be a random vector with joint distribution $F(x, y)$ on $\mathbb{R}^2$. Suppose that $F$ is of absolutely continuous type, i.e., there is a function $f : \mathbb{R}^2 \to \mathbb{R}^+$ such that, for $A \in \mathcal{B}(\mathbb{R}^2)$,*

$$P((X, Y) \in A) = \int_A f(x, y)(dx \otimes dy).$$

*The marginal probability distribution of $X$, say, is*

$$P(X \in B) = P((X, Y) \in B \times \mathbb{R}), \quad \text{for} \quad B \in \mathcal{B}(\mathbb{R}).$$

*An application of Fubini's theorem will show that $X$ is of absolutely continuous type with probability density function*

$$f_X(x) = \int_{-\infty}^{\infty} f(x, y) dy.$$

*Indeed, since $f \geqslant 0$,*

$$P((X, Y) \in B \times \mathbb{R}) = \int_{B \times \mathbb{R}} f(x, y)(dx \otimes dy)$$

$$= \int_B \left( \int_{\mathbb{R}} f(x, y) dy \right) dx = \int_B f_X(x) dx.$$

### 1.2.3   Notes on topological spaces

Later we will need topologies in our discussions of choice under uncertainty. Here is a tutorial. We use the real line $\mathbb{R}$ to be concrete.

For $a, b \in \mathbb{R}$ with $a \leqslant b$, the set $(a, b) = \{x \in \mathbb{R} : a < x < b\}$ is a (bounded) *open* interval. Of course, for $a = b$, we obtain the empty set $\varnothing$. An arbitrary union of open intervals is called an *open set* of $\mathbb{R}$. The collection $\mathcal{T}$ of all open sets has the properties that $\varnothing$ and $\mathbb{R}$ are in $\mathcal{T}$, $\mathcal{T}$ is closed under arbitrary unions, and $\mathcal{T}$ is closed under finite intersections. This leads to the definition of a *topological space*.

**Definition 1.22** *A **topological space** is a set A and a set T of subsets of A such that*

*(i) ∅ and A are in T.*

*(ii) T is closed under arbitrary unions.*

*(iii) T is closed under finite intersections.*

The topological space is denoted as $(A, T)$, the elements of $T$ are called open sets, and $T$ is called a topology on $A$. For example, the pair $(\mathbb{R}, T)$ described above is a topological space. For any topological space, the set complement of an open set is called a *closed set*. By De Morgan's laws, we see that the following dual properties hold for the set $F$ of closed sets.

(a) ∅ and $A$ are in $F$.

(b) $F$ is closed under arbitrary intersections.

(c) $F$ is closed under finite unions.

Other basic concepts are these. For $x \in A$, a set $U$ containing $x$ is called a *neighborhood* of $x$ if there is an open set $T$ containing $x$ and contained in $U$. For the topological space $(\mathbb{R}, T)$, if $x, y \in \mathbb{R}$, and $x \neq y$, then there are disjoint open sets, in fact, disjoint open intervals, one containing $x$, the other containing $y$. These open sets are, of course, neighborhoods of $x$ and $y$, respectively. A topological space having this property, namely that distinct points are contained in distinct disjoint neighborhoods is called a *Hausdorff space*.

The smallest closed set containing a subset $S$ of $A$ is called the *closure* of $S$, and denoted $\bar{S}$. The closure of $S$ is the intersection of all closed subsets of $A$ containing $S$, of which $A$ itself is one. Similarly, there is a largest open set contained in $S$. It is called the *interior* of $S$, and denoted $S^\circ$. The *boundary* of $S$ is the closed set $\bar{S} \setminus S^\circ$, and denoted $\partial(S)$.

If $\mathbb{Q}$ denotes the set of rational numbers, then $\bar{\mathbb{Q}} = \mathbb{R}$. A set with this property is called a *dense subset*. That is, if $D$ is a subset of a topological space $A$, and $\bar{D} = A$, then $D$ is dense in $A$. If $D$ is countable, and $\bar{D} = A$, then the topological space $(A, T)$ is called a *separable space*. For example, the space $(\mathbb{R}, T)$ is separable since $\mathbb{Q}$ is countable. In $\mathbb{R}$, a closed interval $[a, b]$ has the property that any set of open sets whose union contains $[a, b]$ has a finite subset whose union contains $[a, b]$. This is phrased by saying that any open cover of $[a, b]$ has a finite subcover. The set $[a, b]$ is an example of a *compact set*. That is, if $C$ is a subset of topological space $A$ and every open cover of $C$ has a finite subcover, then $C$ is called a compact set. A space is called *locally compact* if every point has a neighborhood that is compact. For example, in $\mathbb{R}$, $[x - \varepsilon, x + \varepsilon]$ is a compact neighborhood of the point $x$.

The function $\rho : \mathbb{R} \times \mathbb{R} \rightarrow \mathbb{R}^+$ defined by $\rho(x, y) = |x - y|$ is called a *metric*, or a distance, and satisfies the following properties.

(i)  $\rho(x, y) = 0$ if and only if $x = y$.

(ii)  $\rho(x, y) = \rho(y, x)$.

(iii)  $\rho(x, z) \leqslant \rho(x, y) + \rho(y, z)$.

The real line $\mathbb{R}$, together with this metric, denoted by $(\mathbb{R}, \rho)$, is a *metric space*. More generally, any set $A$ with a function $\rho : A \times A \rightarrow [0, \infty)$ satisfying (i)–(iii) above is called a metric space. Further, any metric $\rho$ on a set $A$ gives rise to a topology on $A$ by taking as open sets unions of sets of the form $\{y : \rho(x, y) < r\}$, where $r$ ranges over the positive real numbers, and $x$ over the elements of $A$. In the case of $\mathbb{R}$, the metric above gives rise to the topology discussed on $\mathbb{R}$, since the set $\{y : \rho(x, y) < r\}$ are open intervals. If there is a metric on a topological space which gives rise to its topology, then that topological space is called *metrizable*. As just noted, the topological space $\mathbb{R}$ is metrizable.

A sequence $\{x_n, n \geqslant 1\}$ in a metric space is said to converge to the element $x$ if for each $\varepsilon > 0$, there is a positive integer $N$ such that for $n > N$, $\rho(x_n, x) < \varepsilon$. The sequence $\{x_n, n \geqslant 1\}$ is a *Cauchy sequence* if for each $\varepsilon > 0$, there is a positive integer $N$ such that if $m, n > N$, then $\rho(x_n, x_m) < \varepsilon$. If in a metric space, every Cauchy sequence converges, then that metric space is said to be *complete*. The metric space $\mathbb{R}$ is complete. The topological space $\mathbb{R}$ is metrizable, has a countable dense subset and is complete with respect to a metric which gives rise to its topology. Such a topological space is called a *Polish space*. Note that Polish spaces are natural spaces in probability theory. On a topological space, the smallest $\sigma$-field containing the open sets is called the *Borel $\sigma$-field* of that space. The Borel $\sigma$-field of $\mathbb{R}$ will be denoted from now on as $\mathcal{B}(\mathbb{R})$. Elements of $\mathcal{B}(\mathbb{R})$ are called *Borel sets* of $\mathbb{R}$. Examples of Borel sets are singletons $\{a\}$ and all types of intervals such as

$$(a, b), \quad [a, b], \quad [a, b), \quad (-\infty, b), \quad (-\infty, b], \quad (a, \infty), \quad \dots$$

In summary, on a topological space such as $\mathbb{R}$, the collection of events is its Borel $\sigma$-field $\mathcal{B}(\mathbb{R})$, and hence models for random experiments with outcomes in $\mathbb{R}$ are probability spaces of the form $(\mathbb{R}, \mathcal{B}(\mathbb{R}), P)$, where each $P$ is a probability measure on $\mathcal{B}(\mathbb{R})$. Note that, when needed, the extended real line $[-\infty, \infty]$ is topologized appropriately with Borel $\sigma$-field generated by $\{(a, b], -\infty \leqslant a < b \leqslant \infty\}$.

Now, in practice, how can we construct probability measures on $\mathcal{B}(\mathbb{R})$? That is, how can we suggest models from empirical observations? It turns out that the key is in the concept of distribution functions.

## 1.3    Choice under uncertainty

We turn now to *choice under uncertainty*. We follow von Neumann-Morgenstern's model [160]. The simplest case is the choice problem on the set of random variables taking values in a *finite set* $\mathcal{Z}$. Since random variables are characterized by their probability laws, we are led to consider a preference relation $\succ$ on $\mathcal{P}(\mathcal{Z})$, the set of all probability measures on $\mathcal{Z}$, in fact on the power set (as a $\sigma$-field) of $\mathcal{Z}$, or equivalently, that of probability density functions on $\mathcal{Z}$. We will use the notation $\delta_z$ to denote the *Dirac probability measure* at $z \in \mathcal{Z}$ (i.e., the probability measure on the power set of $\mathcal{Z}$ having total mass 1 at $z$: $\delta_z(A) = 1$ or 0 according to $z \in A$ or $z \notin A$).

By abuse of language, each element $p \in \mathcal{P}(\mathcal{Z})$ is called a lottery. Note that $\mathcal{P}(\mathcal{Z})$ is infinitely uncountable, so that results on choice under certainty above are not easy to apply.

We seek a numerical representation of $\succ$ in some specific form, namely by a utility function $U$ on $\mathcal{P}(\mathcal{Z})$ built up from a function $u : \mathcal{Z} \to \mathbb{R}$ as $U(p) = Eu(X)$, where $X$ has $p$ as its distribution, i.e., $U(p) = \sum_{z \in \mathcal{Z}} p(z)u(z)$, called the *expected utility representation* of $\succ$.

This is a convenient form of utility function since you can simply specify your utilities $u$ on $\mathcal{Z}$ (e.g., on money!) and then take its expected value with respect to your "lottery."

The preference relation $\succ$ is said to be *rational* if it satisfies the following two axioms:

**Axiom 1.23 (Continuity (or Archimedean))** *For all $p, q, r \in \mathcal{P}(\mathcal{Z})$, if $p \succ q \succ r$, then there are $\alpha, \beta \in (0, 1)$ such that*

$$\alpha p + (1 - \alpha)r \succ q \succ \beta p + (1 - \beta)r.$$

**Axiom 1.24 (Independence)** *For all $p, q, r \in \mathcal{P}(\mathcal{Z})$ and $\alpha \in (0, 1]$, if $p \succ q$ then $\alpha p + (1 - \alpha)r \succ \alpha q + (1 - \alpha)r$.*

**Theorem 1.25** *Any rational preference relation $\succ$ can be uniquely represented by a function $u$ up to a positive linear transformation, i.e., if $u : \mathcal{Z} \to \mathbb{R}$ is such that*

$$p \succ q \Leftrightarrow \sum_{z \in \mathcal{Z}} p(z)u(z) > \sum_{z \in \mathcal{Z}} q(z)u(z),$$

*then the same holds with $u$ being replaced by $v(.) = au(.) + b$ where $a > 0$, $b \in \mathbb{R}$.*

To prove this result, we need two lemmas.

**Lemma 1.26** *If $\succ$ satisfies the two above axioms, then*

(a) *$p \succ q$ and $0 \le a < b \le 1$ imply $bp + (1 - b)q \succ ap + (1 - a)q$,*

(b) *$p \succeq q \succeq r$ and $p \succ r$ imply that there exists uniquely $a^* \in [0, 1]$ such that $q \approx a^*p + (1 - a^*)r$,*

(c) *$p \approx q$ and $a \in [0, 1]$ imply $ap + (1 - a)r \approx aq + (1 - a)r$, for any $r \in \mathcal{P}(\mathcal{Z})$.*

**Proof.** (a) Put $r = bp + (1 - b)q$. Then Axiom 1.24 implies that $r \succ bq + (1 - b)q = q$, so that (a) holds for $a = 0$. Assume now $a > 0$. Then we have

$$\begin{aligned} r &= (1 - a/b)r + (a/b)r \succ (1 - a/b)q + (a/b)r \\ &= (1 - a/b)q + (a/b)(bp + (1 - b)q) \\ &= ap + (1 - a)q. \end{aligned}$$

(b) If $p \approx q$, take $a^* = 1$. If $q \approx r$, take $a^* = 0$. For the case $p \succ q \succ r$, take

$$a^* = \sup\{a \in [0, 1] : q \succeq ap + (1 - a)r\}.$$

By Axiom 1.23, $bp + (1 - b)r \succ q \succ ap + (1 - a)r \succ r$ for some $a, b$ in $(0, 1)$. Thus, $0 < a^* < 1$.

If $1 \ge a > a^*$, then $ap + (1 - a)r \succ q$ by the definition of $a^*$. On the other hand, if $0 \le a < a^*$, then, by (a), we have

$$q \succeq a'p + (1 - a')r \succ ap + (1 - a)r \tag{1.8}$$

for $a < a' < a^*$.

Case 1: $a^*p + (1-a^*)r \succ q \succ r$. Axiom 1.23 implies that there exists $b \in (0,1)$ such that

$$
\begin{aligned}
ba^*p + (1 - ba^*)R &= b[a^*p + (1 - a^*0r] + (1-b)r \\
&\succ q \succ ba^*p + (1 - ba^*)r
\end{aligned}
$$

by Equation (1.8) since $ba^* < a^*$, a contradiction.

Case 2: $p \succ q \succ a^*p + (1-a^*)r$. Axiom 1.23 implies that there exists $b \in (0,1)$ such that

$$
\begin{aligned}
[1 - b(1-a^*)]p + b(1-a^*)r &= b[a^*p + (1-a^*)r] + (1-b)p \\
&\prec q \prec [1 - b(1-a^*)]p + b(1-a^*)r
\end{aligned}
$$

by Equation (1.8) since $1 - b(1-a^*) > a^*$, a contradiction. Thus, $q \approx a^*p + (1-a^*)r$. The uniqueness of $a^*$ follows from (a).

(c) If $p \approx s$ for any $s \in \mathcal{P}(\mathcal{Z})$, then the result is trivial. Suppose $p \prec s$ for some $s \in \mathcal{P}(\mathcal{Z})$ (the case where $p \succ s$ can be proved analogously). Assume on the contrary that for some $a \in (0,1)$ and some $r \in \mathcal{P}(\mathcal{Z})$,

$$ap + (1-a)r \succ aq + (1-a)r. \tag{1.9}$$

Since $s \succ p$, which implies that $s \succ q$, Axiom 1.24 implies that for any $b$,

$$bs + (1-b)q \succ bq + (1-b)q = q \approx p, \tag{1.10}$$

in particular, when $b = 1/2$,

$$(1/2)s + (1/2)q \succ q.$$

Axiom 1.24 implies

$$a[(1/2)s + (1/2)q] + (1-a)r \succ ap + (1-a)r. \tag{1.11}$$

Axiom 1.23 together with inequalities 1.9 and 1.11 imply that there exists $a^* \in (0,1)$ such that

$$
\begin{aligned}
ap + (1-a)r &\succ a^*[(a/2)s + (a/2)q + (1-a)r] + (1-a^*)[aq + (1-a)r] \\
&= a[(a^*/2)s + (1 - a^*/2)q] + (1-a)r \succ ap + (1-a)r
\end{aligned}
$$

by putting $b = a^*/2$, a contradiction. (The last result follows from Axiom 1.24 and Equation (1.10).) ∎

**Lemma 1.27** *If $\succ$ satisfies the above two axioms, then there exist $z^o$ and $z_o$ in $\mathcal{Z}$ such that $\delta_{z^o} \succeq p \succeq \delta_{z_o}$ for any $p \in \mathcal{P}(\mathcal{Z})$.*

**Proof.** If $\mathcal{Z} = \{z_1\}$ then $\mathcal{P}(\mathcal{Z}) = \{\delta_{z_1}\}$ and the result is trivial. If $\mathcal{Z} = \{z_1, z_2\}$ then $\mathcal{P}(\mathcal{Z}) = \{$convex combinations of $\delta_{z_1}$ and $\delta_{z_2}\}$. If $\delta_{z_1} \approx \delta_{z_2}$ then by (c) of Lemma 1.26, we have

$$a\delta_{z_1} + (1-a)\delta_{z_2} \approx a\delta_{z_2} + (1-a)\delta_{z_2} = \delta_{z_2}$$

implying that $p \approx \delta_{z_2}$ for any $p \in \mathcal{P}(\mathcal{Z})$ and the result follows.
If $\delta_{z_1} \prec \delta_{z_2}$ then

$$a\delta_{z_1} + (1-a)\delta_{z_2} \prec a\delta_{z_2} + (1-a)\delta_{z_2} = \delta_{z_2}$$

and hence $\delta_{z_1} \preceq p$ for any $p \in \mathcal{P}(\mathcal{Z})$.

Now let $\mathcal{Z} = \{z_1, z_2, ..., z_n, z_{n+1}\}$ and suppose we can find $x_o, x^o \in \{z_1, z_2, ..., z_n\}$ such that $\delta_{x_o} \preceq p_n \preceq \delta_{x^o}$ for any $p_n \in \mathcal{P}(\{z_1, z_2, ..., z_n\})$. Let

$$p = \sum_{i=1}^{n+1} a_i \delta_{z_i} \in \mathcal{P}(\mathcal{Z}).$$

Write

$$
\begin{aligned}
p &= (a_1 + \cdots + a_n)[(a_1/a_1 + \cdots + a_n)\delta_{z_1} + \cdots \\
&\quad + (a_n/a_1 + \cdots + a_n)\delta_{z_n}] + a_{n+1}\delta_{z_{n+1}} \\
&= (1 - a_{n+1})p_n + a_{n+1}\delta_{z_{n+1}}.
\end{aligned}
$$

Case 1: $\delta_{x^o} \preceq \delta_{z_{n+1}}$. Then $\delta_{x_o} \preceq p_n \preceq \delta_{x^o} \preceq \delta_{z_{n+1}}$ so that

$$
\begin{aligned}
\delta_{x_o} &= (1 - a_{n+1})\delta_{x_o} + a_{n+1}\delta_{x_o} \\
&\preceq (1 - a_{n+1})p_n + a_{n+1}\delta_{x_o} \\
&\preceq (1 - a_{n+1})p_n + a_{n+1}\delta_{z_{n+1}} \\
&= p \preceq (1 - a_{n+1})\delta_{z_{n+1}} + a_{n+1}\delta_{z_{n+1}} = \delta_{z_{n+1}},
\end{aligned}
$$

implying that, for any $p \in \mathcal{P}(\mathcal{Z})$,

$$\delta_{x_o} \preceq p \preceq \delta_{z_{n+1}}.$$

The following cases are treated similarly:

Case 2: $\delta_{x_o} \preceq \delta_{z_{n+1}} \preceq \delta_{x^o}$.

Case 3: $\delta_{z_{n+1}} \prec \delta_{x_o}$. ∎

Now we are ready to prove Theorem 1.25.

**Proof.** (a) Let $u : \mathcal{Z} \to \mathbb{R}$. Define $\succ$ on $\mathcal{P}(\mathcal{Z})$ by

$$p \succ q \text{ if and only if } U(p) > U(q)$$

where $U(p) = \sum_{z \in \mathcal{Z}} u(z)p(z)$. We are going to show that $\succ$ satisfies the above two axioms of continuity and independence.

Note that $U : \mathcal{P}(\mathcal{Z}) \to \mathbb{R}$ is an affine function, i.e., $U(ap + (1-a)q) = aU(p) + (1-a)U(q)$ for any $a \in [0,1]$ and $p, q \in \mathcal{P}(\mathcal{Z})$. To show the independence axiom, let $p, q, r$ in $\mathcal{P}(\mathcal{Z})$, $a \in (0,1)$ with $p \succ q$. We have

$$
\begin{aligned}
U(ap + (1-a)r) &= aU(p) + (1-a)U(r) \\
&> aU(q) + (1-a)U(r) \\
&= U(aq + (1-a)r)
\end{aligned}
$$

so that

$$ap + (1-a)r \succ aq + (1-a)r.$$

For the continuity axiom, suppose $p \succ q \succ r$. We have $U(p) > U(q) > U(r)$ and we can find, using the convexity of finite intervals in $\mathbb{R}$, $a$ and $b$ in $(0,1)$ such that

$$aU(p) + (1-a)U(r) > U(q)$$

and

$$bU(p) + (1-b)U(r) < U(q),$$

from which we can deduce that

$$ap + (1-a)r \succ q \succ bp + (1-b)r.$$

(b) Let $\succ$ be a binary relation on $\mathcal{P}(\mathcal{Z})$ satisfying the continuity and independence axioms. We are going to show that it can be represented by a utility function. Lemma 1.27 implies that there are $z_o, z^o \in \mathcal{Z}$ such that $\delta_{z_o} \preceq p \preceq \delta_{z^o}$ for any $p \in \mathcal{P}(\mathcal{Z})$.

Case 1: $\delta_{z_o} \approx \delta_{z^o}$. This implies that $p \approx q$ for any $p, q \in \mathcal{P}(\mathcal{Z})$. Take $u(.)$ to be any constant function, say, $u(.) = c$. Since $U(p) = c$ for any $p \in \mathcal{P}(\mathcal{Z})$ showing that

$$p \succ q \Longleftrightarrow U(p) > U(q) \tag{1.12}$$

holds. Moreover, any function $u(.)$ satisfying this equivalence relation must be a constant function. Indeed, for $\delta_z \in \mathcal{P}(\mathcal{Z})$, $z \in \mathcal{Z}$, since here

$\delta_z \approx \delta_{z'}$ for any $z, z'$, we have $u(z) = u(z')$ by Equation (1.12). Thus, $u(.)$ is constant. Finally, for each pair of constants $c, c'$, we can write $c' = ac + d$ where $a = 1$ and $d = c' - c$.

Case 2: $\delta_{z_o} \prec \delta_{z^o}$. For each $p \in \mathcal{P}(\mathcal{Z})$, define $f(p) = a$ where $a$ is the unique number (in c) of Lemma 1.26 in $[0, 1]$ such that $a\delta_{z_o} + (1-a)\delta_{z^o} \approx p$.

Thus, by (a) of Lemma 1.26,

$$f(p) > f(q) \Longleftrightarrow f(p)\delta_{z^o} + (1 - f(p))\delta_{z_o} \qquad (1.13)$$
$$\succ f(q)\delta_{z^o} + (1 - f(q))\delta_{z_o} \Longleftrightarrow p \succ q.$$

The function $f(.)$ so defined is affine. Indeed, by Equation (1.10) we have

$$ap + (1 - a)q$$
$$\approx a[f(p)\delta_{z^o} + (1 - f(p))\delta_{z_o}] + (1 - a)[f(q)\delta_{z^o} + (1 - f(q))\delta_{z_o}]$$
$$= [af(p) + (1 - a)f(q)]\delta_{z^o} + [a(1 - f(p)) + (1 - f(q))]\delta_{z_o}$$
$$= [af(p) + (1 - a)f(q)]\delta_{z^o} + [1 - (af(p) + (1 - a)f(q))]\delta_{z_o},$$

so that, from the definition of $f(.)$, we have

$$f(ap + (1 - a)q) = af(p) + (1 - a)f(q).$$

Now define $u(z) = f(\delta_z)$ on $\mathcal{Z}$. We claim that

$$f(p) = \sum_{z \in \mathcal{Z}} u(z)p(z).$$

If $\mathcal{Z} = \{z\}$ then $\mathcal{P}(\mathcal{Z}) = \{\delta_z\}$ so that $f(\delta_z) = u(z) = u(z)\delta_z(z)$. If $\mathcal{Z} = \{z_1, z_2\}$ then for any $p \in \mathcal{P}(\mathcal{Z})$, $p = a\delta_{z_1} + (1 - a)\delta_{z_2}$ and thus

$$f(p) = af(\delta_{z_1}) + (1 - a)f(\delta_{z_2})$$
$$= au(z_1) + (1 - a)u(z_2) = u(z_1)p(z_1) + u(z_2)p(z_2).$$

Suppose the above holds for $\{z_1, z_2, ..., z_n\}$. Consider

$$\mathcal{Z} = \{z_1, z_2, ..., z_n, z_{n+1}\}.$$

Let $p = \sum_{i=1}^{n+1} a_i\delta_{z_i}$ be a convex combination of the $\delta'_{z_i}$s. Let

$$p_n = (a_1/(1 - a_{n+1}))\delta_{z_1} + \cdots + (a_n/(1 - a_{n+1}))\delta_{z_n}.$$

We can write

$$p = (1 - a_{n+1})p_n + a_{n+1}\delta_{z_{n+1}}.$$

By the induction hypothesis,

$$
\begin{aligned}
f(p) &= (1 - a_{n+1})f(p_n) + a_{n+1}f(\delta_{z_{n+1}}) \\
&= (1 - a_{n+1})[(a_1/1 - a_{n+1})f(\delta_{z_1}) + \cdots \\
&\quad + (a_n/1 - a_{n+1})f(\delta_{z_n})] + a_{n+1}f(\delta_{z_{n+1}}) \\
&= (1 - a_{n+1})[(a_1/1 - a_{n+1})f(\delta_{z_1}) + \cdots \\
&\quad + (a_n/1 - a_{n+1})f(\delta_{z_n})] + a_{n+1}u(z_{n+1}) \\
&= \sum_{i=1}^{n+1} a_i u(z_i) = \sum_{z \in \mathcal{Z}} u(z)p(z).
\end{aligned}
$$

The uniqueness of $u(.)$ can be seen as follows. Suppose both $u(.)$ and $v(.)$ represent $\succ$. We will show that $v(.) = cu(.) + d$ for some constants $c, d$ with $c > 0$. As before, let $U(p) = \sum_{z \in \mathcal{Z}} u(z)p(z)$ and similarly, $V(p) = \sum_{z \in \mathcal{Z}} v(z)p(z)$. Define $f(.)$ and $g(.)$ on $\mathcal{P}(\mathcal{Z})$ as follows:

$$
\begin{aligned}
f(p) &= [U(p) - U(\delta_{z_o})]/[U(\delta_{z^o}) - U(\delta_{z_o})] \\
g(p) &= [V(p) - V(\delta_{z_o})]/[V(\delta_{z^o}) - V(\delta_{z_o})].
\end{aligned}
$$

Since there exists $a \in [0,1]$ such that $p \approx a\delta_{z_o} + (1-a)\delta_{z^o}$ for any $p \in \mathcal{P}(\mathcal{Z})$, we have

$$
f(p) = [aU(\delta_{z^o}) + (1-a)U(\delta_{z_o}) - U(\delta_{z_o})]/[U(\delta_{z^o}) - U(\delta_{z_o})] = a = g(p),
$$

i.e., $f(.) = g(.)$. Thus,

$$
V(p) = f(p)[V(\delta_{z^o}) - V(\delta_{z_o})] + V(\delta_{z_o}) = cU(p) + d
$$

where

$$
c = [V(\delta_{z^o}) - V(\delta_{z_o})]/[U(\delta_{z^o}) - U(\delta_{z_o})] > 0,
$$

and

$$
d = -[[V(\delta_{z^o}) - V(\delta_{z_o})]/[U(\delta_{z^o}) - U(\delta_{z_o})] + V(\delta_{z_o}).
$$

Let $z' \in \mathcal{Z}$. Then we have

$$
\begin{aligned}
v(z') &= v(z')\delta_{z'}(z') - \sum_{z \in \mathcal{Z}} v(z)\delta_{z'}(z) \\
&= V(\delta_{z'}) = cU(\delta_{z'}) + d \\
&= c\sum_{z \in \mathcal{Z}} u(z)\delta_{z'}(z) + d = cu(z') + d
\end{aligned}
$$

so that $v(.) = cu(.) + d$. ∎

We turn now to choice among more general and realistic random quantities of interest such as possible losses in investment portfolios. They are real-valued random variables. The range space $\mathcal{Z}$ is the real line $\mathbb{R}$. More generally, we could consider *random elements*, i.e., random "variables" taking values in arbitrary spaces $\mathcal{Z}$, possibly infinite dimensional such as function spaces. A general structure of $\mathcal{Z}$ as a *complete, separable metric space* is sufficient for applications. In that setting, $\sigma(\mathcal{Z})$ is the Borel $\sigma$-field of $\mathcal{Z}$. Unless $\mathcal{Z}$ is a subset of some Euclidean space $\mathbb{R}^n$, where we could base our analysis on distribution functions, we have to deal directly with probability measures on $\sigma(\mathcal{Z})$.

As a generalization of the finite case, the numerical representation of a preference relation $\succ$ on a class of random variables $\mathcal{X}$ is this. For each decision-maker with her own $\succ$, there exists a utility function (of her own, unique up to a positive linear transformation) $u : \mathcal{Z} \to \mathbb{R}$ such that

$$X, Y \in \mathcal{X}, \; X \succ Y \iff Eu(X) > Eu(Y),$$

provided that the expectations exist.

This is known as the *von Neumann-Morgenstern expected utility representation* of preference relations. When this is the case, decision-makers try to maximize their expected utilities, so that their behavior (in making choices under uncertainty) follows the so-called "expected utility maximization theorem."

The existence of utility functions (and hence of a numerical representation of $\succ$) is restricted to "rational" people. We will detail below what that means. Needless to say whether everybody is rational, in a sense to be specified, is open to debate! In fact, economists (and psychologists) have already raised concerns about the applicability of von Neumann-Morgenstern's model to all real-world problems. This is exemplified by:

(i) *Allais paradox* (1953): a choice problem showing the inconsistency of expected utility theory.

(ii) *Ellsberg paradox* (1961): people's choices in experimental economics violate the expected utility hypotheses, leading to a *generalized expected utility theory* based on *Choquet integral* (see Chapter 5 for background).

(iii) *Non-expected utility theory* (1979): from the work of two psychologists Daniel Kahneman and Amos Tversky leading to the so-called "Prospect Theory."

Interested students should read Fishburn [44] or Kreps [81] for more details.

Now we need to consider the case of *arbitrary probability measures,* e.g., on the real line $\mathbb{R}$, which are probability laws of real-valued random variables (e.g., investment returns or losses). While a general setting could be a separable, complete metric space (as the most general set-up for probability analysis, e.g., a Euclidean space), to be concrete and for stochastic dominance, we take $\mathcal{Z} = \mathbb{R}$.

It turns out that the extension of the expected utility representation for a preference relation $\succ$ on the sets $\mathcal{P}(\mathcal{Z})$ of all probability measures on $\mathcal{Z}$ requires some additional property on $\succ$ of topological nature, namely continuity. By $\mathcal{P}(\mathcal{Z})$ we really mean the set of all Borel probability measures, i.e., on $\mathcal{B}(\mathcal{Z})$, giving the obvious fact that $\mathbb{R} = \mathcal{Z}$ here is a topological space.

The setting of separable, complete metric spaces is general enough for most applications. In fact, if you recall the stochastic calculus in your course on financial economics, the space $C[0,1]$ of continuous real-valued functions, defined on $[0,1]$, with the sup-norm, is a separable and complete metric space (even of infinite dimensions) on which sample paths of Brownian motion live. The separability of the metric space $\mathcal{Z}$ is sufficient for $\mathcal{P}(\mathcal{Z})$ to be metrizable in a topology we are going to explore.

But first, here is the reason for all that! For the subset $\mathcal{S}(\mathcal{Z}) \subseteq \mathcal{P}(\mathcal{Z})$ of probability measures with finite support, a preference relation $\succ$ on it with continuity and independence axioms can be represented by expected utility. On $\mathcal{P}(\mathcal{Z})$, the continuity axiom needs to be strengthened.

We have called this the "continuity axiom," but what does it mean? Look at $\alpha p + (1 - \alpha)r$. If we let $\alpha \to 1$, we get $p$. Thus, for $\alpha$ sufficiently close to 1, we would like $\alpha p + (1 - \alpha)r \succ q$, and similarly for the other side. This is a form of "continuity" on the space of probability measures $\mathcal{S}(\mathcal{Z})$.

On the other hand, we can define the *weak continuity* of $\succ$ on any separable, complete metric space, like our $\mathcal{P}(\mathcal{Z})$, as follows. The following hold for any $p_n, p \in \mathcal{P}(\mathcal{Z})$ for which $p_n \to p$ in the weak topology.

1. If $p \succ q$ for some $q \in \mathcal{P}(\mathcal{Z})$, then $p_n \succ q$ for all sufficiently large $n$, and

2. If $q \succ p$, then $q \succ p_n$ for all sufficiently large $n$.

Note that the weak continuity of $\succ$ in the above sense is stronger than the intuitive idea of continuity expressed in the continuity axiom.

In fact, if $\succ$ is continuous in the above definition then $\mathcal{P}(\mathcal{Z})$ will have a countable $\succ$-order dense subset.

**Remark 1.28** *The continuity of $\succ$ on a separable, metric space, like $\mathcal{P}(\mathcal{Z})$, is equivalent to its continuity at every point $p \in \mathcal{P}(\mathcal{Z})$ in the sense that both $\{q \in \mathcal{P}(\mathcal{Z}) : p \succ q\}$ and $\{q \in \mathcal{P}(\mathcal{Z}) : q \succ p\}$ are open sets of $\mathcal{P}(\mathcal{Z})$.*

So what is the weak topology on $\mathcal{P}(\mathcal{Z})$? The general problem is this. We need to figure out conditions on decision-makers' preference relations on investments (identified as probability measures or distribution functions on $\mathbb{R}$) so that they can be represented by expected utilities, suitable for making decisions. These conditions will define the rationality of investors! On the road to achieve our economic goal, we need a topology for $\mathcal{P}(\mathcal{Z})$. Fortunately, a suitable topology for our purpose is available out there, created much earlier for other purposes, especially for large sample statistics of stochastic processes. For more details, see either [15] or [124].

Here it is. Let $\mathcal{Z} = \mathbb{R}$ (for concreteness) with its usual topology generated by open intervals. The Borel $\sigma$-field of $\mathbb{R}$ is denoted as $\mathcal{B}(\mathbb{R})$ (being the smallest $\sigma$-field containing the open sets of $\mathbb{R}$). The set of all probability measures defined on $\mathcal{B}(\mathbb{R})$ is denoted as $\mathcal{P}(\mathcal{Z})$ or $\mathcal{P}(\mathbb{R})$.

A topology on $\mathcal{P}(\mathbb{R})$ is defined by specifying a concept of convergence in it. We say that $p_n$ *converges weakly to* $p$ in $\mathcal{P}(\mathbb{R})$ if

$$\int_{\mathbb{R}} f(x)dp_n(x) \to \int_{\mathbb{R}} f(x)dp(x) \text{ for any } f \in C_b(\mathbb{R})$$

where $C_b(\mathbb{R})$ denotes the space of bounded and continuous real-valued functions defined on $\mathbb{R}$. The associated topology is generated by the basic neighborhoods of each $p \in \mathcal{P}(\mathbb{R})$ of the form

$$\left\{ q \in \mathcal{P}(\mathbb{R}) : \left| \int f_i dq - \int f_i dp \right| < \varepsilon, \, i = 1, 2, ..., n \right\}.$$

This topology is called the *weak topology* of $\mathcal{P}(\mathbb{R})$.

**Remark 1.29** *There is a "stronger" topology on $\mathcal{P}(\mathbb{R})$ (in the sense that if $p_n \to p$ in this topology then it also converges in the weak topology, but not the converse). For $f \in C_b(\mathbb{R})$, we let*

$$\|f\| = \sup_{x \in \mathbb{R}} |f(x)|$$

*and for $p \in \mathcal{P}(\mathbb{R})$,*

$$\|p\| = \sup\left\{\left|\int_{\mathbb{R}} f(x)dp(x)\right| : \|f\| \leq 1\right\}.$$

*Define $p_n \to p$ when $\|p_n - p\| \to 0$. Let $\delta_{x_n}$ be the Dirac measure at the point $x_n \in \mathbb{R}$. Suppose $x_n \to x$ in $\mathbb{R}$, then $\|\delta_{x_n} - \delta_x\| = 2$, so that $\delta_{x_n}$ does not converge to $\delta_x$ in the "strong" topology, but obviously it converges in the weak topology.*

Now, $\mathbb{R}$ is separable, i.e., having a countable dense subset, namely the set of the rationals. As such, the weak topology on $\mathcal{P}(\mathbb{R})$ is metrizable, i.e., it is generated by a metric on $\mathcal{P}(\mathbb{R})$, namely by the Prohorov metric.

An important fact is this. As a subset of $\mathcal{P}(\mathcal{Z})$, $\mathcal{S}(\mathcal{Z})$ is dense in the weak topology. This is useful for extending the expected utility result for $\mathcal{S}(\mathcal{Z})$ to $\mathcal{P}(\mathcal{Z})$ when we have some appropriate continuity (in the weak topology) property of $\succ$. Specifically, if a preference relation $\succ$ on $\mathcal{P}(\mathbb{R})$ satisfies the independence axiom and is continuous in the weak topology, then $\succ$ can be represented by expected utility, in fact by a continuous and bounded

$$u : \mathbb{R} \to \mathbb{R}.$$

**Remark 1.30** *Clearly, as $\alpha \to 1$, $\alpha p + (1 - \alpha)r$ will converge in the weak topology to $p$, thus this concept of weak continuity for $\succ$ is stronger than the continuity axiom.*

In summary, what we have is this. A "rational" preference relation $\succ$ on a set of random variables $\mathcal{X}$, or equivalently, on the set of their probability measures (or equivalently on the set of their distribution functions, for random variables or vectors) can be represented by expected utility, i.e., there exists a utility function $u : \mathcal{Z} \to \mathbb{R}$ (unique up to a positive affine transformation) such that for, $X, Y \in \mathcal{X}$ with probability laws $p, q \in \mathcal{P}(\mathcal{Z})$, respectively,

$$p \succ q \Longleftrightarrow Eu(X) = \int_{\mathcal{Z}} u(z)dp(z) > Eu(Y) = \int_{\mathcal{Z}} u(z)dq(z)$$

provided, of course, that $E|u(X)| < \infty$, and $E|u(Y)| < \infty$. We also write $E_p(u)$ for $\int_{\mathcal{Z}} u(z)dp(z)$. This is another realization of the von Neumann-Morgenstern expected utility representation of preference relations.

We now provide proofs of the above assertions. First, here are a few observations.

(i) The *weak continuity* of $\succ$ on $\mathcal{P}(\mathbb{R})$ implies the *continuity axiom* of $\succ$. Indeed, let $p, q, r$ in $\mathcal{P}(\mathbb{R})$ such that $p \succ q \succ r$. Let $p_a = ap + (1 - a)r$ for $a \in (0, 1)$. For any $f \in C_b(\mathbb{R})$, $\int_{\mathbb{R}} f(x) dp_a(x)$ converges to $\int_{\mathbb{R}} f(x) dp(x)$ or $\int_{\mathbb{R}} f(x) dr(x)$ when $a \to 1$ or $a \to 0$, respectively, i.e., $p_a$ converges weakly to $p$ or $r$ when $a \to 1$ or $a \to 0$, respectively. Thus, in view of the weak continuity of $\succ$, there exist $a, b \in (0, 1)$ such that $p_a \succ q$ and $p_b \prec q$, i.e., $ap + (1 - a)r \succ q \succ bp(1 - b)r$.

(ii) We leave as an exercise for students to show that if $z_n \to z$ in $\mathbb{R}$ then $\delta_{z_n} \to \delta_z$ weakly.

(iii) The space $\mathcal{S}(\mathbb{R})$ of probability measures on $\mathcal{B}(\mathbb{R})$ with finite support is dense in $\mathcal{P}(\mathbb{R})$ in the weak topology, i.e., for each $p \in \mathcal{P}(\mathbb{R})$, there exists a sequence $\{p_n\}$ of elements of $\mathcal{S}(\mathbb{R})$ such that $p_n \to p$ weakly as $n \to \infty$. For a proof, see, e.g., Billingsley [15, 16].

(iv) Let $\succ$ be a preference relation on $\mathcal{P}(\mathbb{R})$ satisfying the independence and the weak continuity axioms. If $q \prec p$ in $\mathcal{P}(\mathbb{R})$ then there exists $r \in \mathcal{S}(\mathbb{R})$ such that $q \prec r \prec p$. Indeed, choose $p^* \in \mathcal{P}(\mathbb{R})$ such that $q \prec p^* \prec p$, then by denseness of $\mathcal{S}(\mathbb{R})$, there exists $\{p_n\} \subseteq \mathcal{S}(\mathbb{R})$ converging weakly to $p^*$. By the weak continuity of $\succ$, we have $q \prec p_n \prec p$ for all large $n$.

**Theorem 1.31** *A preference relation $\succ$ on $\mathcal{S}(\mathbb{R})$ satisfies the independence and the weak continuity axioms if and only if there exists a bounded and continuous function $u : \mathbb{R} \to \mathbb{R}$ such that*

$$p \succ q \iff E_p(u) > E_q(u). \tag{1.14}$$

*Moreover, this representation is unique up to positive affine transformations.*

**Proof.** Suppose $u$ is a bounded and continuous function satisfying (1.14). As in the finite case, this implies that $\succ$ satisfies the independence axiom. Next, let $p_n \to p$ weakly. Then $E_{P_n}(u) \to E_p(u)$. If $p \succ q$ then $E_{p_n}(u) > E_q(u)$ for all large $n$, implying that $p_n \succ q$ for all large $n$. If $p \prec q$, then $p_n \prec q$ for all large $n$.

Conversely, suppose the preference relation $\succ$ satisfies the independence and weak continuity axioms. Let $\{r_1, r_2, ...\}$ be an enumeration of the rationals $\mathbb{Q}$. Let $\mathcal{P}_n = \mathcal{P}(\{r_1, r_2, ..., r_n\})$. The representation theorem in the finite case implies that there exists $u_n : \{r_1, r_2, ..., r_n\} \to \mathbb{R}$

representing $\succ$ on $\mathcal{P}_n$. For $n > m$, the restriction of $u_n$ to $\{r_1, r_2, ..., r_m\}$ is of the form $cu_m + d$ for some $c, d$ with $c > 0$. Put $u'_n = (1/c)u_n - d/c$, we see that $u'_n$ represents $\succ$ on $\mathcal{P}_n$ as well since it is a positive affine transformation of $u_n$. Moreover $u'_n|\{r_1, r_2, ..., r_m\} = u_m$, this shows that we can find a sequence $\{u_n\}$ such that for each $n$, $u_n : \{r_1, r_2, ..., r_n\} \to \mathbb{R}$ representing $\succ$ on $\mathcal{P}_n$ and

$$u_n|\{r_1, r_2, ..., r_m\} = u_m$$

for $n > m$. In view of this property of the sequence of functions $u_n$, the following function $u$ on $\mathbb{Q}$ is well-defined:

$$u(r_n) = u_n(r_n), \text{ for each } n.$$

Note the following:

(i) $u$ represents $\succ$ on $\mathcal{S}(\mathbb{Q})$. Indeed, $p \succ q$ in $\mathcal{S}(\mathbb{Q}) \iff$ there exists $n$ such that $p \succ q$ in $\mathcal{P}_n \iff$ there exists $n$ such that $E_p(u_n) > E_q(u_n) \iff E_p(u) > E_q(u)$.

(ii) $u$ is continuous on $\mathbb{Q}$, i.e., for each $r_0 \in \mathbb{Q}$,

$$u(r_0) = \liminf_{r \in \mathbb{Q}, r \to r_0} u(r) = \limsup_{r \in \mathbb{Q}, r \to r_0} u(r).$$

Indeed, suppose $u(r_0) > \liminf_{r \in \mathbb{Q}, r \to r_0} u(r)$. Choose a sequence $\{r_n\} \subseteq \mathbb{Q}$ with limit $r_0$ and $\lim u(r_n) = \liminf_{r \in \mathbb{Q}, r \to r_0} u(r)$. We may assume that for some $\varepsilon > 0$, $u(r_n) < u(r_0) - 2\varepsilon$ for all $n$. Choose $a$ sufficiently close to 0 so that

$$(1 - a)u(r_0) \geq u(r_0) - \varepsilon.$$

Since $au(r_1) + (1-a)u(r_0) < u(r_0) - 2a\varepsilon$, we have $a\delta_{r_1} + (1-a)\delta_{r_0} \prec \delta_{r_0}$. As $\delta_{r_n} \to \delta_{r_0}$ weakly, we have $a\delta_{r_1} + (1-a)\delta_{r_0} \prec \delta_{r_n}$ for all large $n$. Since $u$ represents $\succ$ on $\mathcal{S}(\mathbb{Q})$, we see that $E_{\delta_{r_n}}(u) > E_{p_a}(u)$ where $p_a = a\delta_{r_1} + (1-a)\delta_{r_0}$. But this implies

$$u(r_0) - 2\varepsilon > u(r_n) > au(r_1) + (1-a)u(r_0) \geq u(r_0) - \varepsilon,$$

which is impossible.

Now extend $u$ to $\mathbb{R}$ by setting $u_*$ to be the lower envelope of $u$, i.e.,

$$u_*(z) = \sup_{t>0} \inf\{u(r) : |r - z| < t, r \in \mathbb{Q}\}, \; z \in \mathbb{R}.$$

Since $u$ is continuous on $\mathbb{Q}$, $u_*$ is an extension of $u$.

Now $u_*$ represents $\succ$ on $\mathcal{S}(\mathbb{R})$. Indeed, let $p, q$ in $\mathcal{S}(\mathbb{R})$. Note that $A = \mathbb{Q} \cup S_p \cup S_q$ is countable, where $S_p$ denotes the support of $p$. Thus, there exists $v : A \to \mathbb{R}$ which represents $\succ$ on $\mathcal{S}(A)$ in the sense of Equation (1.14). The restriction of $v$ to $\mathbb{Q}$ is of the form $cu + d$ for some $c, d$ with $c > 0$. As before, writing $v' = (1/c)v - d/c$ we see that $v'$ represents $\succ$ on $\mathcal{S}(A)$ and $v'|\mathbb{Q} = u$. Since $v'$ is continuous on $A$, it must be the case that $v' = u_*$ on $A$. Therefore, $u_*$ represents $\succ$ on $\mathcal{S}(A)$, i.e.,

$$p \succ p \Longleftrightarrow E_p(u_*) > E_q(u_*).$$

Consequently, $u_*$ represents $\succ$ on $\mathcal{S}(\mathbb{R})$.

It remains to show that $u_*$ is bounded. By translation, assume $u_*(0) = 0$. Suppose on the contrary that for some strictly increasing sequence $\{z_n\} \subseteq \mathbb{R}$, $u_*(z_n) \geq 2^{n+1}$ for $n \geq 2$, $0 < u_*(z_1) \leq 2$. This can be done since $u_*$ is continuous and $u_*(0) = 0$.

Put $p_n = ((1 - 1/2^n)\delta_0 + (1/2^n)\delta_{z_n}$. For $f \in C_b(\mathbb{R})$,

$$\int_{\mathbb{R}} f(x)dp_n(x) = (1 - 1/2^n)f(0) + (1/2^n)f(z_n)$$
$$\to f(0) = \int_{\mathbb{R}} f(x)d\delta_o(x),$$

since $f$ is bounded. Now $p_1 \succ \delta_o$ as

$$E_{p_1}(u_*) = (1/2)u_*(0) + (1/2)u_*(z_1)$$
$$> 0 = E_{\delta_o}(u_*),$$

and $p_n \succ p_1$ because for any $n \geq 2$,

$$E_{p_n}(u_*) = (1 - 1/2^n)u_*(0) + (1/2^n)u_*(z_n)$$
$$\geq 2 > 1 \geq (1/2)u_*(z_1) = E_{p_1}(u_*).$$

We obtain a contradiction to the weak continuity axiom. $\blacksquare$

**Corollary 1.32** *A preference relation $\succ$ on $\mathcal{P}(\mathbb{R})$ satisfies the independence and weak continuity axioms if and only if there exists a bounded and continuous (utility) function $u : \mathbb{R} \to \mathbb{R}$ such that, for $p, q$ in $\mathcal{P}(\mathbb{R})$,*

$$p \succ q \Longleftrightarrow E_p(u) > E_q(u).$$

**Proof.** Let $p \succ q$ in $\mathcal{P}(\mathbb{R})$. Choose $q_o, p_o$ in $\mathcal{S}(\mathbb{R})$ such that $q \prec q_o \prec p_o \prec p$. Choose $\{q_n\} \subseteq \mathcal{S}(\mathbb{R})$ such that $q_n \to q$ weakly. We have

$$E_q(u) = \lim E_{q_n}(u), E_{q_n}(u) < E_{q_o}(u)$$

for all large $n$. Hence $E_q(u) \leq E_{q_o}(u)$. Note that $E_{q_o}(u) < E_{p_o}(u)$. Similarly, $E_{p_o}(u) \leq E_p(u)$. Therefore, $E_q(u) < E_p(u)$.

Conversely, suppose $E_q(u) < E_p(u)$. Choose sequences $\{q_n\}, \{p_n\}$ in $\mathcal{S}(\mathbb{R})$ such that $q_n \to q$, $p_n \to p$ weakly. For all large $n$, we have $E_{q_n}(u) < E_{p_n}(u)$. Therefore, for those $n$, $q_n \prec p_n$, implying that $q \prec p$.
∎

## 1.4   Utilities and risk attitudes

The buzz word in decision-making, especially in investment, is risk! In the finance context, risk refers to the possibility of a loss in an alternative (e.g., an investment). Like the concept of chance before a quantitative theory of probability was established, the concept of risk is qualitative in nature. Moreover, it is subjective. "How risky a risky situation is" is a matter of subjective judgments of individuals. A qualitative "measure" of risk comes from perception which is based mainly on the magnitude of a loss as well as its probability of occurrence. Risk occurs when uncertainty is present.

We observe that individuals face risk with different attitudes. In other words, decision-makers act according to their risk attitudes.

We have considered utility functions as a means to model preference relations. It turns out that the same concept of utility functions can also be used to model *risk preferences* of individuals, as we will detail next.

### 1.4.1   Qualitative representations of risk attitudes

Let's address first the following question "why do people react differently to risky situations?" Clearly, this is an important question since people do make decisions based upon their own attitudes towards risk, and as such, decision rules to be developed should take risk attitudes of individuals into account. Do not forget what exactly we are trying to do: in the context of investments (in fact for all types of areas where decisions need to be made), not only do we try to model and predict decision-makers' behavior (just like in Game Theory), we seek also to develop tools to advise how to make decisions. For that to be realistic, i.e., applicable to real-world problems, we need to understand any crucial factors in human decision processes so that we can incorporate them into our machinery. Clearly, risk attitudes are among these crucial factors. For example, experiments indicate that people act differently when facing a decision on how to split their investment between a risky lottery and an asset with a fixed return.

Of course, as a rough coarsening of the spectrum of risk, we can say: there are essentially three types of risk attitudes (of people), namely risk neutral, risk averse and risk seeking, as we observe the behavior of gamblers in casinos!

Roughly speaking, an economic agent (or an ordinary person) is:

(i) *risk neutral* if, when facing two risky prospects with the same ex- pected value, will feel indifferent,

(ii) *risk averse* if, when facing two risky prospects with the same ex- pected value, will prefer the less risky one, and

(iii) *risk seeking* if, when facing two risky prospects with the same ex- pected value, will prefer the riskier one.

A decision-maker is risk neutral if she cares only about her expected gains or losses. She is risk averse if, confronted with two choices with the same expected value, she would prefer the smaller and more certain of the options.

For example, suppose if you are asked to choose between getting $100 for sure, and play the game: "Flip a coin. If heads comes up, you get $200, if tails comes up, you get nothing." If you say: "I don't care," then you are risk neutral. If you say you prefer to get $100 for sure, you are risk averse; whereas if you prefer to play the game, you are risk seeking.

More specifically, in contrast to risk neutral people, a risk averse person cares not only about the expected value of losses, but also about the possible magnitude of losses. For example, a risk averse person will find a situation involving a 5% chance of losing $20,000 worse than that involving a 10% chance of losing $10,000, which in turn is worse than that involving a sure loss of $1,000 (note that each situation has the same expected loss of $1,000). A risk neutral would not find any of these situations worse than any other. In other words, risk averse people dislike uncertainty about the size of losses per se.

Now, suppose people make decisions based upon their own utility functions. The above risk attitudes are related to the utility they attach to their wealth. While one's utility increases with the level of her wealth, it does so at a decreasing rate (i.e., if the utility function $u$ is assumed to be twice differentiable, the rate is $u'$, and "decreasing rate" means $u'' < 0$, so that $u$ is a concave function).

Thus, if the utility of wealth has the shape of a concave function, like $u(x) = \log x$, this person will dislike bearing the risk of large losses, noting that each person evaluates a risky prospect by measuring its effect on her expected utility.

More formally, let the random variable $X$ be a risky prospect, and $u$ is the utility function of a decision-maker. When she has to choose between the mean $EX$ and $Eu(X)$, she is

(i) Risk neutral when her utility function $u$ is such that $u(EX) = Eu(X)$, for example $u(x) = x$ (linear),

(ii) Risk averse when $u(EX) > Eu(X)$, for example $u(x) = \log x$ (concave), and

(iii) Risk seeking when $u(EX) < Eu(X)$, for example $u(x) = x^2$ (convex).

A statement like (ii) means she prefers the mean of the prospect with certainty than the prospect itself.

### Note on risk reduction

One way to reduce risk in, say, portfolio selection is *diversification*, according to the old saying "don't put all your eggs in one basket," meaning that we should allocate our resources to different activities whose outcomes are not closely related.

We now proceed to show the connection between risk attitudes of individuals and the shapes of their utility functions. For simplicity, we consider utility functions $u : R \to R$. We can characterize risk attitudes in terms of lotteries.

**(i) Risk neutral**   In the finance context, as investors obviously prefer "more money than less" (!), so they assign higher utility values to higher money amounts than smaller money amounts. And as utility functions are functions of money, it seems reasonable to postulate that in general, i.e., without specific type of risk attitude, utility functions are *nondecreasing*. In fact, this follows from at least two reasons:

*Where do utility functions come from?* They come from decision-makers' preference relations. Recall that, making decisions is making choices among alternatives. Decision-makers have "something" on their mind when ranking alternatives for selections. That "something" is what we call "utility." The theory of utility consists of showing that, for rational decision-makers, their preference relations can be expressed in terms of utility functions. And hence, utility functions are the building blocks of decision theory.

*Why nondecreasing?* This qualitative characteristics of utility functions is *context dependent*. As we talk about money in finance, we "prefer more than less!" and hence we assign higher utility values on larger amounts of money (the "attribute" of interest), so that utility functions should be nondecreasing.

Thus, for all rational decision-makers, without further information about their risk attitudes, i.e., considered as "neutral," we assign nondecreasing utility functions to them, and use such utility functions to predict their behavior in the face of risk or uncertainty.

Thus, we consider only nondecreasing utility functions in this work. Note however, in other contexts, utility functions could be different, such as "nonincreasing," e.g., if we have to choose response times when calling for ambulance service, then clearly we prefer shorter times, and hence utility functions in this case are nonincreasing.

**(ii) Risk averse** Roughly speaking, a risk-averse person is a person who prefers to behave conservatively.

*How can we quantify such a statement?* In other words, how can we find out the risk attitude of a person? The actual device for this purpose is *lotteries*, a concept familiar to all, even for non-gamblers! A lottery is just a random variable with *known* probability distribution. Gambling is a problem of *decision-making under risk*.

To make this more specific, consider the situation where a person is facing a lottery with two outcomes $x$ and $y$, with equal probability. To find out her attitude toward risk, you ask her whether she will play this lottery (where she has to pay $(x + y)/2$ dollars) or not. If she does not play this lottery, then she will "receive" the amount $(x + y)/2$ for certain (it's her money!), whereas if she plays the lottery, she could lose money if the outcome is $x$ (assuming that $x < y$) or she could gain money if the outcome is $y$ (since $(x + y)/2 < \max(x, y)$). The price to pay for getting more money is risk!

To put it differently, you ask her to state her preference between receiving the expected consequence $(x + y)/2$ for certain and the lottery $((x, 1/2), (y, 1/2))$. If she prefers $(x + y)/2$ for certain to the lottery, then her attitude is to avoid the risks associated with the lottery. This type of attitude is called risk aversion.

We can formulate this risk attitude as follows.

**Definition 1.33** *A decision-maker is **risk averse** if she prefers the ex-*

*pected consequence of any nondegenerate lottery to that lottery. Specifi-
cally, a decision-maker is risk averse if her utility function u is such that
$u(EX) > E(u(X))$ for any nondegenerate random variable X for which
$E(|X|) < \infty$.*

Note that a nondegenerate lottery is one where no single outcome
has a probability one of occurring. A lottery is a random variable. As
such, in general, it can be a continuous type random variable, although in
practice, a lottery has a finite, or countable number of possible outcomes.

**(iii) Risk seeking**   This is the opposite of the risk aversion concept,
i.e., when a decision-maker does not behave conservatively. This reflects
the fact that the decision-maker is more than willing to accept the risks
associated with lotteries. More formally,

**Definition 1.34**   *A decision-maker is* **risk seeking** *if she prefers any
nondegenerate lottery to the expected consequence of that lottery. Specifically, a decision-maker is risk seeking if her utility function u is such
that $u(EX) < E(u(X))$ for any nondegenerate random variable for which
$E(|X|) < \infty$.*

Now we need to translate the above definitions into qualitative properties of associated utility functions for decision analysis. Such transla-
tions provide classifications of decision-makers in terms of their utility
functions.

**Theorem 1.35**   *A decision-maker is risk averse if and only if her (nondecreasing) utility function is concave.*

**Proof.**   *Necessity*: Let $u : \mathbb{R} \to \mathbb{R}$ be the utility function of a risk averse
person. Consider a lottery X with $P(X = x) = p = 1 - P(X = y)$, with
$x \neq y$, and $0 < p < 1$. By definition of risk aversion, this decision-maker
will assign a higher utility value on $EX$ than $E(u(X))$, i.e., $u(EX) >
E(u(X))$ so that $u(px + (1 - p)y) > pu(x) + (1 - p)u(y)$ which means
that u is (strictly) concave.
   *Sufficiency:* Let u be a (strictly) concave utility function. Let X
be a nondegenerate lottery such that $E(|X|) < \infty$. Then it follows by
Jensen's inequality (see Section 1.4.3) that $u(EX) > E(u(X))$. ∎

### 1.4.2  Notes on convex functions

Recall that concave and convex functions are dual in the sense that if $u$ is concave then $-u$ is convex. Thus, technically, it suffices to look at, say, convex functions.

A real-valued function $u$, defined on an interval (convex subset of $\mathbb{R}$) $A$, is said to be *convex* if for any $x, y \in A$ and $\alpha \in [0, 1]$, we have

$$u(\alpha x + (1 - \alpha)y) \leq \alpha u(x) + (1 - \alpha)u(y)$$

The function $u$ is said to be *strictly convex* if the above inequality is strict whenever $x \neq y$.

Here are some useful facts about convex functions:

(i) By induction, $u$ is convex if and only if $u\left(\sum_{i=1}^{n} \alpha_i x_x\right) \leq \sum \alpha_i u(x_i)$ for any $x_1, ..., x_n \in A$ and $\alpha_i \geq 0$ such that $\sum_{i=1}^{n} \alpha_i = 1$.

(ii) A convex function $u : A \to \mathbb{R}$ is continuous in the interior $A^o$ of $A$.

(iii) A convex function $u : A \to \mathbb{R}$ has left and right derivatives (both nondecreasing) on $A^o$, denoted respectively as

$$D^-u(x) \quad = \quad \lim_{y \nearrow x} [u(y) - u(x)]/(y - x)$$
$$D^+u(x) \quad = \quad \lim_{y \searrow x} [u(y) - u(x)]/(y - x).$$

Moreover, if $u$ is strictly convex, then these derivatives are increasing.

(iv) Let $u$ be a diffrentiable function, defined on an open interval $A$ of $\mathbb{R}$. Then

    (a) $u$ is convex on $A$ if and only if its derivative $f'$ is nondecreasing on $A$, and

    (b) $u$ is strictly convex on $A$ if and only if $f'$ is increasing on $A$.

We need the following lemma to prove Jensen's inequality.

**Lemma 1.36** *Let $u$ be convex on $\mathbb{R}$. Then for each $x_0 \in \mathbb{R}$, there exists a number $\lambda(x_0)$ such that, for each $x \in \mathbb{R}$,*

$$u(x) \geq u(x_0) + (x - x_0)\lambda(x_0).$$

*If $u$ is strictly convex, then this inequality is strict for $x \neq x_0$.*

**Proof.** By fact (iii) above, $u$ has left and right derivatives on $\mathbb{R}$. If $x < x_0$ then

$$[u(x_0) - u(x)]/(x_0 - x) \leq D^- u(x_0).$$

Thus,

$$u(x) \geq u(x_0) + (x - x_0)D^- u(x_0).$$

Similarly, if $x > x_0$, then

$$[u(x) - u(x_0)]/(x - x_0) \geq D^+ u(x_0),$$

so that

$$u(x) \geq u(x_0) + (x - x_0)D^+ u(x_0).$$

But $D^- u(x_0) \leq D^+ u(x_0)$, so we have that

$$u(x) \geq u(x_0) + (x - x_0)D^+ u(x_0)$$

for all $x$.

Suppose that $u$ is strictly convex, then for $h > 0$,

$$\varphi(h) = \frac{u(x + h) - u(x)}{h}$$

is (strictly) increasing, and hence, for $x \neq x_0$, we have

$$u(x) > u(x_0) + (x - x_0)D^+ u(x_0).$$

∎

### 1.4.3 Jensen's inequality

Let $u$ be a convex function on an open interval $A$. Let $X$ be a random variable taking values almost surely in $A$ with $E(|X|) < \infty$. Then

$$E(u(X)) \geq u(EX).$$

Furthermore, this inequality is strict if $u$ is strictly convex and $X$ is nondegenerate.

Taking $x_0 = EX$ in the above lemma, we get

$$u(x) \geq u(x_0) + (x - x_0)\lambda(x_0) \text{for all } x \in A.$$

Then

$$E(u(X)) \geq u(x_0) + E(X - x_0)\lambda(x_0) = u(x_0) = u(EX).$$

If $u$ is strictly convex, then

$$u(x) > u(x_0) + (x - x_0)\lambda(x_0)$$

for all $x \in A$ with $x \neq x_0$ and hence $E(u(X)) > u(x_0)$ since $P(X = x_0) < 1$.

The proof of the following theorem is obvious by duality.

**Theorem 1.37** *A decision-maker is risk seeking (or is a risk lover) if and only if her (nondecreasing) utility function is convex.*

If $u$ is convex, then, for any $x, y$ in its domain, we have

$$u\left(\frac{x+y}{2}\right) \leq \frac{u(x) + u(y)}{2}. \tag{1.15}$$

This condition (1.15) was used by Jensen [74] to investigate convex functions. Jensen made major contributions to the foundations of the theory of convex functions.

Note that (1.15) is not equivalent to the convexity of $u$ since there exist discontinuous functions, on an open interval, that satisfy (1.15), recalling that a convex function defined on an open interval must be continuous on it. However, (1.15) is just a little weaker than convexity. Specifically, by induction, $u$ satisfies (1.15) if and only if

$$u\left(\sum_{i=1}^{n} \alpha_i x_i\right) \leq \sum \alpha_i u(x_i) \tag{1.16}$$

for any $x_1, ..., x_n \in A$ and $\alpha_i \geq 0$, *rationals*, such that $\sum_{i=1}^{n} \alpha_i = 1$.

From that we can see that if $u$ is continuous and satisfies (1.15) then it must be convex, since continuity will force (1.16) to be true also with all convex combinations. A sufficient condition for a function $u$ satisfying (1.15) to be continuous is that it is bounded from above on a neighborhood of just one point in its domain.

The point is this: for continuous functions, (1.15) is equivalent to convexity.

## 1.5   Exercises

1. Let $\succeq$ be a binary relation on a finite set $\mathcal{Z}$. Show that if $\succeq$ is represented by a utility function, i.e., there is a function

$$u : \mathcal{Z} \to \mathbb{R}$$

such that for all $x, y \in \mathcal{Z}$, we have $x \succeq y \iff u(x) \geq u(y)$, then $\succeq$ must be complete and transitive.

2. Let $\mathcal{Z} = \{z_1, z_2, ..., z_n\}$ be a finite set, and $\mathcal{S}(\mathcal{Z})$ be the set of all lotteries with prizes in $\mathcal{Z}$, i.e., $\mathcal{S}(\mathcal{Z})$ is the set of functions $p : \mathcal{Z} \to [0, 1]$, such that $\sum_{i=1}^{n} p(z_i) = 1$.

   A utility function $U : \mathcal{S}(\mathcal{Z}) \to \mathbb{R}$ is said to have an expected utility form if there exists a function $u : \mathcal{Z} \to \mathbb{R}$ such that

   $$U(p) = \sum_{i=1}^{n} u(z_i)p(z_i).$$

   The function $U$ is said to be linear on $\mathcal{S}(\mathcal{Z})$ if for any $p_i \in \mathcal{S}(\mathcal{Z})$, $\alpha_i \in [0, 1]$ with $\sum_{i=1}^{n} \alpha_i = 1$, we have

   $$U\left(\sum_{i=1}^{n} \alpha_i p_i\right) = \sum_{i=1}^{n} \alpha_i U(p_i).$$

   Show that a utility function $U : \mathcal{S}(\mathcal{Z}) \to \mathbb{R}$ has an expected utility form if and only if it is linear.

3. Consider the setting of Exercise 2.

   (a) Verify that if $U$ has an expected utility form via $u$, then it has also an expected utility form via the function $v(.) = au(.) + b$ where $a > 0$ and $b \in \mathbb{R}$ (such a transformation of $u(.)$ is called a positive linear (or affine) transformation).

   (b) If we transform $u$ by a monotone transformation (like in the case of consumer choice theory), for example, $w(.) = u^2(.)$, can $U$ still have an expected utility form via $w$?

   To answer this, examine the following example, with $u(z) = \sqrt{z}$. Let $\mathcal{Z} = \{0, 240, 600, 1000\}$ and $p, q \in \mathcal{S}(\mathcal{Z})$ with

   $$p(0) = 0.5, \; p(240) = p(600) = 0, \; p(1000) = 0.5,$$
   $$q(0) = q(1000) = 0, \; q(240) = 1/3, \; q(600) = 2/3.$$

4. Show directly that if the utility function $U : \mathcal{S}(\mathcal{Z}) \to \mathbb{R}$ has an expected utility form, then the preference relation on $\mathcal{S}(\mathcal{Z})$ which represents it must satisfy the independence axiom.

5. Show that if $\int_A |X(\omega)|dP(\omega) < \infty$, then $P(\omega \in A : |X(\omega)| = \infty) = 0$.

6. Show that if $P(A) = 0$, then $\int_A X(\omega)dP(\omega) = 0$.

7. (Fatou's Lemma). If $\{X_n, n \geq 1\}$ is a sequence of nonnegative random variables such that $\liminf_{n\to\infty} X_n = X$ almost surely (a.s.), then

$$\int X(\omega)dP(\omega) \leq \lim_{n\to\infty} \inf \int X_n(\omega)dP(\omega).$$

8. Two balls are drawn, without replacement, from a box containing $n$ balls, labeled from 1 to $n$. Let $X, Y$ denote the number shown on the first and second ball, respectively. Compute $E(XY)$.

9. Let $P$ be a probability measure on $\mathcal{B}(\mathbb{R})$. Let $F : \mathbb{R} \to [0, 1]$ be defined as $F(x) = P((-\infty, x])$. Show that $F$ is a distribution function.

10. A class $\mathcal{K}$ of subsets of a set $\Omega$ is called a *compact class* if every sequence $\{K_n, n \geq 1\}$, $K_n \in \mathcal{K}$, such that $\cap_{n\geq 1} K_n = \varnothing$ implies the existence of an integer $N$ such that $\cap_{n=1}^{N} K_n = \varnothing$.

    Let $\mathcal{K}$ be a compact class of a field $\mathcal{A}$ of subsets of a set $\Omega$. Let $P : \mathcal{A} \to [0, 1]$ with $P(\Omega) = 1$, be an additive map. Show that, if

$$P(A) = \sup\{P(K) : K \in \mathcal{K}, K \subseteq A\}$$

    for any $A \in \mathcal{A}$, then $P$ is continuous at $\varnothing$ (and hence $\sigma$-additive).

11. Let $\succ$ be a preference relation on a set $\mathcal{Z}$. Show that

    (a) $z \succ x$ and $x \approx y$ imply $z \succ y$.
    (b) $z \succ x$ and $z \approx y$ imply $y \succ x$.

12. Let $\succ$ be a preference relation on $\mathcal{P}(\mathcal{Z})$ satisfying the continuity and independence axioms. Show that $q \approx q'$, $p \approx p'$ and $q \prec p$ imply $q' \prec p'$.

13. Let $\succ$ be a binary relation on a set $\mathcal{Z}$. Suppose $\succ$ is assumed to be asymmetric and transitive. Let $\approx$ be defined as $x \approx y$ if and only if "not $x \succ y$" and "not $y \succ x$." Construct an example to show that $\approx$ so defined need not be transitive.

14. Let $\succ$ be a preference relation on $\mathcal{P}(\mathbb{R})$ (space of all Borel probability measures on $\mathbb{R}$) satisfying the weak continuity and independence axioms. Construct a countable $\succ$-order dense subset of $\mathcal{P}(\mathbb{R})$.

15. Let $\succ$ be a preference relation on $\mathcal{P}(\mathbb{R})$ satisfying the weak continuity and independence axioms. Show that for $z_n$, $z$ in $\mathbb{R}$ and $p, q$ in $\mathcal{P}(\mathbb{R})$,

    (a) $\delta_{z_n} \Longrightarrow \delta_z$ (weak convergence) if $z_n \to z$, as $n \to \infty$.

    (b) $ap + (1-a)q \Longrightarrow p$ as $a \to 1$, and $ap + (1-a)q \Longrightarrow q$ as $a \to 0$.

16. Let $\succ$ be a preference relation on $\mathcal{P}(\mathbb{R})$ satisfying the weak continuity and independence axioms. If $q \prec p$ in $\mathcal{P}(\mathbb{R})$, show that $q \prec q_1 \prec p_1 \prec p$ for some $q_1, p_1$ in $\mathcal{P}(\mathbb{R})$ with finite support.

17. Let $\succ$ be a preference relation on $\mathcal{P}(\mathbb{R})$ satisfying the weak continuity and independence axioms. Show that if $p_n \Longrightarrow p$, $q_n \Longrightarrow q$ and $q_n \prec p_n$ for all $n$ then $q \preceq p$. Show by an example that $p \not\succ q$.

18. Let $\succ$ be a preference relation on $\mathcal{P}(\mathbb{R})$ satisfying the weak continuity and independence axioms. Show that the function $u(.)$ in the proof of Theorem 1.31 is continuous.

19. Let $\succ$ be a preference relation on $\mathcal{P}(\mathbb{R})$ satisfying the weak continuity and independence axioms. Show why, for any function $v$ representing $\succ$ with $v|\mathbb{Q} = u$, we must have $v = u_*$ (lower envelope of $u$).

20. Let $(U, \rho)$ be a metric space, and $\mathcal{P}(U)$ be the space of all probability measures on the Borel $\sigma$-field $\mathcal{B}(U)$ of $U$. For $P, Q$ in $\mathcal{P}(U)$, let $d(P, Q) = \inf\{\varepsilon > 0 : P(A) \leq Q(A^\varepsilon) + \varepsilon,\ A$ closed set of $U\}$, where
    $$A^\varepsilon = \{x \in U : \rho(x, A) < \varepsilon\}.$$

    (a) Verify that $d(\cdot, \cdot)$ is a metric on $\mathcal{P}(U)$. The metric is called the *Prohorov metric*.

    (b) Show that $d(P_n, P) \to 0$ if and only if $P_n$ converges weakly to $P$, when $U$ is separable.

# Chapter 2

# Foundations of Stochastic Dominance

*This chapter lays down the foundations of stochastic dominance rules from a rigorous and general setting. We choose to present the theory from first principles.*

## 2.1 Some preliminary mathematics

In Chapter 1 we obtained quantitative presentations for preference relations of individuals in terms of their utility functions. However, in practice, it is hard to specify utility functions of individuals. Thus, it seems natural to ask whether we could go a step further to represent the total order defined by utility functions by other stochastic orders which depend only on distributions of random variables involved, but not on utility functions. As we will see, this is possible only locally, i.e., when we consider a coarsening of risk attitudes of decision-makers and address the problem in each category of risk attitude. As explained in Chapter 1, each type of risk attitude is characterized by the shape of utility functions. Thus, the derivation of a quantitative representation of preference relations for an individual belonging to a given class of risk attitude depends on the shape of utility functions of individuals in that class. Moreover, the quantitative representations obtained lose the power of a total order, in other words, they are only partial orders. The practical aspect of this "finer" representation of preference relations is that, within each class of risk attitude, decisions could be based only on distributions of random variables (e.g., future returns on investment prospects) which, while unknown, can be estimated by "historical data." In Chapter 6, we will discuss reliable estimation based on such data.

In order to derive presentations of preferences, in terms of distribution functions, in each category of risk attitude, we need some results on approximations of appropriate functions as well as a lemma involving the Stieltjes integral.

## 2.1.1  Approximation of utility functions

In Chapter 1, we considered a coarsening of risk attitude of decision-makers consisting of the three main subclasses, namely risk neutral, risk averse and risk seeking. The utility functions in each class have similar shapes expressed as nondecreasing, nondecreasing and concave, and non-decreasing and convex, respectively. This subsection is a preparation for deriving representations, in terms of distribution functions, of preference relations.

### Approximation of nonnegative monotone functions

Let $u : \mathbb{R} \to \mathbb{R}^+$ be a nonnegative monotone function. As a measurable function, $u$ is approximated by the standard procedure in measure theory as follows. Let

$$
u_n(x) = \begin{cases} \dfrac{i}{2^n} & \text{if } \dfrac{i}{2^n} \leq u(x) < \dfrac{i+1}{2^n}, \ i = 0, 1, ..., n2^n - 1 \\[2mm] n & \text{if } u(x) \geq n. \end{cases}
$$

Then $u_n(x) \nearrow u(x)$, for each $x \in \mathbb{R}$, as $n \to \infty$.

For each $n \geq 1$, let

$$
A_i = \{x \in \mathbb{R} : u(x) \geq i/2^n\}.
$$

Clearly the $A_i$'s are nested,

$$
A_1 \supseteq A_2 \supseteq ... \supseteq A_{n2^n},
$$

with possible equality for some. Moreover, since $u$ is monotone, these $A_i$, $i = 1, 2, ..., n2^n$, are *intervals*.

We distinguish two cases.

a. *Nonnegative and nondecreasing functions*

   In this case, $A_i$ is of the form $(a_i, \infty)$. Thus we can write each $u_n$ as a positive linear combination of indicator functions of the form

$1_{(a_i,\infty)}$. Specifically, we have

$$u_n(x) = 1/2^n \sum_{i=1}^{n2^n} 1_{(a_i,\infty)}(x).$$

b. *Nonnegative and nonincreasing functions*

In this case, the intervals $A_i$ are of the form $(-\infty, b_i]$, and each $u_n$ is a positive linear combination of indicator functions of the form $1_{(-\infty,b_i]}$, namely

$$u_n(x) = 1/2^n \sum_{i=1}^{n2^n} 1_{(-\infty,b_i]}(x).$$

## Approximation of nondecreasing concave functions

Let $u : \mathbb{R} \to \mathbb{R}^+$ be a nonnegative concave function such that $u(0) = 0$. It is well known that there exists a nonnegative and nonincreasing function $g$ such that, for any $x \in \mathbb{R}$, $u(x) = \int_0^x g(t)dt$. To approximate $u$, we first approximate $g$ using the above (Section 2.1.1.b), replacing $u_n$ by $g_n$. Now observe that

$$\xi_b(x) = \begin{cases} \int_0^x 1_{(-\infty,b]}(t)dt = x \wedge b = (x-b)1_{(-\infty,b]}(x) + b & \text{if } b \geq 0 \\ (x-b)1_{(-\infty,b]}(x) & \text{if } b < 0. \end{cases}$$

Thus, each $u_n(x) = \int_0^x g_n(t)dt$ is a positive linear combination of nondecreasing concave functions of the forms

$$(x-b)1_{(-\infty,b]}(x) + b.$$

Moreover as $g_n \nearrow g$, we have $u_n \nearrow u$.

## Approximation of functions in class $\mathcal{U}_3$

We denote by $\mathcal{U}_3$ the class of functions $f : \mathbb{R} \to \mathbb{R}$ of the form $f(x) = \int_0^x h(y)dy$ where $h$ is nonnegative, nonincreasing and convex. Such functions are nondecreasing and concave. When such $f$ are differentiable, up to third order, they are characterized by $f' \geq 0$, $f'' \leq 0$, and $f''' \geq 0$. If we let $\mathcal{U}_2$ be the class of utility functions of risk averse individuals, i.e., the class of nondecreasing and concave functions, then $\mathcal{U}_3$ is a subclass of $\mathcal{U}_2$.

Now, $h$ can be written as

$$h(x) = \int_0^x p(y)dy + h(0),$$

where $p$ is nonpositive and nondecreasing.

Let $q(x) = -p(x)$ then $q$ is nonnegative and nonincreasing. Moreover, since $h(x)$ is nonnegative,

$$h(0) \geq -\int_0^x p(y)dy \qquad \text{i.e.,} \qquad h(0) \geq \int_0^x q(y)dy.$$

Let $A_i = \{x \in \mathbb{R} : q(x) \geq \frac{i}{2^n}\}$, for each $n \geq 1$. Since $q$ is nonincreasing, $A_i$ is of the form $(-\infty, b_i]$. Moreover, clearly the $A_i$'s are nested,

$$A_1 \supseteq A_2 \supseteq \cdots \supseteq A_{n2^n},$$

so $b_1 \geq b_2 \geq \cdots \geq b_{n2^n}$.

Then $q$ can be approximated by $1_{(-\infty, b_i]}$'s, namely,

$$q_n(x) = \frac{1}{2^n} \sum_{i=1}^{n2^n} 1_{(-\infty, b_i]}(x),$$

and $q_n(x) \nearrow q(x)$. Let $m$ be the index at which $b_m \geq 0$ and $b_{m+1} < 0$. Then we have

$$
\begin{aligned}
h(0) &\geq \int_0^{b_1} q(y)dy = \int_0^{b_m} q(y)dy + \int_{b_m}^{b_{m-1}} q(y)dy + \int_{b_{m-1}}^{b_{m-2}} q(y)dy + \cdots \\
&\quad + \int_{b_3}^{b_2} q(y)dy + \int_{b_2}^{b_1} q(y)dy \\
&\geq \frac{m}{2^n} b_m + \frac{m-1}{2^n}(b_{m-1} - b_m) + \frac{m-2}{2^n}(b_{m-2} - b_{m-1}) + \cdots \\
&\quad + \frac{2}{2^n}(b_2 - b_3) + \frac{1}{2^n}(b_1 - b_2) \\
&= \frac{1}{2^n}(b_m + b_{m-1} + \cdots + b_2 + b_1).
\end{aligned}
$$

Hence,

$$h(0) - \frac{1}{2^n} \sum_{i=1}^m b_i \geq 0.$$

Let

$$p_n(x) = -q_n(x) = -\frac{1}{2^n} \sum_{i=1}^{n2^n} 1_{(-\infty, b_i]}(x),$$

then $p_n(x) \searrow p(x)$. Let

$$h_n(x) = \int_0^x p_n(y)dy + h(0).$$

Then by the monotone convergence theorem,

$$h_n(x) \searrow h(x) \text{ for } x \geq 0 \text{ and } h_n(x) \nearrow h(x) \text{ for } x < 0.$$

It is easy to see that

$$\int_0^x 1_{(-\infty,b_i)}(y)dy = \begin{cases} (x - b_i)1_{(-\infty,b_i)}(x) & \text{if } b_i < 0 \\ (x - b_i)1_{(-\infty,b_i)}(x) + b_i & \text{if } b_i \geq 0 \end{cases}$$

so we have

$$h_n(x) = -\frac{1}{2^n}\sum_{i=m+1}^{n2^n}(x-b_i)1_{(-\infty,b_i)}(x) - \frac{1}{2^n}\sum_{i=1}^{m}[(x-b_i)1_{(-\infty,b_i)}(x)+b_i]+h(0).$$

Let $f_n(x) = \int_0^x h_n(y)dy$, then $f_n(x) \searrow f(x)$ for $x \geq 0$ and $f_n(x) \nearrow f(x)$ for $x < 0$. Let's compute:

1. $\int_0^x (y - b_i)1_{(-\infty,b_i)}(y)dy$ for $b_i < 0$,

2. $\int_0^x [(y - b_i)1_{(-\infty,b_i)}(y) + b_i]dy$ for $b_i \geq 0$, and

3. $\int_0^x h(0)dy = xh(0)$.

First, for $b_i < 0$,

$$\int_0^x (y - b_i)1_{(-\infty,b_i)}(y)dy = \begin{cases} 0 & \text{if } x > b_i \\ \int_{b_i}^x (y - b_i)dy & \text{if } x \leq b_i \end{cases}$$

$$= \begin{cases} 0 & \text{if } x > b_i \\ [\frac{1}{2}y^2 - b_iy]_{y=b_i}^x & \text{if } x \leq b_i \end{cases}$$

$$= \begin{cases} 0 & \text{if } x > b_i \\ \frac{1}{2}x^2 - b_ix - \frac{1}{2}b_i^2 + b_i^2 & \text{if } x \leq b_i \end{cases}$$

$$= \begin{cases} 0 & \text{if } x > b_i \\ \frac{1}{2}(x^2 - 2b_ix + b_i^2) & \text{if } x \leq b_i \end{cases}$$

$$= \begin{cases} 0 & \text{if } x > b_i \\ \frac{1}{2}(x - b_i)^2 & \text{if } x \leq b_i \end{cases}$$

$$= \frac{1}{2}(x - b_i)^2 1_{(-\infty,b_i]}(x).$$

Next, for $b_i \geq 0$,

$$
\int_0^x [(y - b_i)1_{(-\infty, b_i)}(y) + b_i] dy =
\begin{cases}
\int_0^{b_i} y\, dy + \int_{b_i}^x b_i dy & \text{if } x > b_i \\
\int_0^x y\, dy & \text{if } x \leq b_i
\end{cases}
$$

$$
=
\begin{cases}
\frac{1}{2} b_i^2 + b_i(x - b_i) & \text{if } x > b_i \\
\frac{1}{2} x^2 & \text{if } x \leq b_i
\end{cases}
$$

$$
=
\begin{cases}
\frac{1}{2}(2b_i x - b_i^2) & \text{if } x > b_i \\
\frac{1}{2} x^2 & \text{if } x \leq b_i.
\end{cases}
$$

Let

$$
I_{b_i}(x) =
\begin{cases}
-\frac{1}{2}(2b_i x - b_i^2) & \text{if } x > b_i \\
-\frac{1}{2} x^2 & \text{if } x \leq b_i
\end{cases}
$$

for $b_i \geq 0$. Then

$$
I_{b_i}(x) + x b_i =
\begin{cases}
\frac{1}{2} b_i^2 & \text{if } x > b_i \\
-\frac{1}{2}(x - b_i)^2 + \frac{1}{2} b_i^2 & \text{if } x \leq b_i
\end{cases}
$$

$$
= -\frac{1}{2}(x - b_i)^2 1_{(-\infty, b_i]}(x) + \frac{1}{2} b_i^2,
$$

and hence,

$$
f_n(x) = \frac{1}{2^n} \sum_{i=m+1}^{n2^n} -\frac{1}{2}(x - b_i)^2 1_{(-\infty, b_i)}(x) + \frac{1}{2^n} \sum_{i=1}^{m} I_{b_i}(x) + x h(0)
$$

$$
= \frac{1}{2^n} \sum_{i=m+1}^{n2^n} -\frac{1}{2}(x - b_i)^2 1_{(-\infty, b_i)}(x) + \frac{1}{2^n} \sum_{i=1}^{m} [I_{b_i}(x) + x b_i]
$$

$$
+ x \left[ h(0) - \frac{1}{2^n} \sum_{i=1}^{m} b_i \right]
$$

$$
= \frac{1}{2^n} \sum_{i=m+1}^{n2^n} -\frac{1}{2}(x - b_i)^2 1_{(-\infty, b_i)}(x)
$$

$$
+ \frac{1}{2^n} \sum_{i=1}^{m} \left[ -\frac{1}{2}(x - b_i)^2 1_{(-\infty, b_i]}(x) + \frac{1}{2} b_i^2 \right]
$$

$$
+ x \left[ h(0) - \frac{1}{2^n} \sum_{i=1}^{m} b_i \right].
$$

## 2.1.2    A fundamental lemma

The following simple lemma is useful for our analysis.

**Lemma 2.1** *Let* $u : \mathbb{R} \to \mathbb{R}$ *be measurable and nondecreasing. For random variables* $X, Y$ *(with distribution functions* $F, G$, *respectively) such that* $E|u(X)|, E|u(Y)|$ *are finite, we have*

$$\Delta u(X, Y) = Eu(X) - Eu(Y) = \int_{-\infty}^{\infty} [G(t) - F(t)] du(t).$$

**Proof.** Recall that $du(t)$ denotes the Stieltjes measure on $\mathcal{B}(\mathbb{R})$ associated with the monotone nondecreasing function $u$. We have

$$
\begin{aligned}
\Delta u(X, Y) &= Eu(X) - Eu(Y) \\
&= \int_{-\infty}^{\infty} u(x) dF(x) - \int_{-\infty}^{\infty} u(x) dG(x).
\end{aligned}
$$

Now using integration by parts for Stieltjes integral, we get

$$\int_{-\infty}^{\infty} u(x) dF(x) = \int_{-\infty}^{\infty} d[u(x) F(x)] - \int_{-\infty}^{\infty} F(x) du(x),$$

so that

$$
\begin{aligned}
\Delta u(X, Y) &= \int_{-\infty}^{\infty} d[u(x)(F(x) - G(x))] + \int_{-\infty}^{\infty} [G(x) - F(x)] du(x) \\
&= \lim_{x \to \infty} u(x)[F(x) - G(x)] - \lim_{x \to -\infty} u(x)[F(x) - G(x)] \\
&\quad + \int_{-\infty}^{\infty} [G(x) - F(x)] du(x).
\end{aligned}
$$

Thus it suffices to show that both limits

$$\lim_{x \to \infty} u(x)[F(x) - G(x)]$$
$$\lim_{x \to -\infty} u(x)\lfloor F(x) - G(x)]$$

are zero, since $E\,|u(X)| < \infty$, $E\,|u(Y)| < \infty$.

This can be seen as follows. Let $\nu(A) = \int_A u(t) dF(t)$ be a signed measure, absolutely continuous with respect to $dF(t)$. We have

$$\nu((x, \infty)) = \int_x^{\infty} u(t) dF(t) \ge u(x) dF((x, \infty))$$

for any $x \in \mathbb{R}$. But $\lim_{x \to \infty} dF((x, \infty)) = \lim_{x \to \infty} [1 - F(x)] = 0$ implying that $\lim_{x \to \infty} \nu((x, \infty)) = 0$ by absolute continuity, which in turn, implies that

$$\lim_{x \to \infty} u(x) dF((x, \infty)) = \lim_{x \to \infty} u(x)[1 - F(x)] = 0.$$

We have the same result when replacing $F$ by $G$, and hence

$$\lim_{x \to \infty} u(x)[F(x) - G(x)] = \lim_{x \to \infty} u(x)[(1 - G(x)) - (1 - F(x))] = 0.$$

Similarly, $\nu((-\infty, x]) = \int_{-\infty}^{x} u(t) dF(t) \leq u(x) dF((-\infty, x])$. By absolute continuity, $\lim_{x \to -\infty} dF((-\infty, x]) = \lim_{x \to -\infty} F(x) = 0$, implying

$$\lim_{x \to -\infty} \nu((-\infty, x]) = 0.$$

Since $u$ is nondecreasing, for $x < 0$, we have $u(x)F(x) \leq u(0)F(x)$, so that

$$0 = \lim_{x \to -\infty} \nu((-\infty, x]) \leq \lim_{x \to -\infty} u(x)F(x) \leq \lim_{x \to -\infty} u(0)F(x) = 0,$$

and hence $\lim_{x \to -\infty} u(x)F(x) = 0$. Applying this also to $G$ we get

$$\lim_{x \to -\infty} u(x)[F(x) - G(x)] = 0.$$

■

## 2.2 Deriving representations of preferences

As stated above, we seek necessary and sufficient conditions for "rational" choices among random variables of decision-makers with various risk attitudes, in terms of distribution functions.

### 2.2.1 Representation for risk neutral individuals

Consider the space $\mathcal{U}_1$ of all nondecreasing functions $u : \mathbb{R} \to \mathbb{R}$, i.e., the space of utility functions of risk neutral individuals. We seek a necessary and sufficient condition for

$$Eu(X) \geq Eu(Y) \text{ for any } u \in \mathcal{U}_1$$

in terms of the distribution functions $F, G$ (of $X, Y$, respectively).

**Theorem 2.2** *Let $X, Y$ be random variables with distribution functions $F, G$, respectively. Then $Eu(X) \geq Eu(Y)$ for any $u \in \mathcal{U}_1$ if and only if $F(.) \leq G(.)$.*

**Proof.** Suppose $Eu(X) \geq Eu(Y)$ for any $u \in \mathcal{U}_1$. Since $u_a(.) = 1_{(a,\infty)}(.) \in \mathcal{U}_1$ for all $a \in \mathbb{R}$, we have

$$
\begin{aligned}
Eu_a(X) &= \int_a^\infty dF(x) = 1 - F(a) \\
&\geq Eu_a(Y) = \int_a^\infty dG(x) = 1 - G(a),
\end{aligned}
$$

resulting in

$$F(a) \leq G(a) \text{ for all } a \in \mathbb{R}. \tag{2.1}$$

Conversely, suppose $F(.) \leq G(.)$ on $\mathbb{R}$. Let $\mathcal{C}_1$ be the subset of $\mathcal{U}_1$ consisting of functions of the form $1_{(a,\infty)}$ for $a \in \mathbb{R}$. Since

$$
\begin{aligned}
E(1_{(a,\infty)}(X)) &= \int_a^\infty dF(x) = 1 - F(a) \\
&\geq 1 - G(a) = \int_a^\infty dG(x) - E(1_{(a,\infty)}(Y)),
\end{aligned}
$$

we have

$$Eu(X) \geq Eu(Y) \text{ for any } u \in \mathcal{C}_1.$$

For $u^+$ we use the approximation of nonnegative, nondecreasing functions in: $u_n^+ \nearrow u^+$, with $Eu_n^+(X) \geq Eu_n^+(Y)$ for each $n$, we have, by the monotone convergence theorem,

$$Eu^+(X) = \lim_{n \to \infty} Eu_n^+(X) \geq \lim_{n \to \infty} Eu_n^+(Y) = Eu^+(Y).$$

For $u^-$ we use the approximation of nonnegative, nonincreasing functions: $u_n^- \nearrow u^-$ with $Eu_n^-(X) \leq Eu_n^-(Y)$ for each $n$, by noting that $1_{(-\infty,b]}(x) = 1 - 1_{(b,\infty)}(x)$, so that

$$E(1_{(-\infty,b]}(X)) = 1 - E(1_{(b,\infty)}(X)) \leq 1 - E(1_{(b,\infty)}(Y)) = E(1_{(-\infty,b]}(X)).$$

Again, by the monotone convergence theorem,

$$Eu^-(X) = \lim_{n \to \infty} Eu_n^-(X) \leq \lim_{n \to \infty} Eu_n^-(Y) = Eu^-(Y),$$

and hence

$$Eu(X) = Eu^+(X) - Eu^-(X) \geq Eu^+(Y) - Eu^-(Y) = Eu(Y).$$

∎

## 2.2.2   Representation for risk averse individuals

Consider the space $\mathcal{U}_2$ consisting of nondecreasing and concave functions, i.e., the space of utility functions of risk averse individuals. We seek a necessary and sufficient condition for $Eu(X) \geq Eu(Y)$ for any $u \in \mathcal{U}_2$.

**Theorem 2.3** *Let $X, Y$ be random variables with distribution functions $F, G$, respectively. Then $Eu(X) \geq Eu(Y)$ for any $u \in \mathcal{U}_2$ if and only if*

$$\int_{-\infty}^{a} [G(t) - F(t)]dt \geq 0 \text{ for all } a \in \mathbb{R}.$$

**Proof.** Suppose $Eu(X) \geq Eu(Y)$ for any $u \in \mathcal{U}_2$. Let $\mathcal{C}_2$ denote the subclass of $\mathcal{U}_2$ consisting of functions of the form

$$\varphi_a(x) = (x - a)1_{(-\infty,a]}(x) \text{ for } a \in \mathbb{R}.$$

Then

$$E\varphi_a(X) - E\varphi_a(Y) \geq 0 \text{ for all } a \in \mathbb{R}$$

or

$$\int_{-\infty}^{a} (x - a)d[F - G](x) \geq 0.$$

Now, by integration by parts,

$$\int_{-\infty}^{a} (x - a)d[F - G](x) = \int_{-\infty}^{a} [G(t) - F(t)]dt,$$

since $\lim_{x \to -\infty}(x - a)[F(x) - G(x)] = 0$ as $E|X|$ and $E|Y|$ are finite. Thus, a *necessary condition* is

$$\int_{-\infty}^{a} [G(t) - F(t)]dt \geq 0 \text{ for all } a \in \mathbb{R}.$$

Conversely, suppose

$$\int_{-\infty}^{a} [G(t) - F(t)]dt \geq 0 \text{ for all } a \in \mathbb{R}. \qquad (2.2)$$

Recall that Equation (2.2) means $Eu(X) \geq Eu(Y)$ for any $u \in \mathcal{C}_2$. Let $u \in \mathcal{U}_2$. Without loss of generality, we can assume that (as in the approximation subsection above) $u(0) = 0$. This is so, since otherwise, take $v(x) = u(x) - u(0)$, then $v \in \mathcal{U}_2$ and $v(0) = 0$, and moreover $Eu(X) = Ev(X) - u(0) \geq Ev(Y) - u(0) = Eu(Y)$ if and only if $Ev(X) \geq$

$Ev(Y)$. Thus, we just apply the approximation of $u$ as in the above subsection. Now $\xi_a\,(.) \in \mathcal{C}_2$ for all $a < 0$, where

$$\xi_a(x) = \varphi_a(x) = (x - a)1_{(-\infty,a]}(x) \text{ if } a < 0.$$

But for $a \geq 0$

$$\xi_a(x) = (x - a)1_{(-\infty,a]}(x) + a$$

so that

$$E\xi_a(X) = E\varphi_a(X) + a \geq E\xi_a(Y) = E\varphi_a(Y) + a$$

by Equation (2.2). It follows by the monotone convergence theorem that $Eu(X) \geq Eu(Y)$, i.e., the necessary condition is also sufficient. ∎

### 2.2.3  Representation for risk seeking individuals

Let $\mathcal{U}'_2$ be the space of nondecreasing and convex functions, i.e., the space of utility functions of risk seeking individuals.

**Theorem 2.4** *Let $X, Y$ be random variables with distribution functions $F, G$, respectively. Then $Eu(X) \geq Eu(Y)$ for any $u \in \mathcal{U}'_2$ if and only if*

$$\int_a^\infty [G(t) - F(t)]dt \geq 0 \text{ for each } a \in \mathbb{R}.$$

**Proof.** Suppose $Eu(X) \geq Eu(Y)$ for any $u \in \mathcal{U}'_2$. Let $\mathcal{C}'_2$ denote the subclass of $\mathcal{U}'_2$ consisting of functions of the form $\psi_a(x) = (x-a)1_{[a,\infty)}(x)$, $a \in \mathbb{R}$. We arrive at the necessary condition

$$\int_a^\infty [G(t) - F(t)]dt \geq 0 \text{ for each } a \in \mathbb{R} \tag{2.3}$$

since

$$
\begin{aligned}
E\psi_a(X) - E\psi_a(Y) &= \int_a^\infty (x - a)d[F - G](t)dt \\
&= \int_a^\infty [G(t) - F(t)]dt + \lim_{x\to\infty}(x - a)[F(x) - G(x)]
\end{aligned}
$$

and

$$
\begin{aligned}
\lim_{x\to\infty}(x - a)[F(x) - G(x)] &= \lim_{x\to\infty}(x - a)[(1 - G(x)) - (1 - F(x))] \\
&= 0.
\end{aligned}
$$

To show that Equation (2.3) is also *sufficient*, we proceed similarly as follows. Let $u \in \mathcal{U}_2'$ with $u(0) = 0$. By duality between concave and convex functions, $u$ can be represented as

$$u(x) = \int_0^x g(t)dt,$$

for all $x \in \mathbb{R}$, for some nonnegative and nondecreasing function $g$ on $\mathbb{R}$. Now observe that

$$v_a(x) = \int_0^x 1_{[a,\infty)}(t)dt = \begin{cases} (x-a)1_{[a,\infty)}(x) & \text{if } a \geq 0 \\ x \vee a & \text{if } a < 0 \end{cases}$$

where $x \vee a = (x-a)1_{[a,\infty)}(x) + a$ so that $Ev_a(X) \geq Ev_a(Y)$, and hence by the monotone convergence theorem, we get $Eu(X) \geq Eu(Y)$. ∎

### 2.2.4  A subclass of risk averse individuals

We present here a representation for an interesting subclass of risk averse individuals, the so-called "decreasing absolute risk averse" (DARA) individuals.

Recall from the approximations in the previous section that $\mathcal{U}_3$ denotes the set of functions $u$ such that there exists a convex decreasing nonnegative function $h$ such that

$$u(x) = \int_0^x h(y)dy + u(0).$$

Let $\mathcal{C}_3$ denote the subset of $\mathcal{U}_3$ consisting of functions of the form

$$\varphi_a(x) = -(x-a)^2 1_{(-\infty,a]}(x), \text{ for } a \in \mathbb{R}.$$

Note that $E\varphi_a(X)$ exists since $\varphi_a(.) \leq 0$.

**Theorem 2.5** $Eu(X) \geq Eu(Y)$ *for any* $u \in \mathcal{U}_3$ *with* $Eu^+(X) < \infty$ *and* $Eu^+(Y) < \infty$ *if and only if* $E\varphi_a(X) \geq E\varphi_a(Y)$ *for any* $a \in \mathbb{R}$ *and* $E(X) \geq E(Y)$.

**Proof.** Let $u$ be a function in class $\mathcal{U}_3$ such that $u(x) = \int_0^x h(y)dy$ where $h$ is nonnegative, nonincreasing and convex. By the approximation for functions in class $\mathcal{U}_3$, there exists $\{u_n\}$ such that $u_n(x) \searrow u(x)$ for $x \geq 0$ and $u_n(x) \nearrow u(x)$ for $x < 0$. In particular,

$$u_n(x) = \frac{1}{2^n} \sum_{i=m+1}^{n2^n} -\frac{1}{2}(x - a_i)^2 1_{(-\infty, a_i)}(x)$$

$$+ \frac{1}{2^n} \sum_{i=1}^{m} [-\frac{1}{2}(x - a_i)^2 1_{(-\infty, a_i]}(x)$$

$$+ \frac{1}{2} a_i^2] + x[h(0) - \frac{1}{2^n} \sum_{i=1}^{m} a_i] + x$$

with $h(0) - \frac{1}{2^n} \sum_{i=1}^{m} a_i \geq 0$. Hence, $E_X(u_n) \geq E_Y(u_n)$ (recall that $E(X) \geq E(Y)$).

Since $u_n(x) \searrow u(x)$ for $x \geq 0$ and $u_n(x) \nearrow u(x)$ for $x < 0$, by the monotone convergence theorem, we get

$$E_X(u) \geq E_Y(u).$$

(Note: $u(x) \geq 0$, for $x \geq 0$ and $u(x) < 0$ for $x < 0$.)

The other direction is trivial since $\mathcal{C}_3 \subset \mathcal{U}_3$ and $f(x) = x$ is in the class $\mathcal{U}_3$. ∎

**Special cases**

(1) If $X$ and $Y$ are bounded below by $b$, then

$$\Delta \varphi_a = E\varphi_a(X) - E\varphi_a(Y) = \int_b^a -(x - a)^2 d[F - G](x)$$

$$= -(x - a)^2 (F(x) - G(x)) |_{x=b}^a$$

$$+ \int_b^a (x - a)[F - G]dx \text{ (by integration by parts)}$$

$$= \int_b^a (x - a)[F - G]dx.$$

(Note: $F(b) = G(b) = 0$.) Hence, the condition

$$E\varphi_a(X) \geq E\varphi_a(Y) \text{ for all } a \in \mathbb{R}$$

is equivalent to

$$\int_b^a (x - a)[F - G]dx \geq 0 \text{ for all } a \in \mathbb{R},$$

which implies the sufficient condition

$$\int_b^a (x-a)[F-G]dx \geq 0 \text{ for all } a \in \mathbb{R}$$

and $E(X) \geq E(Y)$.

(2) If the second moments $E(X^2)$ and $E(Y^2)$ are finite, a *necessary condition* is derived by restricting to $C_3$ resulting in

$$E\varphi_a(X) - E\varphi_a(Y) \geq 0 \text{ for all } a \in \mathbb{R},$$

or equivalently,

$$\int_{-\infty}^a -(x-a)^2 d[F-G](x) \geq 0.$$

Now, by integration by parts,

$$\int_{-\infty}^a -(x-a)^2 d[F-G](x) = 2\int_{-\infty}^a (x-a)[F(x)-G(x)]dx$$

since $\lim_{x\to-\infty}(x-a)^2[F(x)-G(x)] = 0$ (because of finiteness of the second moments). Applying integration by parts one more time, we have

$$\begin{aligned}
\int_{-\infty}^a (x-a)[F(x)-G(x)]dx &= (x-a)\int_{-\infty}^x [F(t)-G(t)]dt\big|_{x=-\infty}^a \\
&+ \int_{-\infty}^a \int_{-\infty}^x [G(t)-F(t)]dtdx.
\end{aligned}$$

Notice that $\lim_{x\to-\infty} x\int_{-\infty}^x F(t)dt = 0$. Indeed, let

$$F_2(x) = \int_{-\infty}^x F(y)dy$$

then $F_2$ is nondecreasing and continuous. Moreover,

$$\int_{-\infty}^\infty x dF_2(x) = \int_{-\infty}^\infty x F(x)dx < \infty \text{ (since } E(X^2) < \infty)$$

so we can let $\upsilon(A) = \int_A x dF_2(x)$, and then $\upsilon$ is a signed-measure. We have

$$\upsilon((-\infty, x]) = \int_{-\infty}^x t dF_2(t) \leq x\int_{-\infty}^x dF_2(t) = xF_2(x) \leq 0 \text{ for } x < 0.$$

If we let $x \to -\infty$, then $v((-\infty, x]) \to 0$, which implies $\lim_{x \to -\infty} x F_2(x) = 0$. Therefore,

$$\int_{-\infty}^{a} (x - a)[F(x) - G(x)]dx = \int_{-\infty}^{a} \int_{-\infty}^{x} (G - F)(t)dtdx.$$

Moreover, $\int_{-\infty}^{a} \int_{-\infty}^{t} (G - F)(y)dydt \geq 0$ for all $a$ implies $E(X) \geq E(Y)$ in the case where the second moments of $X$ and $Y$ are finite (see a proof below).

Hence, if the second moments of $X$ and $Y$ are finite, then the necessary and sufficient condition is

$$\int_{-\infty}^{a} \int_{-\infty}^{x} (G - F)(t)dtdx \geq 0 \text{ for all } a \in \mathbb{R}.$$

**Example 2.6** *Here is an example showing that, in general,*

$$\int_{-\infty}^{x} \int_{-\infty}^{y} (G - F)(t)dtdy \geq 0$$

*for any $x$ does not imply $E(X) \geq E(Y)$. Consider*

$$f(x) = \frac{24}{x^4}\bigg|_{(-\infty, -2)}.$$

*Then*

$$F(x) = \begin{cases} -\frac{8}{x^3} & \text{if } x \leq -2 \\ 1 & \text{if } x > -2, \end{cases}$$

*and if*

$$g(x) = \frac{-2}{x^3}\bigg|_{(-\infty, -1)}$$

*then*

$$G(x) = \begin{cases} \frac{1}{x^2} & \text{if } x \leq -1 \\ 1 & \text{if } x > -1. \end{cases}$$

*Then*

$$E(X) = \int_{-\infty}^{-2} xf(x)dx = \int_{-\infty}^{-2} \frac{24}{x^3}dx = \left[\frac{-12}{x^2}\right]_{x=-\infty}^{-2} = -3,$$

$$E(Y) = \int_{-\infty}^{-1} xg(x)dx = \int_{-\infty}^{-1} \frac{-2}{x^2}dx = \frac{2}{x}\bigg|_{x=-\infty}^{-1} = -2.$$

*Note that $E(Y^2) = \infty$.*

*For $x \leq -2$, let's compute $I(x) = \int_{-\infty}^{x} \int_{-\infty}^{y} [G(t) - F(t)] \, dt dy$. First,*

$$\int_{-\infty}^{y} [G(t) - F(t)] \, dt = \int_{-\infty}^{y} \left[ \frac{1}{t^2} + \frac{8}{t^3} \right] dt$$

$$= \left[ -\frac{1}{t} - \frac{4}{t^2} \right]_{t=-\infty}^{y} = -\frac{1}{y} - \frac{4}{y^2}.$$

*Then*

$$I(x) = \int_{-\infty}^{x} \left( -\frac{1}{y} - \frac{4}{y^2} \right) dy = \left[ -\ln|y| + \frac{4}{y} \right]_{y=-\infty}^{x} = +\infty.$$

*Now for $x > -2$, we can write*

$$I(x) = \int_{-\infty}^{-2} \int_{-\infty}^{y} (G - F)(t) dt dy + \int_{-2}^{x} \int_{-\infty}^{y} (G - F)(t) dt dy.$$

*It is obvious that $\int_{-2}^{x} \int_{-\infty}^{y} (G - F)(t) dt dy < \infty$, so $I(x) = +\infty$. In summary, $\int_{-\infty}^{x} \int_{-\infty}^{y} (G - F)(t) dt dy = +\infty$ for any $x$; but $E(X) < E(Y)$.*

Thus, the condition $I(x) = \int_{-\infty}^{x} \int_{-\infty}^{y} (G - F)(t) dt dy \geq 0$ for any $x$ will imply $E(X) \geq E(Y)$ if and only if $I(x) < \infty$ (e.g., $E(X^2) < \infty, E(Y^2) < \infty$). Indeed, assuming

$$\int_{-\infty}^{a} \int_{-\infty}^{x} (G - F)(t) dt dx \geq 0 \text{ for all } a \in \mathbb{R}, \qquad (2.4)$$

let

$$h(y) = \int_{-\infty}^{y} (G - F)(t) dt.$$

By the fundamental lemma (applied to $u(t) = t$),

$$\lim_{y \to \infty} h(y) = \int_{-\infty}^{\infty} (G - F)(t) dt = EX - EY.$$

Thus, it suffices to show that

$$\lim_{y \to \infty} \int_{-\infty}^{y} (G - F)(t) dt \geq 0.$$

Suppose $\lim_{y \to \infty} \int_{-\infty}^{y} (G-F)(t) dt = a < 0$, i.e., $\lim_{y \to \infty} h(y) = a < 0$. Then, for $x$ sufficiently large, $\int_{-\infty}^{x} h(y) dy < 0$, contradicting (2.4).

Indeed, let $\varepsilon > 0$, and $\varepsilon < |a|$, then for $x \geq x(\varepsilon)$, $a - \varepsilon \leq h(y) \leq a + \varepsilon$. Let

$$A = \left| \int_{-\infty}^{x(\varepsilon)} h(y) dy \right|$$

and take $x > x(\varepsilon) - A/(a + \varepsilon)$. We have that $\int_{-\infty}^{x} h(y) dy < 0$, since

$$\int_{-\infty}^{x} h(y) dy = \int_{-\infty}^{x(\varepsilon)} h(y) dy + \int_{x(\varepsilon)}^{x} h(y) dy < A + (a + \varepsilon) \int_{x(\varepsilon)}^{x} dy$$

$$= A + (a + \varepsilon)(x - x(\varepsilon)) < 0$$

for $x > x(\varepsilon) - A/(a + \varepsilon)$ (noting that $(a + \varepsilon) < 0$). Of course, we need to assume that $A < \infty$ which is equivalent to $EX^2 < \infty$, $EY^2 < \infty$.

**Remark 2.7** *For completeness, we provide here the proof (due to Schmid [139]) that when $X, Y$ have finite second moments, the additional condition $EX \geq EY$ in the theorem is not needed.*

First, recall that when $EX^2 < \infty$ and $EY^2 < \infty$, the necessary condition

$$E\varphi_a(X) - E\varphi_a(Y) \geq 0 \text{ for all } a \in \mathbb{R}$$

becomes

$$\int_{-\infty}^{a} \int_{-\infty}^{x} (G - F')(t) dt dx \geq 0 \text{ for all } a \in \mathbb{R}$$

or

$$\int_{-\infty}^{a} \int_{-\infty}^{x} (F - G)(t) dt dx \leq 0 \text{ for all } a \in \mathbb{R}.$$

Now this necessary condition implies that $EX \geq EY$. This can be seen as follows. Observe, by integration by parts, that

$$\int_{-\infty}^{x} (F - G)(t) dt = x(F(x) - G(x)) + \int_{-\infty}^{x} t dG(t) - \int_{-\infty}^{x} t dF(t).$$

Since the second moments of $X, Y$ are finite, we have

$$\lim_{x \to \infty} \int_{-\infty}^{x} (F - G)(t) dt = \int_{-\infty}^{\infty} t dG(t) - \int_{-\infty}^{\infty} t dF(t) = EY - EX.$$

Thus, if we assume $EX < EY$, i.e., $EY - EX > 0$, then for $\varepsilon = (EY - EX)/2 > 0$, we have

$$\int_{-\infty}^{x} (F - G)(t) dt > \varepsilon/2$$

for $x$ sufficiently large, say $x \geq x_0$. But then, for $a \geq x_0$,

$$
\begin{aligned}
\int_{-\infty}^{a} \int_{-\infty}^{x} (F - G)(t) dt dx \; &= \; \int_{-\infty}^{x_0} \int_{-\infty}^{x} (F - G)(t) dt dx \\
&+ \; \int_{x_0}^{a} \int_{-\infty}^{x} (F - G)(t) dt dx \\
&\geq \; c + (\varepsilon/2)(a - x_0)
\end{aligned}
$$

where

$$
c = \int_{-\infty}^{x_0} \int_{-\infty}^{x} (F - G)(t) dt dx.
$$

Let $a_0$ be such that $c + (\varepsilon/2)(a_0 - x_0) > 0$. We have, for $a \geq a_0$,

$$
\int_{-\infty}^{a} \int_{-\infty}^{x} (F - G)(t) dt dx > 0,
$$

contradicting

$$
\int_{-\infty}^{a} \int_{-\infty}^{x} (F - G)(t) dt dx \leq 0 \text{ for all } a \in \mathbb{R}.
$$

## 2.3   Stochastic dominance

Stochastic dominance rules refer to various partial orders on the space of distribution functions of random variables. The very first stochastic order was introduced by Lehmann in 1955 [87].

In economics literature, some special stochastic orders of interest are termed *stochastic dominance rules*. The rationale of these rules lies in the principle of maximization of expected utility as detailed in the previous section.

Note that in descriptive statistics, an ordering of distribution functions of random variables has been used to make precise the idea that one distribution has less of the characteristic than the other, where characteristics of a distribution refer to things such as location, dispersion, kurtosis, skewness and so forth.

Stochastic dominance rules refer to criteria to order distributions of random variables modeling uncertain quantities such as future returns of investments. They form a repertoire of tools for making financial decisions such as portfolio selections. Specifically, stochastic dominance refers to a set of rules to choose uncertain prospects, mostly in financial economics, based solely on distributions of future returns viewed as random variables.

## 2.3.1 First-order stochastic dominance

If $X, Y$ are future returns in two investment prospects, then clearly $X$ will be preferred to $Y$ is $X \geq Y$ (almost surely). Unfortunately, as it is typical when modeling uncertain quantities as random variables, we cannot reason on the variables themselves (!) but have to rely only on their distributions. Now if $X \geq Y$ then clearly $F(.) \leq G(.)$ where, $F, G$ are distribution functions of $X, Y$, respectively. What happens if we observe that $F(.) \leq G(.)$? Well, we can construct a pair of random variables, $X', Y'$, possibly defined on a different probability space, such that $X', Y'$ have distributions, $F, G$, respectively, and moreover $X' \leq Y'$ (a.s.). This is the method of coupling. Indeed, let $X^* = F^{-1}(U)$, $Y^* = G^{-1}(U)$, where $U : \Omega \to (0, 1)$ is a random variable defined on some $(\Omega, \mathcal{A}, P)$, uniformly distributed on $[0, 1]$, and $F^{-1}$ is the quantile function of $F$, i.e., its *left-continuous inverse* defined by

$$F^{-1} \; : \; (0, 1) \to \mathbb{R}$$
$$F^{-1}(\alpha) \; = \; \inf\{x \in \mathbb{R} : F(x) \geq \alpha\}.$$

Then $X^* = X$ and $Y^* = Y$ in distribution, i.e., having the same distribution functions: $P(X^* \leq x) = F(x)$ for any $x \in \mathbb{R}$. Moreover, since $F(.) \leq G(.)$ is equivalent to $G^{-1}(.) \leq F^{-1}(.)$ so that $Y^* = G^{-1}(U) \leq F^{-1}(U) = X^*$ $P$-a.s.

Note also that in fact $F^{-1}(\alpha) = \min\{x \in \mathbb{R} : F(x) \geq \alpha\}$ since the infimum is attained, i.e., $P(X \leq F^{-1}(\alpha)) \geq \alpha$. This can be seen as follows. For each $n \geq 1$, by definition of infimum, there is $x_n \leq F^{-1}(\alpha) + 1/n$ such that $F(x_n) \geq \alpha$. Since $F$ is non-decreasing, we have

$$F(F^{-1}(\alpha) + 1/n)) \geq F(x_n) \geq \alpha.$$

But, by right continuity of $F$, we have

$$\lim_{n \to \infty} F(F^{-1}(\alpha) + 1/n) = F(F^{-1}(\alpha)),$$

implying that $F(F^{-1}(\alpha)) \geq \alpha$ since, for each $n$, $F(F^{-1}(\alpha) + 1/n)) \geq \alpha$.

**Remark 2.8** *The right continuous inverse of $F$ is*

$$F_+^{-1}(\alpha) = \inf\{x \in \mathbb{R} : F(x) > \alpha\}.$$

Thus, in a sense, we should prefer $X$ to $Y$ when $F(.) \leq G(.)$. But, at the logical level, this stochastic ordering is strengthened by decision theory in which risk neutral decision-makers do behave this way, as demonstrated in the previous subsection.

**Definition 2.9** *Let $X, Y$ be two random variables with distribution functions $F, G$, respectively. Then $X$ is said to dominate $Y$ in the first-order stochastic dominance (FSD), denoted as $X \succeq_1 Y$, if and only if $F(.) \leq G(.)$.*

**Remark 2.10** *We can add to the above definition the requirement that there exists some $x_0 \in \mathbb{R}$ such that $F(x_0) < G(x_0)$ which is equivalent to there exists $u_0 \in \mathcal{U}_1$ such that $Eu_0(X) > Eu_0(Y)$. Indeed, if $F(x_0) < G(x_0)$, then consider $u_0 = 1_{(x_0, \infty)} \in \mathcal{U}_1$. We have $Eu_0(X) = \int u_0(X)dP = P(X > x_0) > P(Y > x_0) = Eu_0(Y)$. Conversely, suppose $F(.) = G(.)$, then $Eu(X) = Eu(Y) \; \forall u \in \mathcal{U}_1$.*

**Remark 2.11** *Clearly the FSD induces a partial order relation $\preceq$ on the space of distribution functions: $F \preceq G$ if and only if $F(.) \geq G(.)$. Moreover, equipped with this partial order, the space of distribution functions is a lattice, the join of $F, G$ is $\min\{F(.).G(.)\}$ and the meet of $F, G$ is $\max\{F(.), G(.)\}$.*

Now several observations are in order.

(i)   Clearly if $X \succeq_1 Y$ then $EX \geq EY$, by noting that

$$EX = \int_0^\infty (1 - F(t))dt - \int_{-\infty}^0 F(t)dt$$

but clearly the converse is not true.

(ii)   If $X$ and $Y + a$ have the same distribution, for some $a \geq 0$, then $X \succeq_1 Y$ (exercise).

(iii)   Note the difference in ranking random variables by their distribution functions with the *Lorenz ordering* for inequalities in income distributions. Recall that, the Lorenz curve of an income distribution $F$, as a representation of inequality, is defined by

$$L_F : [0, 1] \to [0, 1] \text{ where } L_F(t) = \frac{\int_0^t F^{-1}(u)du}{\int_0^1 F^{-1}(u)du]}$$

(of course where the mean of income $\int_0^1 F^{-1}(u)du \in (0, \infty)$).

The first degree of dominance in Lorenz ordering sense between two income variables $X$ and $Y$, with distribution functions $F$ and $G$, respectively, is defined consistently with the fact that the higher Lorenz curve displays less inequality than the lower one: $X$ dominates $Y$ when $L_F(.) \geq L_G(.)$.

(iv)   Take $X, Y$ with distribution functions $F, G$, respectively. If $E(X) = E(Y)$ and $F(.) \leq G(.)$ then necessarily $F(.) = G(.)$. To see this, let $u(x) = x$ (a nondecreasing utility function), $du(x) = dx$. We have

$$\Delta u(X, Y) = EX - EY = 0 = \int_{\mathbb{R}} [G(x) - F(x)] dx.$$

But $G(.) - F(.) \geq 0$ and right continuous (as a difference of two right continuous functions), so its integral being zero will imply that $G(.) - F(.)$ is identically zero. Indeed, let $g(x) = G(x) - F(x)$ and choose $x_0$ such that $g(x_0) > 0$ (noting that $g \geq 0$ and $\int_{\mathbb{R}} g(x) dx = 0$ imply that $g - 0$ almost everywhere). By right continuity of $g$ at $x_0$, for $\varepsilon = g(x_0)/2 > 0$, there is $\delta > 0$ such that $|g(x) - g(x_0)| < \varepsilon$ for all $x \in [x_0, x_0 + \delta)$, so that

$$\int_{x_0}^{x_0+\delta} g(x) dx \geq \varepsilon \delta > 0,$$

a contradiction.

(v)   As preference relations on the space of distribution functions, they are essentially stochastic orders introduced in statistics much earlier. It was in the context of investment that the concept of stochastic dominance became known in economics, due to the pioneering work of J. Hadar and W. Russell [65], G. Hanoch and H. Levy [67] and G.A. Whitmore [167].

(vi)   How can we rank distributions if they intersect, i.e., when FSD cannot be applied? This might be often the case in applied economics. Well, we have to seek weaker criteria, yet consistent with FSD (i.e., orderings which are implied by FSD), for reaching decisions! The weaker orderings which will be explained next serve this purpose. Like FSD, they can be justified rigorously from the principle of utility maximization.

   In another practical note, since distribution functions of future returns are unknown, they need to be estimated from data. As such, to apply FSD, we need to carry out tests. We will address this issue in Chapter 6.

(vii)   From a common sense viewpoint, we could consider various types of *risk measures* on which we base our decisions. Risk measures will be detailed in Chapter 4. For example, in finance, the concept of *Value-at-Risk* (VaR) is this: In the context of investment, let $W_0$ be the initial investment (to a prospect), and let $X$ be its future return. For $\alpha \in (0, 1)$,

$$VaR_X(\alpha) = \sup\{x \in \mathbb{R} : P(W_0 - X \geq x) \geq \alpha\}.$$

*Interpretation:* $W_0 - X$ is the loss and $VaR_X(\alpha)$ is the value such that the probability of loss more than it is strictly less than $\alpha$. Specifically, if $x > VaR_X(\alpha)$, then $P(W_0 - X \geq x) < \alpha$. (Note that we write $VaR_X(\alpha)$, with the deterministic quantity $W_0$ fixed, in the comparisons with other future returns $Y$ invested on other prospects.) For each level $\alpha$, $VaR_X(\alpha)$ is a characteristic value for $X$ in the above sense. Thus $X$ is preferred to $Y$ if $VaR_X(\alpha) \leq VaR_Y(\alpha)$, for all $\alpha \in (0, 1)$.

It turns out that this concept is subsumed by the concept of stochastic dominance. This is so for the following reasons. First, let $F$ be the distribution function of $X$. Then

$$VaR_X(\alpha) = W_0 - F^{-1}(\alpha).$$

*Indeed,*

$$
\begin{aligned}
W_0 - F^{-1}(\alpha) &= W_0 - \inf\{x \in \mathbb{R} : F(x) \geq \alpha\} \\
&= -\inf\{x - W_0 : F(x) \geq \alpha\} \\
&= \sup\{W_0 - x : F(x) \geq \alpha\} \\
&= \sup\{y \in \mathbb{R} : F(W_0 - y) \geq \alpha\} \\
&= \sup\{y \in \mathbb{R} : P(X \leq W_0 - y) \geq \alpha\} \\
&= VaR_X(\alpha).
\end{aligned}
$$

**Theorem 2.12** $F(.) \leq G(.)$ *if and only if* $VaR_X(\alpha) \leq VaR_Y(\alpha)$, *for all* $\alpha \in (0, 1)$.

**Proof.** Observe that $F \leq G$ is equivalent to $F^{-1} \geq G^{-1}$. Indeed, if $F \leq G$ then clearly $F^{-1} \geq G^{-1}$. Conversely, if $F^{-1} \geq G^{-1}$, then $G^{-1}(F(x)) \leq F^{-1}(F(x)) \leq x$ but this is equivalent to $F(x) \leq G(x)$. Next, clearly,

$$VaR_X(\alpha) = W_0 - F^{-1}(\alpha) \leq W_0 - G^{-1}(\alpha) = VaR_Y(\alpha)$$

if and only if $F^{-1} \geq G^{-1}$. ■

## 2.3.2 Second-order stochastic dominance

As observed above, when FSD cannot be applied, we seek other weaker orderings to choose among risky prospects. Since $\mathcal{U}_2 \subseteq \mathcal{U}_1$, a stochastic order can be justified by using the necessary and sufficient condition for preferences among risk averse decision-makers.

**Definition 2.13** *The random variable $X$ with distribution function $F$ is said to dominate the random variable $Y$ with distribution function $G$, in second-order stochastic dominance (SSD), denoted as $X \succeq_2 Y$, if and only if $\int_{-\infty}^{x} F(y)dy \leq \int_{-\infty}^{x} G(y)dy$, for all $x \in \mathbb{R}$.*

Note that the criterion for SSD is more conveniently written as

$$\int_{-\infty}^{x} [G(t) - F(t)]dt \geq 0 \text{ for all } x \in \mathbb{R}.$$

**Remark 2.14** *Since $X \succeq_2 Y \Leftrightarrow Eu(X) \geq Eu(Y) \; \forall \, u \in \mathcal{U}_2$, we see that the additional condition "there exists $u_0 \in \mathcal{U}_2$ such that $Eu_0(X) > Eu_0(Y)$" is equivalent to "there exists $x_0 \in \mathbb{R}$ such that $\int_{-\infty}^{x_0} (G-F)(t)dt > 0$." Indeed, simply take $u_0(t) = (t - x_0)1_{(-\infty,x_0]}(t)$, we have*

$$\int_{-\infty}^{x_0} (G - F)(t)dt = \int_{-\infty}^{\infty} (G - F)(t)du_0(t)$$
$$= Eu_0(X) - Eu_0(Y).$$

*Conversely, if $\int_{-\infty}^{x}(G - F)(t)dt = 0 \; \forall \, x \in \mathbb{R}$, then $G(.) = F(.)$ which implies that $Eu(X) = Eu(Y) \; \forall \, u \in \mathcal{U}_2$.*

We can put the above in the following equivalent forms.

(a) $\Delta u(X,Y) \geq 0$ for all $u \in \mathcal{U}_2$ and there exists $u_0 \in \mathcal{U}_2$ such that $\Delta u_0(X,Y) > 0$.

(b) $\int_{-\infty}^{x}[G(t) - F(t)]dt \geq 0$ for all $x \in \mathbb{R}$ and there exists $x_0 \in \mathbb{R}$ such that $G(x_0) \neq F(x_0)$.

Indeed, if $G(.) = F(.)$ then $\Delta u(X,Y) = 0$ for all $u \in \mathcal{U}_2$. Thus, (a) $\Rightarrow$ (b).

For the other direction, in view of the above remark, it suffices to show that (b) implies

$$\int_{-\infty}^{x} [G(t) - F(t)]dt \geq 0 \text{ for all } x \in \mathbb{R}$$

and there exists $x_0 \in \mathbb{R}$ such that

$$\int_{-\infty}^{x_0} (G - F)(t)dt > 0.$$

But this follows by observing that

$$\int_{-\infty}^{x} [G(t) - F(t)]dt = 0 \text{ for all } x \in \mathbb{R}$$

implies that $G(.) = F(.)$.

Here are some useful observations.

(i)  Clearly FSD implies SSD. Also SSD implies that $EX \geq EY$.

(ii)  If $X \succeq_2 Y$ and $EX = EY$ then the variance of $X$ is less than the variance of $Y$ (exercise). What happens if we have two random variables $X, Y$ having the same mean and same variance? Do we expect that $X \succeq_2 Y$? Of course the answer is no in general. From the decision analysis viewpoint, it suffices to find a nondecreasing and concave utility function with different expected utilities for such distribution functions. In other words, risk averse decision-makers might not be indifferent between two such risky prospects. Thus, the standard deviation of a distribution function is not a good criterion for measuring risk.

(iii)  Since the negative of a concave function is a convex function, second stochastic dominance for risk seeking decision-makers is dual to the SSD as defined for risk averse decision-makers. Specifically, a risk seeking individual will prefer $X$ to $Y$ (with distribution functions $F, G$, respectively), denoted as $X \succeq_{2'} Y$ if

$$\int_{x}^{\infty} (G(t) - F(t))dt \geq 0 \text{ for all } x \in \mathbb{R}.$$

This stochastic order is referred to as *Risk Seeking Stochastic Dominance* (RSSD).

(iv)  If we require only $Eu(X) \geq Eu(Y)$ for $u$ convex (not necessary nondecreasing), then the associated stochastic order is called "convex order." Note that $X \succeq_{2'} Y$ and $EX = EY$ is equivalent to $X$ dominates $Y$ in the convex order sense. Our RSSD order is in fact equivalent to the well-known *stop-loss* order in actuarial science, namely "$X$ dominates $Y$ in the stop-loss order" if

$$E(X sgn)^+ \geq E(Y - t)^+ \text{ for all } t \in \mathbb{R},$$

where $(X - t)^+$ is the positive part of $(X - t)$, i.e., $(X - t)^+ = \max(X - t, 0)$. Note that

$$E(X - t)^+ = \int_t^\infty P(X > x)dx.$$

## 2.3.3 Third-order stochastic dominance

A yet weaker stochastic order for a subclass of risk averse individuals is termed the *third-order stochastic dominance* (TSD). The criterion is derived from utility theory. Note that we do not assume differentiability of utility functions.

**Definition 2.15** *$X$ (with distribution function $F$) is said to dominate $Y$ (with distribution function $G$) in the third-order stochastic dominance (TSD), denoted as $X \succeq_3 Y$ if and only if*

$$\int_{-\infty}^x \int_{-\infty}^y (G - F)(t)dtdy \geq 0 \text{ for all } x \in \mathbb{R}.$$

**Remark 2.16** *In all generalities, if $X, Y$ are assumed to have means (finite or not) only, then as mentioned before, we need to add the condition $EX \geq EY$. However, if both $X$ and $Y$ have finite variances, this additional condition is automatically satisfied as proved previously. Note also that the above definition of TSD is consistent with expected utility theory for random variables with finite second moments. And as such, like FSD and SSD, if $X \succeq_3 Y$ then $EX \geq EY$.*

(i) For differentiable utility functions, the TSD is applied to utility functions $u$ such that $u'(.) \geq 0$, $u''(.) \leq 0$ and $u'''(.) \geq 0$.

(ii) The additional condition: "there exists at least one $x_0 \in \mathbb{R}$ such that $I(x_0) = \int_{-\infty}^{x_0} \int_{-\infty}^y [G(t) - F(t)]dtdy > 0$" is equivalent to "there exists $u_0 \in \mathcal{U}_3$ such that $Eu_0(X) > Eu_0(Y)$." This can be seen as follows.

First, using integration by parts, we have

$$
\begin{aligned}
I(x) &= \int_{-\infty}^x \int_{-\infty}^y (G - F)(t)dtd \\
&= \int_{-\infty}^x (x - t)(G - F)(t)dt \\
&= \frac{1}{2} \int_{-\infty}^\infty (G - F)(t)du_x(t) \\
&\quad - \frac{1}{2}(Eu_x(X) - Eu_x(Y))
\end{aligned}
$$

where $u_x(t) = -(t-x)^2 1_{(-\infty,x]}(t)$. Thus,

$$I(x_0) = \frac{1}{2}(Eu_0(X) - Eu_0(Y))$$

where $u_0(t) = -(t-x_0)^2 1_{(-\infty,x_0]}(t)$.

Now if there exists $x_0$ such that $I(x_0) > 0$, then take

$$u_0(t) = -(t-x_0)^2 1_{(-\infty,x_0]}(t)$$

and we have

$$Eu_o(X) > Eu_0(Y).$$

Conversely, if

$$I(x) = \int_{-\infty}^{x} \int_{-\infty}^{y} (G-F)(t)dtdy = 0, \text{ for all } x \in \mathbb{R}$$

then

$$\int_{-\infty}^{x} (G-F)(t)dt = 0, \text{ for all } x \in \mathbb{R},$$

so that

$$G(x) = F(x), \text{ for all } x \in \mathbb{R}$$

implying that

$$Eu(X) = Eu(Y), \text{ for all } u \in \mathcal{U}_3.$$

As in the case of SSD, we also observe that the following are equivalent:

(a) $\Delta u(X,Y) \geq 0$ for all $u \in \mathcal{U}_3$, and there exists $u_0 \in \mathcal{U}_3$ such that $\Delta u_0(X,Y) > 0$.

(b) $\int_{-\infty}^{x} \int_{-\infty}^{y} (G-F)(t)dtdy \geq 0$, for all $x \in \mathbb{R}$, and there exists $x_0 \in \mathbb{R}$ such that $G(x_0) \neq F(x_0)$.

(c) $\int_{-\infty}^{x} \int_{-\infty}^{y} (G-F)(t)dtdy \geq 0$, for all $x \in \mathbb{R}$, and there exists $x_0 \in \mathbb{R}$ such that

$$\int_{-\infty}^{x_0} \int_{-\infty}^{y} (G-F)(t)dtd\dot{y} > 0.$$

(iii) For TSD, as mentioned in the section on derivation of representations, the condition $EX \geq EY$ is not needed when variances of random variables involved are finite. Thus, when risky prospects $X, Y$ have finite variances, that additional condition is superfluous. So, the condition

$$I(x) = \int_{-\infty}^{x} \int_{-\infty}^{y} [G(t) - F(t)]\, dt dy \geq 0$$

for any $x$ will imply $E(X) \geq E(Y)$ if and only if $I(x) < \infty$ (e.g., $E(X^2) < \infty, E(Y^2) < \infty$).

## 2.4   Exercises

1. Let $X, Y$ be nonnegative random variables, defined on $(\Omega, \mathcal{A}, P)$, and $u : \mathbb{R} \to \mathbb{R}$ be a convex, nondecreasing function, twice differentiable. Show directly that

$$\int_{y}^{\infty} P(X > x)dx \geq \int_{y}^{\infty} P(Y > x)dx \text{ for all } y \geq 0$$

implies

$$Eu(X) \geq Eu(Y).$$

2. Let $X, Y$ be two random variables taking values in an interval $[a, b]$. Show that

    (a) If $X \succeq_1 Y$ then $X \succeq_2 Y$.
    (b) If $X \succeq_2 Y$ then $EX \geq EY$.
    (c) If $X \succeq_2 Y$ and $EX = EY$, then $Eu(X) \geq Eu(Y)$ for all concave (and twice differentiable) function $u$ (increasing or not).
    (d) If $X \succeq_2 Y$ and $EX = EY$, then Variance$(X) \leq$ Variance$(Y)$.

3. Consider differentiable utility functions $u$. Show directly that if $X, Y$ have distribution functions $F, G$, then $F(.) \leq G(.)$ if and only if $Eu(X) \geq Eu(Y)$ for all $u$ nondecreasing.

4. This example shows that the standard deviation is not a good risk measure. Let $X, Y$ such that

$$P(X = 2) = 1 - P(X = 12) = 1/5$$
$$P(Y = 8) = 1 - P(Y = 18) = 4/5.$$

(a) Verify that $EX = EY, V(X) = V(Y)$.

(b) Consider the nondecreasing and concave utility function $u(x) = \log x$ (for $x > 0$). Does any risk averse individual feel indifferent between the two risky prospects $X$ and $Y$? In other words, is $Eu(X) = Eu(Y)$?

5. Let $X$ be a random variable with distribution function $F$. Suppose $E|X| = \int_{-\infty}^{\infty} |x| dF(x) < \infty$. Use Fubini's theorem to show that $EX = \int_0^\infty (1 - F(t)) dt - \int_{-\infty}^0 F(t) dt$.

6. Show that if $X$ and $Y + a$ (for some $a \geq 0$) have the same distribution, then $X \succeq_1 Y$.

7. Let nonnegative $X$ have distribution function $F$. Show that the Lorenz curve of $X$ (or of $F$) is $L_F(F(t)) = (1/\mu) \int_0^t x dF(x)$, $t \in \mathbb{R}^+$, where $\mu = E_F(X) = \int_0^\infty x dF(x)$.

8. Let $F$ and $G$ be two distribution functions such that $F(.) \leq G(.)$. Using integration by parts, show that for all differentiable and nondecreasing functions $u : \mathbb{R} \to \mathbb{R}$, $\int_{\mathbb{R}} u(x) dF(x) \geq \int_{\mathbb{R}} u(x) dG(x)$.

9. Construct two distribution functions $F, G$ having the same mean and the same variance and a utility function $u$, nondecreasing and concave such that $E_F(u) \neq E_G(u)$.

10. Let $X$ be a nonnegative random variable with distribution and density functions $F, f$, respectively. The failure (or hazard) rate function of $X$ is defined to be

$$\lambda_X : \mathbb{R}^+ \to \mathbb{R}^+, \lambda_X(t) = f(t)/[1 - F(t)].$$

We say that "$X$ is larger than $Y$ in the hazard rate sense" if $\lambda_X(.) \leq \lambda_Y(.)$. Show that $\lambda_X(.) \leq \lambda_Y(.)$ if and only if the function $t \to [1 - G(t)]/[1 - F(t)]$ is nonincreasing, where $G$ is the distribution function of $Y$.

11. Let nonnegative random variables $X, Y$ have density functions $f, g$, respectively. We say that "$X$ is larger than $Y$ in the likelihood ratio sense" if $f(.)/g(.)$ is nondecreasing. Show that the likelihood order is stronger than the hazard rate order (Exercise 10).

12. Continuing with Exercise 10, show that the hazard rate order is stronger than FSD.

13. Let $X$ be a random variable, with distribution function $F$, such that $E|X| < \infty$. Show that

    (a) $\lim_{x \to \infty} xF(x) = 0$.
    (b) $\lim_{x \to \infty} x[1 - F(x)] = 0$.

14. Let $I$ be an interval in $\mathbb{R}$. A function $f : I \to \mathbb{R}$ is *convex* if, for all $x, y \in I$, and $\lambda \in (0, 1)$, we have

$$f(\lambda x + (1 - \lambda)y) \leq \lambda f(x) + (1 - \lambda)f(y).$$

If $-f$ is convex, then $f$ is *concave*.

Show that $f : (a, b) \to \mathbb{R}$ is convex if and only if there exist a nondecreasing function $g : (a, b) \to \mathbb{R}$ and a point $c \in (a, b)$ such that

$$f(x) - f(c) = \int_c^x g(t)dt \text{ for all } x \in (a, b).$$

Deduce that if $f$ is concave and nondecreasing, with $0 \in (a, b)$ and $f(0) = 0$, then

$$f(x) = \int_0^x g(t)dt$$

with $g$ being nonnegative and nonincreasing.

15. Show that $X \succeq_{2'} Y$ and $EX = EY \iff X \geq Y$ in the convex order sense.

16. Show that $X \succeq_{2'} Y \iff X \geq Y$ in the stop-loss order sense.

# Chapter 3

# Issues in Stochastic Dominance

*This chapter is a continuation of Chapter 2, in which we addressed issues related to stochastic dominance.*

## 3.1 A closer look at the mean-variance rule

As we will see, stochastic dominance is developed as a set of tools for investment decisions. A systematic investigation into investment decisions has been carried out by Markowitz [104] using his Mean-Variance (MV) rule.

Heuristically, when comparing two random variables $X, Y$ (with distribution functions $F, G$, respectively), we are concerned with "utilities" and "risk" involved. For utility, of course we would like to be able to see whether, say $X \geq Y$. But that is impossible since they are random! So let's "predict" their values (say, at the end of an investment period) by their expected values $EX, EY$. Thus, as far as utilities are concerned, we prefer $X$ to $Y$ if $EX \geq EY$. To reach a final decision, we need to compare risks. Here, a quantitative risk is taken as the standard deviation, or, mathematically equivalent, of the variance. Thus $Y$ is riskier than $X$ if $\sigma^2(X) \leq \sigma^2(Y)$.

MV rule: $X$ is preferred to $Y$ when $EX \geq EY$ and $\sigma^2(X) \leq \sigma^2(Y)$.

The MV rule is for risk averse individuals and is consistent with expected utility theory for *normally distributed returns*. This can be seen as follows.

**Theorem 3.1 (MV rule)** *Let* $X, Y$ *be two random variables, normally distributed with means* $\mu_1, \mu_2$ *and variances* $\sigma_1^2$, $\sigma_2^2$, *respectively. Then the following are equivalent:*

(i) $Eu(X) \geq Eu(Y)$ *for any* $u : \mathbb{R} \to \mathbb{R}$ *nondecreasing and concave.*

(ii) $\mu_1 \geq \mu_2$ *and* $\sigma_1^2 \leq \sigma_2^2$.

**Proof.** $[(i) \Rightarrow (ii)]$ Obviously, $(i)$ implies $\mu_X \geq \mu_Y$. We prove, by contradiction, that $(i)$ implies also $\sigma_1^2 \leq \sigma_2^2$. Thus, suppose $\sigma_X > \sigma_Y$. Let $t_0 = \frac{\mu_Y - \mu_X}{\sigma_X - \sigma_Y}$. Observe that $t_0 \sigma_X + \mu_X = t_0 \sigma_Y + \mu_Y$. Call this common value $t_1$. Hence

$$t\sigma_X + \mu_X \leq t_1 \Leftrightarrow t \leq t_0 \Leftrightarrow t\sigma_Y + \mu_Y \leq t_1.$$

Consider the nondecreasing and concave function $u(x) = (x - t_1)\chi_{(-\infty, t_1]}$. Then

$$
\begin{aligned}
Eu(X) &= (1/\sigma_X\sqrt{2\pi}) \int_{-\infty}^{t_1} (x - t_1) \exp(-1/2)(x - \mu_X)^2/\sigma_X^2 \, dx \\
&= \frac{1}{\sqrt{2\pi}} \int_{-\infty}^{t_0} (t\sigma_X + \mu_X - t_1) e^{\frac{-t^2}{2}} \, dt \\
&< \frac{1}{\sqrt{2\pi}} \int_{-\infty}^{t_0} (t\sigma_Y + \mu_Y - t_1) e^{\frac{-t^2}{2}} \, dt \\
&= Eu(Y).
\end{aligned}
$$

This is so because for $t < t_0$ we have $t\sigma_X + \mu_X < t\sigma_Y + \mu_Y$, resulting in

$$(t\sigma_X + \mu_X - t_1) e^{\frac{-t^2}{2}} < (t\sigma_Y + \mu_Y - t_1) e^{\frac{-t^2}{2}}.$$

The inequality is strict since otherwise there are (in fact, almost everywhere $dt$) $t < t_o$ such that $t\sigma_X + \mu_X = t\sigma_Y + \mu_Y$ which is not possible. We thus get a contradiction. Therefore we must have $\sigma_X \geq \sigma_Y$ as desired.

$[(ii) \Rightarrow (i)]$ If $\sigma_X = \sigma_Y = \sigma$ then $F_X(x) = P(X \leq x) = P(Z \leq (x - \mu_X)/\sigma) \leq P(Z \leq (x - \mu_Y)/\sigma) = F_Y(x)$, for all $x \in \mathbb{R}$ (noting that $\mu_X \geq \mu_Y$), where $Z$ denotes the standard normal random variable $N(0, 1)$.

That $F_X(.) \leq F_Y(.)$ implies $Eu(X) \geq Eu(Y)$ for any nondecreasing function $u$ (and hence a fortiori implies $(ii)$) can be seen as follows. By coupling, we have $X^*$, $Y^*$ having distribution functions as $F_X$, $F_Y$, respectively and $X^* \geq Y^*$ with probability one. Since $u$ is nondecreasing, $u(X^*) \geq u(Y^*)$ with probability one, and hence $F_{u(X*)} \leq F_{u(Y*)}$. Since

$$P(u(X) \leq t) = \int_{\{x:u(x)\leq t\}} dF_X(x) = P(u(X^*) \leq t),$$

the distribution functions of $u(X^*), u(Y^*)$ are the same as those of $u(X), u(Y)$.

The above can be also seen as follows. First, $F_{u(X)} \leq F_{u(Y)}$. Indeed, we have

$$P(u(X) > t) = P(X > u^{-1}(t)) \geq P(Y > u^{-1}(t)) = P(u(Y) > t),$$

where $u^{-1}(t) = \inf\{x : u(x) > t\}$.

Second, note that for any two random variables $X, Y$, with $F_X(.) \leq F_Y(.)$, we have $EX \geq EY$, $X = X^+ - X^-$, where

$$X^+ = \begin{cases} X & \text{if } X \geq 0 \\ 0 & \text{if } X < 0 \end{cases} \quad \text{and} \quad X^- = \begin{cases} 0 & \text{if } X \geq 0 \\ -X & \text{if } X < 0, \end{cases}$$

then $F_X(.) \leq F_Y(.)$ implies $F_{X^+}(.) \leq F_{Y^+}(.)$ and $F_{Y^-}(.) \leq F_{X^-}(.)$.

Now, for nonnegative random variables, we have

$$EX^+ = \int_0^\infty P(X^+ > t)dt \geq \int_0^\infty P(Y^+ > t)dt = EY^+$$

and similarly

$$EY^- \geq EX^-.$$

Hence, $EX = EX^+ - EX^- \geq EY^+ - EY^- = EY$.

Assume now that $\sigma_X < \sigma_Y$, and put $t_0 = \frac{\mu_Y - \mu_X}{\sigma_X - \sigma_Y} \geq 0$, and recall that $t_1 = t_0\sigma_X + \mu_X(= t_0\sigma_Y + \mu_Y)$. Let $s = t_0 - t$ we get

$$\int_{-\infty}^{t_0} [u(t\sigma_X + \mu_X) - u(t\sigma_Y + \mu_Y)] e^{\frac{-t^2}{2}} dt$$

$$= \int_0^\infty [u(-s\sigma_X + t_1) - u(-s\sigma_Y + t_1)] e^{-\frac{(s-t_0)^2}{2}} ds$$

$$\geq \int_0^\infty [u(-s\sigma_X + t_1) - u(-s\sigma_Y + t_1)] e^{-\frac{(s+t_0)^2}{2}} ds,$$

and for $s = t - t_0$

$$\int_{t_0}^\infty [u(t\sigma_X + \mu_X) - u(t\sigma_Y + \mu_Y)] e^{\frac{-t^2}{2}} dt$$

$$= \int_0^\infty [u(s\sigma_X + t_1) - u(s\sigma_Y + t_1)] e^{-\frac{(s+t_0)^2}{2}} ds.$$

Hence

$$
\begin{aligned}
E(u(X) - u(Y)) \;&=\; \int_{-\infty}^{\infty} [u(t\sigma_X + \mu_X) - u(t\sigma_Y + \mu_Y)]\, e^{\frac{-t^2}{2}}\, dt \\
&\geq\; \int_{0}^{\infty} ([u(-t\sigma_X + t_1) + u(t\sigma_X + t_1)] \\
&\qquad - [u(-t\sigma_Y + t_1) + u(t\sigma_Y + t_1)]) e^{-(t+t_0)^2/2}\, dt \\
&\geq\; 0
\end{aligned}
$$

since

$$
[u(-t\sigma_X + t_1) + u(t\sigma_X + t_1)] - [u(-t\sigma_Y + t_1) + u(t\sigma_Y + t_1)] \geq 0.
$$

This can be seen as follows. For $t \geq 0$, let

$$
a = -t\sigma_Y + t_1 < b = -t\sigma_X + t_1 < c = t\sigma_X + t_1 < d = t\sigma_Y + t_1,
$$

then $a + d = b + c$, and the result follows from the following lemma which is of interest in its own right.

**Lemma 3.2** *Let* $u : \mathbb{R} \to \mathbb{R}$ *be a nondecreasing concave function,* $a < b < c < d$ *and* $a + d = b + c$. *Then*

$$
u(a) + u(d) \leq u(b) + u(c).
$$

**Proof.**  Let $e = (a + d)/2$ and write $b = \alpha a + (1 - \alpha)e$ for some $\alpha \in (0, 1)$, so that $c = \alpha d + (1 - \alpha)e$. Also let $A = \alpha u(a) + (1 - \alpha)u(e)$ and $B = \alpha u(d) + (1 - \alpha)u(e)$. By concavity, we have

$$
u(b) \geq \alpha u(a) + (1 - \alpha)u(e) = A
$$

and

$$
u(c) \geq \alpha u(d) + (1 - \alpha)u(e) = B.
$$

Note that, also by concavity $(u(e) = u((a+d)/2) \geq u(a)/2 + u(d)/2)$, and we have

$$
A \geq \frac{1 + \alpha}{2} u(a) + \frac{1 - \alpha}{2} u(d),
$$

and

$$
B \geq \frac{1 + \alpha}{2} u(d) + \frac{1 - \alpha}{2} u(a).
$$

Thus

$$
u(b) + u(c) \geq A + B \geq u(a) + u(b).
$$

∎

## 3.2 Multivariate stochastic dominance

It seems logical to mention at least the concept of stochastic dominance for random vectors. In the *univariate* case, i.e., for random variables (with values in the real line $\mathbb{R}$), we have developed the complete theory for FSD, SSD and TSD. They are to be applied to decisions concerning the choice among various investments with returns being considered as random variables.

Recall the usual partial order relation on $\mathbb{R}^n$: for $\mathbf{x} = (x_1, x_2, ..., x_n)$ and $\mathbf{y} = (y_1, y_2, ..., y_n)$ in $\mathbb{R}^n$,

$$\mathbf{x} \leq \mathbf{y} \text{ means } x_i \leq y_i \text{ for all } i = 1, 2, ..., n.$$

A function $u : \mathbb{R}^n \to \mathbb{R}$ is said to be *nondecreasing* if

$$\mathbf{x} \leq \mathbf{y} \text{ implies } u(\mathbf{x}) \leq u(\mathbf{y}).$$

In the univariate case, the FSD rule can be written as

$$X \succeq_1 Y \text{ if and only if } P(X > t) \geq P(Y > t) \text{ for all } t \in \mathbb{R},$$

or equivalently,

$$P(X \in A) \geq P(Y \in A), \text{ for any } A \subseteq \mathbb{R} \text{ of the form } A = (a, \infty).$$

Clearly, such subsets $A$ have nondecreasing indicator functions. A subset of the form $(-\infty, a)$ has a nonincreasing indicator function.

A subset of $\mathbb{R}^n$ is called an *upper set* (*lower set*) if its indicator function is nondecreasing (nonincreasing). Note that $A \subseteq \mathbb{R}^n$ is an upper set if and only if $\mathbf{x} \in A$ and $\mathbf{y} \geq \mathbf{x}$ imply $\mathbf{y} \in A$.

As in the univariate case, the expected utility representation of preference relations on random vectors of risk neutral individuals is this. The random vector $X$ is preferred to the random vector $Y$ if and only if

$$Eu(X) \geq Eu(Y) \text{ for any } u \in \mathcal{U}_1$$

where $\mathcal{U}_1$ is the space of all nondecreasing functions $u : \mathbb{R}^d \to \mathbb{R}$, and $X, Y$ are random elements with values in $\mathbb{R}^d$.

This first-order stochastic dominance (FSD) on random vectors can be characterized by probability laws of random vectors involved.

**Theorem 3.3** *A necessary and sufficient condition for*

$$Eu(X) \geq Eu(Y) \text{ for any } u \in \mathcal{U}_1$$

*is*

$$P(X \in A) \leq P(Y \in A)$$

*for any upper (Borel) set $A$ of $\mathbb{R}^d$.*

**Proof.** We prove this theorem by using approximations of functions as follows. Let $u : \mathbb{R}^d \to \mathbb{R}^+$ be a nonnegative increasing function. As a measurable function, $u$ is approximated by the standard procedure in measure theory, namely as follows. Let

$$u_n(x) = \begin{cases} \dfrac{i}{2^n} & \text{for } i/2^n \leq u(x) < (i+1)/2^n, \ i = 0, 1, ..., n2^n - 1 \\ n & \text{for } u(x) \geq n. \end{cases}$$

Then $u_n(x) \nearrow u(x)$, for each $x \in \mathbb{R}^d$, as $n \to \infty$.

Let $A_i = \{x \in \mathbb{R}^d : u(x) \geq i/2^n\}$, for each $n \geq 1$. Clearly the $A_i$'s are nested: $A_1 \supseteq A_2 \supseteq ... \supseteq A_{n2^n}$, with possible equality for some, and each $A_i$ is an (Borel) upper set. Thus we can write each $u_n$ as a positive linear combination of indicator functions of $A_i$'s as follows:

$$u_n(x) = 1/2^n \sum_{i=1}^{n2^n} 1_{A_i}(x).$$

Now if $u$ is a nonnegative decreasing function, then the $A_i$'s are still nested, but they are lower sets. Let $\mathcal{C}_1$ be the subset of $\mathcal{U}_1$ consisting of functions of the form $1_A$ where $A$ is an upper set in $\mathbb{R}^d$. ∎

To figure out a practical criterion for FSD, we observe that a *necessary condition* for FSD is this. Since $\mathcal{C}_1 \subseteq \mathcal{U}_1$, FSD implies

$$\Delta u(X, Y) = Eu(X) - Eu(Y) \geq 0 \text{ for } u \in \mathcal{C}_1.$$

Specifically, for each upper set $A$ in $\mathbb{R}^d$, take $u = 1_A$, and we have

$$E_X(1_A) \geq E_Y(1_A) \text{ or } P(X \in A) \geq P(Y \in A).$$

The simplest upper set in $\mathbb{R}^d$ is of the form $(a, \infty) = (a_1, \infty) \times (a_2, \infty) \times \cdots \times (a_d, \infty)$ where $a = (a_1, a_2, ..., a_d) \in \mathbb{R}^d$. Then we have

$$E_X(1_{(a,\infty)}) \geq E_Y(1_{(a,\infty)}) \text{ or } \int_a^\infty dF(x) \geq \int_a^\infty dG(x),$$

or equivalently, $\overline{F}(a) \leq \overline{G}(a)$ where $\overline{F}$ and $\overline{G}$ are survival distributions of $X$ and $Y$, i.e., $\overline{F}(a) = P(X \geq a)$.

Similarly, let $\mathcal{C}_1'$ be the subset of $\mathcal{U}_1$ consisting of functions of the form $-1_A$ where $A$ is a lower set in $\mathbb{R}^d$, then $\mathcal{C}_1' \subset \mathcal{U}_1$. Specifically, for each lower set $A$ in $\mathbb{R}^d$, take $u = -1_A$, if $X \succeq_1 Y$, we have

$$E_X(-1_A) \geq E_Y(-1_A) \text{ or } P(X \in A) \leq P(Y \in A).$$

The simplest lower set in $\mathbb{R}^d$ is of the form $(-\infty, a) = (-\infty, a_1) \times (-\infty, a_2) \times \cdots \times (-\infty, a_d)$ where $a = (a_1, a_2, ..., a_d) \in \mathbb{R}^d$, then we have

$$E_X(-1_{(-\infty,a)}) \geq E_Y(-1_{(-\infty,a)}) \text{ or } \int_{-\infty}^{a} dF(x) \geq \int_{-\infty}^{a} dG(x)$$

i.e., $F(x) \leq G(x)$.

We are going to show that the conditions

$$P(X \in A) \geq P(Y \in A) \text{ for all upper sets } A \text{ in } \mathbb{R}^d$$
$$P(X \in A) \leq P(Y \in A) \text{ for all lower sets } A \text{ in } \mathbb{R}^d$$

are also *sufficient* for FSD.

Let $u \in \mathcal{U}_1$. Let $u^+ = \max(u, 0)$ and $u^- = -\min(u, 0)$, the positive and negative parts of $u$, noting that they are both nonnegative functions, and $u^+$ is nondecreasing, while $u^-$ is nonincreasing.

For $u^+$ we use the approximation above: $u_n^+ \nearrow u^+$, with $Eu_n^+(X) \geq Eu_n^+(Y)$ for each $n$, we have, by the monotone convergence theorem,

$$Eu^+(X) = \lim_{n\to\infty} Eu_n^+(X) \geq \lim_{n\to\infty} Eu_n^+(Y) = Eu^+(Y).$$

For $u^-$ we can obtain a sequence $u_n(x) = 1/2^n \sum_{i=1}^{n2^n} 1_{A_i}(x)$ where each $A_i$'s is a lower set in $\mathbb{R}^d$, so that $u_n^- \nearrow u^-$. Then $Eu_n^-(X) \leq Eu_n^-(Y)$ for each $n$. Again, by the monotone convergence theorem,

$$Eu^-(X) = \lim_{n\to\infty} Eu_n^-(X) \leq \lim_{n\to\infty} Eu_n^-(Y) = Eu^-(Y)$$

and hence

$$Eu(X) = Eu^+(X) - Eu^-(X) \geq Eu^+(Y) - Eu^-(Y) = Eu(Y).$$

**Remark 3.4** *In portfolio optimization, there is a random vector of returns* $\mathbf{X} = (X_1, X_2, ..., X_n)$ *where, say, $X_i$ is the future return for stock $i$ in our portfolio consisting of $n$ different stocks. The investor needs to find an optimal allocation of her budget $a$ to the $n$ stocks.*

*Suppose $u$ is her utility function, then the problem is to find an optimal allocation $(b_1^*, b_2^*, ..., b_n^*)$ maximizing*

$$Eu\left(\sum_{i=1}^{n} b_i X_i\right)$$

*subject to*

$$\sum_{i=1}^{n} b_i = a.$$

*Thus we are facing the univariate case: comparing stochastically the random variables (not vectors) $\sum_{i=1}^{n} b_i X_i$ for different allocations $(b_1, b_2, ..., b_n)$.*

## 3.3   Stochastic dominance via quantile functions

Although the distribution function of a random variable contains all information about the random evolution of the variable, its quantile function plays a useful role in many aspects of statistical analysis, such as in quantile regression problems in econometrics. Here we merely mention that stochastic dominance rules can be expressed also in terms of quantile functions of random variables.

**Definition 3.5** *The* quantile function *or* inverse distribution *of a distribution $F$ is defined for $p \in (0, 1)$ as*

$$F^{-1}(p) = \inf\{x : F(x) \geq p\}.$$

**Remark 3.6** *Some useful facts about quantile functions are*

1. *$F^{-1}(F(x)) \leq x, \ x \in \mathbb{R}$.*

2. *$F(F^{-1}(p)) \geq p, \ p \in (0, 1)$.*

3. *$F(x) \geq t$ if and only if $x \geq F^{-1}(t)$.*

4. *$F^{-1} \circ F \circ F^{-1} = F^{-1}, \ F \circ F^{-1} \circ F = F$, and $(F^{-1})^{-1} = F$.*

5. *$F(x) \geq G(x), \forall x \in \mathbb{R}$ if and only if $G^{-1}(p) \geq F^{-1}(p), \forall p \in (0, 1)$.*

6. *$F^{-1}$ is increasing and left continuous and hence the set of discontinuity points of $F^{-1}$ is at most countable. Moreover, $F^{-1}$ is discontinuous at $p$ if and only if $F(x) \equiv p$ for all $x$ in some interval $I$.*

The following lemma is useful for computational purposes.

**Lemma 3.7** *For any continuous function* $g$,

$$E_X(g) = \int_{-\infty}^{\infty} g(x)dF(x) = \int_0^1 g(F^{-1}(y))dy$$

*in the sense that if one side exists, then so does the other and the two are equal.*

**Proof.** If $F$ is continuous and strictly increasing then $F^{-1}$ is the ordinary inverse of $F$. Hence, by simple substitution $y = F(x)$, we get the result. In fact, if $F$ is continuous and strictly increasing on some open interval $(a,b)$, then we also have

$$\int_I g(x)dF(x) = \int_{F(a^+)}^{F(b^-)} g(F^{-1}(y))dy.$$

Moreover, the formula above is still valid if $F$ is only continuous. Suppose, for example, that $F$ is constant on each $[b_i, a_{i+1}]$ then

$$\int_I g(x)dF(x) = \int_{(a,b_1)} g(x)dF(x) + \int_{(a_2,b_2)} g(x)dF(x)$$

$$+ \int_{(a_3,b_3)} g(x)dF(x) + \cdots$$

On each interval $[a_i, b_i]$, $F$ is strictly increasing (and continuous), so

$$\int_{(a_i,b_i)} g(x)dF(x) = \int_{F(a_i^+)}^{F(b_i^-)} g(F^{-1}(y))dy.$$

Hence,

$$\int_I g(x)dF(x) = \int_{F(a^+)}^{F(b_1^-)} g(F^{-1}(y))dy + \int_{F(a_2^+)}^{F(b_2^-)} g(F^{-1}(y))dy$$

$$+ \int_{F(a_3^+)}^{F(b_3^-)} g(F^{-1}(y))dy + \cdots$$

But $F(b_i^-) = F(b_i) = F(a_{i+1}) = F(a_{i+1}^+)$, and

$$\int_I g(x)dF(x) = \int_{F(a^+)}^{F(b^-)} g(F^{-1}(y))dy.$$

Now, suppose the set of discontinuities of $F$ is $\{c_1, c_2, ...\}$. Then

$$
\begin{aligned}
E_X(g) &= \int_{(-\infty,c_1)} g(x)dF(x) + \int_{[c_1,c_1]} g(x)dF(x) + \int_{(c_1,c_2)} g(x)dF(x) \\
&\quad + \int_{[c_2,c_2]} g(x)dF(x) + \int_{(c_2,c_3)} g(x)dF(x) + \cdots \\
&= \int_{(-\infty,c_1)} g(x)dF(x) + g(c_1)(F(c_1^+) - F(c_1^-)) \\
&\quad + \int_{(c_1,c_2)} g(x)dF(x) + g(c_2)(F(c_2^+) - F(c_2^-)) \\
&\quad + \int_{(c_2,c_3)} g(x)dF(x) + \cdots
\end{aligned}
$$

(Note: $F$ is continuous on each $(c_i, c_{i+1})$.)

$$
\begin{aligned}
&= \int_0^{F(c_1^-)} g(F^{-1}(y))dy + g(c_1)(F(c_1^+) - F(c_1^-)) \\
&\quad + \int_{F(c_1^+)}^{F(c_2^-)} g(F^{-1}(y))dy + g(c_2)(F(c_2^+) - F(c_2^-)) \\
&\quad + \int_{F(c_2^+)}^{F(c_3^-)} g(F^{-1}(y))dy + \cdots
\end{aligned}
$$

On the other hand,

$$
\int_{F(c_i^-)}^{F(c_i^+)} g(F^{-1}(y))dy = \int_{F(c_i^-)}^{F(c_i^+)} g(c_i)dy = g(c_i)(F(c_i^+) - F(c_i^-)).
$$

Therefore,

$$
\begin{aligned}
E_X(g) &= \int_0^{F(c_1^-)} g(F^{-1}(y))dy + \int_{F(c_1^-)}^{F(c_1^+)} g(F^{-1}(y))dy \\
&\quad + \int_{F(c_1^+)}^{F(c_2^-)} g(F^{-1}(y))dy + \int_{F(c_2^-)}^{F(c_2^+)} g(F^{-1}(y))dy \\
&\quad + \int_{F(c_2^+)}^{F(c_3^-)} g(F^{-1}(y))dy + \cdots \\
&= \int_0^1 g(F^{-1}(y))dy.
\end{aligned}
$$

∎

### 3.3.1  First-order stochastic dominance

Recall that, with $\mathcal{U}_1$ being the space of all nondecreasing functions $u :$ $\mathbb{R} \to \mathbb{R}$, the random variable $X$, with distribution function $F$, dominates the random variable $Y$, with distribution $G$, in FSD, denoted as $X \succeq_1 Y$ if

$$Eu(X) \geq Eu(Y) \text{ for any } u \in \mathcal{U}_1,$$

which is equivalent to

$$F(.) \leq G(.)$$

which, in turn, is equivalent to $F^{-1}(.) \geq G^{-1}(.)$ (by Remark 3.6, part 5).

### 3.3.2  Second-order stochastic dominance

Consider the space $\mathcal{U}_2$ consisting of nondecreasing and concave functions. Recall that by SSD we mean

$$Eu(X) \geq Eu(Y) \text{ for any } u \in \mathcal{U}_2,$$

and we write $X \succeq_2 Y$.

Let $\mathcal{C}_2$ denote the subset of $\mathcal{U}_2$ consisting of functions of the form

$$\varphi_a(x) = (x-a)1_{(-\infty,a]}(x), \text{ for } a \in \mathbb{R},$$

then $\mathcal{C}_2$ spans $\mathcal{U}_2$ in the sense that

$$Eu(X) \geq Eu(Y), \forall u \in \mathcal{U}_2 \text{ if and only if } Eu(X) \geq Eu(Y), \forall u \in \mathcal{C}_2,$$

resulting in

$$E\varphi_a(X) \geq E\varphi_a(Y) \text{ for all } a \in \mathbb{R},$$

or equivalently,

$$\int_{-\infty}^{a} (x-a)dF(x) \geq \int_{-\infty}^{a} (x-a)dG(x).$$

By a formal substitution $y = F(x)$, we have

$$\int_{-\infty}^{a} (x-a)dF(x) = \int_{0}^{F(a)} (F^{-1}(y) - a)dy.$$

Thus, $X \succeq_2 Y$ if and only if

$$\int_{0}^{F(a)} (F^{-1}(y) - a)dy \geq \int_{0}^{G(a)} (G^{-1}(y) - a)dy \text{ for all } a \in \mathbb{R}$$

$$\Leftrightarrow$$

$$\int_{0}^{G(a)} (a - G^{-1}(y))dy \geq \int_{0}^{F(a)} (a - F^{-1}(y))dy \text{ for all } a \in \mathbb{R}.$$

Note that $G^{-1}(y) \leq a$ for $y \leq G(a)$.

**Theorem 3.8** $X \succeq_2 Y$ *if and only if* $\int_0^p (F^{-1}(y) - G^{-1}(y))dy \geq 0$ *for all* $p \in (0, 1)$. *(Note that* $\int_0^1 F^{-1}(y)dy = E(X)$.*)*

**Proof.** ($\Rightarrow$) For all $p \in (0, 1)$, let $a = G^{-1}(p)$ then we have

$$\int_0^{G(a)} (a - G^{-1}(y))dy \geq \int_0^{F(a)} (a - F^{-1}(y))dy.$$

But since $a - F^{-1}(y) \geq 0$ if and only if $y \leq F(a)$, then

$$\int_0^{F(a)} (a - F^{-1}(y))dy \geq \int_0^p (a - F^{-1}(y))dy,$$

whether $p \geq F(a)$ or $p < F(a)$.

In addition,

$$\int_0^{G(a)} (a - G^{-1}(y))dy = \int_0^{G(G^{-1}(p))} (a - G^{-1}(y))dy$$

$$= \int_0^p (a - G^{-1}(y))dy + \int_p^{G(G^{-1}(p))} (a - G^{-1}(y))dy.$$

(Recall that $G(G^{-1}(p)) \geq p$.)

However, for all $y$,

$$p \leq y \leq G(G^{-1}(p)) \Rightarrow G^{-1}(p) \leq G^{-1}(y) \leq G^{-1}G(G^{-1}(p)) = G^{-1}(p).$$

That means $G^{-1}(y) = G^{-1}(p) = a$ for all $y$ such that $p \leq y \leq G(G^{-1}(p))$.

Hence,

$$\int_0^{G(a)} (a - G^{-1}(y))dy = \int_0^p (a - G^{-1}(y))dy.$$

Finally, we get

$$\int_0^p (a - G^{-1}(y))dy \geq \int_0^p (a - F^{-1}(y))dy,$$

so

$$\int_0^p (F^{-1}(y) - G^{-1}(y))dy \geq 0.$$

($\Leftarrow$) For all $a \in \mathbb{R}$, let $p = F(a)$ then we have

$$\int_0^{F(a)} (F^{-1}(y) - G^{-1}(y))dy \geq 0$$

or

$$\int_0^{F(a)} (a - G^{-1}(y))dy \geq \int_0^{F(a)} (a - F^{-1}(y))dy.$$

By the same argument we have

$$\int_0^{G(a)} (a - G^{-1}(y))dy \geq \int_0^{F(a)} (a - G^{-1}(y))dy$$

(whether $G(a) \geq F(a)$ or not). Hence,

$$\int_0^{G(a)} (a - G^{-1}(y))dy \geq \int_0^{F(a)} (a - F^{-1}(y))dy.$$

∎

### 3.3.3 Stochastic dominance rule for risk seekers

Let $\mathcal{U}_2'$ be the space of nondecreasing and convex functions, and $\mathcal{C}_2'$ its subset consisting of elements of the form $\psi_a(x) = (x - a)1_{[a,\infty)}(x)$, $a \in \mathbb{R}$.

Recall that $X$ dominates $Y$ in RSSD (Risk Seeking Stochastic Dominance), denoted as $X \succeq_{2'} Y$, if $Eu(X) \geq Eu(Y)$ for any $u \in \mathcal{U}_2'$.

We already showed that $\mathcal{C}_2'$ spans $\mathcal{U}_2'$ in the sense that

$$Eu(X) \geq Eu(Y), \forall u \in \mathcal{U}_2' \text{ if and only if } Eu(X) \geq Eu(Y), \forall u \in \mathcal{C}_2',$$

resulting in

$$E\psi_a(X) \geq E\psi_a(Y) \text{ for all } a \in \mathbb{R},$$

or equivalently,

$$\int_a^\infty (x - a)dF(x) \geq \int_a^\infty (x - a)dG(x).$$

By a formal substitution $y = F(x)$, we have

$$\int_a^\infty (x - a)dF(x) = \int_{F(a)}^1 (F^{-1}(y) - a)dy.$$

(Note that the integrand $F^{-1}(y) - a \geq 0$ if and only if $y \geq F(a)$.)

Then we get

$$X \succeq_2 Y$$

if and only if

$$\int_{F(a)}^1 (F^{-1}(y) - a)dy \geq \int_{G(a)}^1 (G^{-1}(y) - a)dy, \forall a \in \mathbb{R}.$$

**Theorem 3.9** $X \succeq_2' Y$ *if and only if* $\int_p^1 (F^{-1}(y) - G^{-1}(y))dy \geq 0$ *for all* $p \in (0, 1)$.

**Proof.** ($\Rightarrow$) For any $p \in (0, 1)$, let $a = F^{-1}(p)$ then by assumption

$$\int_{F(a)}^1 (F^{-1}(y) - a)dy \geq \int_{G(a)}^1 (G^{-1}(y) - a)dy. \qquad (3.1)$$

Since

$$F(a) = F(F^{-1}(p)) \geq p,$$

we get

$$\int_{F(a)}^1 (F^{-1}(y) - a)dy = \int_p^1 (F^{-1}(y) - a)dy - \int_p^{F(a)} (F^{-1}(y) - a)dy.$$

But for any $y$ such that $p \leq y \leq FF^{-1}(p)$, we have

$$F^{-1}(p) \leq F^{-1}(y) \leq F^{-1}FF^{-1}(p) \equiv F^{-1}(p).$$

Thus,

$$\int_p^{F(a)} (F^{-1}(y) - a)dy = 0.$$

Hence,

$$\int_{F(a)}^1 (F^{-1}(y) - a)dy = \int_p^1 (F^{-1}(y) - a)dy. \qquad (3.2)$$

On the other hand, if $p > G(a)$, then

$$\int_{G(a)}^1 (G^{-1}(y) - a)dy \geq \int_p^1 (G^{-1}(y) - a)dy. \qquad (3.3)$$

(Note that the integrand $G^{-1}(y) - a \geq 0$ *for* $y \geq G(a)$.) And if $p \leq G(a)$, then

$$\int_{G(a)}^1 (G^{-1}(y) - a)dy = \int_p^1 (G^{-1}(y) - a)dy - \int_p^{G(a)} (G^{-1}(y) - a)dy.$$

But for $y \leq G(a)$, $G^{-1}(y) \leq a$, so

$$\int_p^{G(a)} (G^{-1}(y) - a)dy \leq 0.$$

Hence,

$$\int_{G(a)}^{1} (G^{-1}(y) - a)dy \geq \int_{p}^{1} (G^{-1}(y) - a)dy. \tag{3.4}$$

By Equations (3.1), (3.2), (3.3) and (3.4), we get

$$\int_{p}^{1} (F^{-1}(y) - a)dy \geq \int_{p}^{1} (G^{-1}(y) - a)dy$$

which implies

$$\int_{p}^{1} [F^{-1}(y) - G^{-1}(y)]dy \geq 0.$$

( $\Longleftarrow$ ) For any $a \in \mathbb{R}$, let $p = G(a)$. Then by the assumption we have

$$\int_{G(a)}^{1} [F^{-1}(y) - G^{-1}(y)]dy$$

or

$$\int_{G(a)}^{1} (F^{-1}(y) - a)dy \geq \int_{G(a)}^{1} (G^{-1}(y) - a)dy.$$

Now, if $F(a) \geq G(a)$ then

$$\int_{G(a)}^{1} (F^{-1}(y) - a)dy = \int_{F(a)}^{1} (F^{-1}(y) - a)dy + \int_{G(a)}^{F(a)} (F^{-1}(y) - a)dy.$$

But the last integral is nonpositive since $F^{-1}(y) - a \leq 0$ for $y \leq F(a)$. Hence,

$$\int_{F(a)}^{1} (F^{-1}(y) - a)dy \geq \int_{G(a)}^{1} (F^{-1}(y) - a)dy.$$

And if $F(a) < G(a)$, then

$$\int_{F(a)}^{1} (F^{-1}(y) - a)dy \geq \int_{G(a)}^{1} (F^{-1}(y) - a)dy.$$

Therefore,

$$\int_{F(a)}^{1} (F^{-1}(y) - a)dy \geq \int_{G(a)}^{1} (G^{-1}(y) - a)dy,$$

which ends the proof. ∎

## 3.4　Exercises

1. Provide an example to show that the MV rule might not be consistent with expected utility criterion for risk averse decision-makers. Specifically, find two random variables $X, Y$ with distribution functions $F, G$, respectively, with $EX \geq EY$ and $\sigma^2(X) \leq \sigma^2(Y)$, and a utility function $u$ (nondecreasing and concave) such that $Eu(X) < Eu(Y)$.

2. Let $X, Y$ be normally distributed with mean $\mu_1, \mu_2$ and variance $\sigma_1^2, \sigma_2^2$, respectively. Show that if the distribution functions $F, G$ of $X$ and $Y$ are distinct, they can only intersect each other at most once, i.e.,

   (a) There exists exactly one point $x_0$ such that $F(x_0) = G(x_0)$ if $\sigma_1^2 \neq \sigma_2^2$.

   (b) $\sigma_1^2 = \sigma_2^2$ if and only if $F(x) \neq G(x)$ for all $x \in \mathbb{R}$.

3. Let $X, Y$ be normally distributed with the same mean and variances $\sigma_X^2 \neq \sigma_Y^2$. Can either $X$ or $Y$ dominate the other in FSD sense?

4. If $X \succeq_2 Y$ and $EX = EY$, show that $Var(X) \leq Var(Y)$.

5. Let $X$ be a real-valued random variable, defined on $(\Omega, \mathcal{A}, P)$. The distribution function of $X$ is $F : \mathbb{R} \rightarrow [0, 1]$, $F(x) = P(X \leq x)$. The left-continuous inverse of $F$ is $F^{-1} : (0, 1) \rightarrow \mathbb{R}$, $F^{-1}(t) = \inf\{x \in \mathbb{R} : F(x) \geq t\}$. Show the following:

   (a) If $U$ is a random variable, uniformly distributed on $(0, 1)$, then $F^{-1}(U)$ is equal to $X$ in distribution.

   (b) $F \circ F^{-1}$ is the identical function if $F$ is continuous.

   (c) $F(X)$ is uniformly distributed on $(0, 1)$ if $F$ is continuous.

   (d) $F$ is continuous if and only if $F^{-1}$ is strictly increasing.

   (e) Let $F, G$ be two continuous distribution functions. Let $X$ have $F$ as its distribution, then $G^{-1}(F(X))$ has $G$ as its distribution.

6. Let $X$ be a random variable with distribution function $F$.

(a) Consider the problem of minimizing $E|X - a|$ over $a \in \mathbb{R}$. Write

$$E|X - a| = \int_{-\infty}^{a} (a - x)dF(x) + \int_{a}^{\infty} (x - a)dF(x)$$

and show that the solution is the median of $F$, i.e., $F^{-1}(1/2)$.

(b) Consider the objective function, for $\alpha \in (0, 1)$,

$$(1 - \alpha) \int_{-\infty}^{a} (a - x)dF(x) + \alpha \int_{a}^{\infty} (x - a)dF(x).$$

Show that the solution of the minimization of this objective function, over $a \in \mathbb{R}$, is the $\alpha$-quantile of $F$, i.e., $F^{-1}(\alpha)$.

7. Let $X$ be a normal random variable with distribution function $F$. Compute $F^{-1}$, in terms of the distribution function of the standard normal random variable, when

(a) $X$ is normally distributed with mean $\mu$ and variance $\sigma^2$.

(b) $X$ is lognormal, i.e., $X = e^Y$ where $Y$ is normally distributed with mean $\mu$ and variance $\sigma^2$.

# Chapter 4

# Financial Risk Measures

*This chapter presents the fundamentals of an important related concept in investment and actuarial sciences, namely risk. We discuss mainly the problem of risk modeling and that of consistency with respect to stochastic dominance.*

## 4.1   The problem of risk modeling

Assuming that individuals are capable of ranking random variables in some rational fashion, we have introduced the concept of utility functions to *model preference relations* in Chapter 1. Of course, preference relations or their utility representations encode implicitly the risk factor. Once an alternative is chosen according to a preference relation (equivalently, according to its utility representation), we face a new problem: how to "manage" the follow-up activities. For example, banks need to figure out required capitals to back their trading activities.

It's about risk! Indeed, the so-called *financial risk management* is about "managing" exposure to risk, such as market risk and credit risk. For these purposes, we need a *quantitative concept of risk* in its own right.

Managing risk in all sectors of an economy is viewed as the key to making organizations successful in delivering their objectives while projecting the interests of their stakeholders. Specifically, risk management allows an organization to have increased confidence in achieving its desired outcomes, effectively constrain threats to acceptable levels and take informed decisions about exploiting opportunities.

It is all about risk! In Chapter 1, we have seen that utility theory (as a rational model for preferences among alternatives) incorporates risk factor via *shapes* of utility functions, exhibiting *risk attitudes* of individuals. These are qualitative concepts.

But risk is a concept by itself! It is related to any chosen alternative. Of course, we can proceed to consider risk first, then suggest preference relations for choosing alternatives according to risk's characteristics, namely the probability of a possible loss and the amount of the possible loss (as an operator on random variables which is a function of these two main characteristics of risk).

In the finance context it is clear that choices among uncertain quantities involve risk. The stochastic dominance rules which are derived from the expected utility principle implicitly take risk into account, but rather at a qualitative level. This is manifested in the fact that stochastic dominance is derived for different shapes of utility functions which, in turn, represent different risk attitudes of individuals.

Just like the concept of chance, risk needs to be modeled quantitatively for applications. While risk seems to be understood from a common sense viewpoint, its numerical modeling seems much more delicate than chance. Perhaps this is due to the fact that, for the concept of chance, we can rely on concrete templates to propose general axioms for probability measures, namely games of chance where we know the game structure.

Moreover, risk is context dependent. Thus it is unlikely that we can achieve an axiomatic theory of risk. As we will see, the best we can achieve is listing a set of desirable and natural properties that a risk measure should satisfy. Of course, such a list depends on our perception of risk in a given context.

As statistical decision theory, while restricted to specific problems of statistical inference, addresses choices under uncertainty, no stone is left unturned on its paths. For example, in the search for good estimators of population parameters, we define risk as a measure of the error we might commit when using an estimator. Specifically, we specify a loss function and define the risk of an estimator, with respect to that loss function, as the expected value of the loss. This numerical (quantitative) measure of risk provides us with a guideline to search for best estimators, i.e., solving the problem of choice among various possible estimators.

Of course, the choice of a loss function remains a matter of debate. For unbiased estimation, with the square loss function, the variance of an estimator is used as a measure of risk. This symmetric risk measure is appropriate in the context of statistical estimation.

We turn now to the question, "how can we assess the risk, say, of a financial position?" While stochastic dominance rules shed light on which future prospects are riskier than others, they do not provide us with numerical values of risk which we need to manage further our financial positions. The need for quantitative risk modeling appears to be obvious. Note however that the qualitative choice of uncertain (risky) prospects and the assessment of quantitative risks to these prospects are related. This will be the problem of consistency of risk assessment with respect to stochastic dominance.

Basically, the problem of risk modeling is this. A financial position is captured by a random variable $X$. Information about $X$ is in its distribution function $F$. In this financial context where risk is about risk of losing money, we wish to define numerical risk for $X$ as some appropriate number $\rho(X)$, and we call the functional $\rho(.)$ a risk measure. Clearly, the problem is how we are going to define $\rho(.)$ in such a way as to capture our risk perception and be suitable for financial management purposes.

Note that the risky prospect $X$ is the risk itself! Thus we are talking about measuring risk of a risk!, and yet, we are unable to give a *definition of risk*! This is somewhat similar to the concept of *shape*, say, of geometric objects, where we can give examples but not a formal definition. However, from a common sense viewpoint, the concept of risk is understood. For the needs of economic decisions, what we can do is to provide a *quantitative theory for risk behavior*. Quoting R. Engle (Nobel lecture 2003), "The advantage of knowing about risks is that we can change our behavior to avoid them."

In this chapter we will present the state-of-the-art of risk modeling, provide the mathematical foundations for risk measures, investigate their properties, including their consistency with stochastic dominance rules. The problem of consistency of risk measures with stochastic dominance is an important issue in financial decision-making since risk modeling should be conducted within rational decision theory from which stochastic dominance is derived.

As we will see, the situation for risk modeling, i.e., how to suggest risk measures, is somewhat similar to the beginning of statistical estimation. People suggested plausible risk measures based only on intuitive grounds, then a theory of risk began taking shape, in which desirable conditions are listed to guide the search for realistic risk measures. Apparently, there is still no consensus in the economic community about which risk measures to use; the research continues.

Since both stochastic dominance rules and risk measures are just mathematical formulae containing unknown distributions, we need to use

statistics for estimation and testing. That will be discussed in Chapters
6 and 7.

## 4.2   Some popular risk measures

Random variables involved represent uncertain quantities such as future
returns or losses of investment portfolios, or payments in actuarial science. A risk measure is a mapping from some collection of random vari-
ables $\mathcal{X}$ of interest to the real line $\mathbb{R}$.   All random variables are defined on some probability space $(\Omega, \mathcal{A}, P)$ and taking values in $\mathbb{R}$, i.e.,
$\rho(.) : \mathcal{X} \to \mathbb{R}$. As stated above, for $X \in \mathcal{X}$, the risk of $X$, denoted as
$\rho(X)$, is in fact a function of $X$, through its distribution function $F$. Us-
ing statistical language, $\rho(X)$ is a population parameter. In other words,
$\rho(X)$ depends only on $F$, so that two random variables $X, Y$ having the
same distribution will have the same risk. This is referred in the literature
as the *law invariance* for risk modeling, where "law" refers to probabilistic distributions of random variables, by analogy with dynamical physical
systems.

   We survey in this section several well-known risk measures which
were proposed from different perspectives. Systematic constructions of
risk measures can take place after we have a reasonable framework to
work with.  The evaluation of each proposed risk measure is based on
how intuitive it captures the concept of financial risk.

### 4.2.1   Variance

In the early theory of portfolio selections, Markowitz proposed the use of
variance to quantify risk of portfolios,

$$\rho(X) \;=\; Var(X) = E(X - EX)^2$$
$$=\; \int_{-\infty}^{\infty} \int_{-\infty}^{\infty} (1/2)(x - y)^2 dF(x)dF(y)$$

in the sense that the portfolio with small variance is considered as less
risky.

   However, this symmetric measure of risk seems to have the draw-
back of treating equally the desirable upside and undesirable downside
variations of return distributions. Moreover, as it can be verified easily, the variance as a risk measure is not consistent with first stochastic dominance rule, i.e., if $X \succeq_1 Y$, it does not follow in general that
$Var(X) \leq Var(Y)$.  In fact the contrary could happen.  Indeed, if

$X \succeq_1 Y$, then $Eu(X) \geq Eu(Y)$ for all nondecreasing utility function $u$, in particular for $u$ nondecreasing and convex. This is an indication that "$X$ is more variable than $Y$." For example, for nonnegative random variables, $X \succeq_1 Y$ will imply $Eu(X) \geq Eu(Y)$ for all convex $u$ when $EX = EY$, so that, taking $u(x) = x^2$, we see that $Var(X) \geq Var(Y)$.

In view of the above reasons, economists search for risk measures which should be based on the downside tail distribution of returns.

## 4.2.2 Value-at-risk

For a *loss* random variable $X$ with distribution function $F$ (where $F(x) = P(X \leq x)$), the Value-at-Risk of $X$, denoted from now on by $VaR_\alpha(X)$, or $VaR_\alpha(F)$, is taken to be the $\alpha$-quantile of $F$, i.e.,

$$VaR_\alpha(X) = F^{-1}(\alpha) = \inf\{x \in \mathbb{R} : F(x) \geq \alpha\}$$

for $\alpha \in (0,1)$. Note that the above infimum is in fact a minimum, i.e., the infimum is attained: the $\alpha$-quantile $F^{-1}(\alpha)$ is the position such that $P(X \leq F^{-1}(\alpha)) \geq \alpha$. Thus, $P(X > F^{-1}(\alpha)) \leq 1 - \alpha$.

**Remark 4.1** *We have termed $F^{-1}(.)$ the quantile function of $X$ (or of $F$). In fact it is the lower quantile if we consider an $\alpha$-quantile as any value $q(\alpha)$ in the interval*

$$[\inf\{x \in \mathbb{R} : F(x) \geq \alpha\}, \inf\{x \in \mathbb{R} : F(x) > \alpha\}].$$

*The left endpoint of the above interval is the lower quantile $F^{-1}(\alpha)$, which is also equal to $\sup\{x \in R : F(x) < \alpha\}$, and is called the left-continuous inverse of $F$, noting that $F^{-1}(.)$ is nondecreasing and left continuous. The right endpoint is the upper quantile function of $F$, namely $F^+(\alpha) = \inf\{x \in R : F(x) > \alpha\} = \sup\{x \in R : F(x) \leq \alpha\}$. The function $F^{-1}(.)$ is nondecreasing and right continuous. So a quantile function is any function $q(.)$ such that $F^{-1}(.) \leq q(.) \leq F^+(.)$. Note that any quantile function is nondecreasing. In fact all quantile functions coincide almost everywhere on $(0,1)$. Note also that $q(.)$ is a quantile function of $F$ if and only if*

$$F(q(\alpha)-) \leq \alpha \leq F(q(\alpha)) \text{ for all } \alpha \in (0,1)$$

*where $F(q(\alpha)-)$ denotes the left limit of $F$ at $q(\alpha)$, noting that $F(q(\alpha)) = F(q(\alpha)+)$.*

Viewing $X$ as a loss variable, $VaR_\alpha(X)$ is the maximum possible loss at level $\alpha$ in the sense that the probability that the loss will exceed $VaR_\alpha(X)$ is less than $1 - \alpha$. The incorporation of $\alpha$ in the definition of

$VaR_\alpha(X)$ bears the flavor of statistical confidence intervals. $VaR_\alpha(X)$ assesses the risk at some confidence level. Here it sounds like risk is a matter of degree.

The rationale for this proposed risk measure can be elaborated upon as follows. Being unable to assert with certainty a maximum possible loss, we are content with replacing it with a bound with high probability. As a practice in statistics or uncertainty analysis, we specify the linguistic term "high probability" with a number $\alpha \in (0, 1)$, called our confidence level, say, $\alpha = 0.95$, and take the plausible quantity $F^{-1}(\alpha) = \min\{x \in \mathbb{R} : F(x) \geq \alpha\}$, which is the $\alpha$-quantile of $F$, to be our risk measure $\rho_\alpha(X)$. Whether or not this is a "good" measure of the risk we intend to quantify remains to be seen. It brings out, however, the road map for formulating appropriate risk measures in each given context. In this example, the upper bound $F^{-1}(\alpha)$ is such that $P(X > a) \leq 1 - \alpha$, for any $a \geq F^{-1}(\alpha)$, so that $F^{-1}(\alpha)$ serves as an upper bound for $X$ at high confident level $\alpha$.

If the loss variable $X$ is the payment to future claims of an insurance company, the above $\rho_\alpha(X) = F^{-1}(\alpha)$ is the capital the company should hold to avoid insolvency with a high degree $\alpha$ of confidence. Note that we need to choose $\alpha$ (!) just like in confidence interval estimation.

Another setting is this. The variables of interest are future net worths of financial positions. In this case, we are concerned with the risk of having an "unacceptable" position. To avoid such situations, we would like to look at risk measures as some *capital requirement*, namely as an amount of money $\rho(X)$ needed to add to $X$, by some investment strategy, to make the position acceptable.

The point is this. In trying to quantify risks, we need to place ourself in a specific economic context, and spell out the meaning of the quantitative risk we intend to model.

It is interesting, in fact useful, to observe that while the mean of $X$ is a Lebesgue integral, the quantiles of $X$ can be written as another type of integral generalizing the Lebesgue integral. This can be seen as follows.

$$
\begin{aligned}
F^{-1}(\alpha) &= \int_0^{F^{-1}(\alpha)} dt = \int_{\{t : t < F^{-1}(\alpha)\}} dt \\
&= \int_{\{t : F(t) < \alpha\}} dt = \int_{\{t : 1 - F(t) > 1 - \alpha\}} dt \\
&= \int_0^\infty 1_{(1-\alpha, 1]}(1 - F(t)) dt.
\end{aligned}
$$

Thus, if we let

$$g_\alpha : [0,1] \to [0,1] \text{ where } g_\alpha(x) = 1_{(1-\alpha,1]}(x)$$

then

$$
\begin{aligned}
F^{-1}(\alpha) &= \int_0^\infty g_\alpha(1 - F(t))dt \\
&= \int_0^\infty (g_\alpha \circ P)(X > t)dt.
\end{aligned}
$$

Now observe that the set function $v = g_\alpha \circ P$ in the above integral, defined on the $\sigma$-field $\mathcal{A}$, is no longer additive (let alone a probability measure). Since $g_\alpha(.)$ is nondecreasing and $g_\alpha(0) = 0$, $g_\alpha(1) = 1$, it follows that the set function $v$ is monotone increasing, i.e., $A \subseteq B \Longrightarrow v(A) \le v(B)$, and $v(\varnothing) = 0, v(\Omega) = 1$. We will call such a set function a *capacity*. If we recall, say for $X \ge 0$, that

$$EX = \int_\Omega X(\omega)dP(\omega) = \int_0^\infty P(X > t)dt,$$

then, replacing (or generalizing) the probability measure $P$ by the capacity $v$, we obtain a quantile value

$$\Gamma^{-1}(\alpha) = \int_0^\infty v(X > t)dt.$$

We refer to this as a *Choquet integral representation* of $VaR_\alpha(.)$.

For an arbitrary capacity $v$, the last integral is viewed as an integral of $X$ with respect to $v$, and is called a *Choquet integral* in honor of Gustave Choquet who considered it in his work on theory of capacities [26]. As we will see, Choquet integral representations of risk measures are essential in risk modeling. For example, using the above Choquet integral representation for $VaR_\alpha(.)$, it follows readily that $VaR_\alpha(.)$ has several desirable properties such as

(i) If $X \ge Y$ (a.s.), then $VaR_\alpha(X) \ge VaR_\alpha(Y)$,

(ii) For any $\lambda > 0, VaR_\alpha(\lambda X) = \lambda VaR_\alpha(X)$, and

(iii) For any $t \in \mathbb{R}$, $VaR_\alpha(X + t) = VaR_\alpha(X) + t$.

Details of these results will be explained after an entire section on the theory of the Choquet integral.

However, the risk measure $VaR_\alpha(.)$ lacks a basic desirable property in investment science, namely *diversification*. Diversification should reduce risk. In terms of risk operators, this is expressed by saying that a risk measure $\rho(.)$ is compatible with diversification if it is *subadditive*, i.e.,

$$\rho(X + Y) \le \rho(X) + \rho(Y).$$

This fact can be also examined by using the Choquet integral representation of $VaR_\alpha(.)$, but here is an example.

**Example 4.2** *Let $X, Y$ be two independent and identically distributed Bernoulli random variables with parameter $\theta$, i.e., $P(X = 1) = 1 - P(X = 0) = \theta$. Then for any $\alpha$ such that $(1 - \theta)^2 < \alpha < 1 - \theta$, we have $F_X^{-1}(\alpha) = F_Y^{-1}(\alpha) = 0$, whereas $F_{X+Y}^{-1}(\alpha) > 0$. Thus $\rho_\alpha(X + Y) \le \rho_\alpha(X) + \rho_\alpha(Y)$ does not hold for such $\alpha$.*

### 4.2.3  Tail value-at-risk

More conservative than the risk measure $VaR_\alpha(.)$, another risk measure is defined by considering the average of $VaR_\alpha(.)$ over the $\alpha$-upper tail. We call it the *Tail Value-at-Risk*, in symbols $TVaR_\alpha(.)$. For $\alpha \in (0, 1)$, and $X$ with distribution function $F$, the tail value-at-risk at level $\alpha$ is defined as

$$TVaR_\alpha(X) = (1/(1 - \alpha)) \int_\alpha^1 F^{-1}(t)dt.$$

Clearly $TVaR_\alpha(X) \ge VaR_\alpha(X)$. In fact, we have

$$TVaR_\alpha(X) = VaR_\alpha(X) + E(X - F^{-1}(\alpha))^+/(1 - \alpha),$$

where $(X - a)^+$ denotes the positive part of $(X - a)$, i.e., $\max\{X - a, 0\}$.

In actuarial science, $E(X - a)^+$ is the *stop-loss* with retention $a$. The expectation $E(X - F^{-1}(\alpha))^+$ is referred to as the *expected shortfall* at level $\alpha$, which is the expected loss for a stop-loss contract with retention $F^{-1}(\alpha)$.

Note that when $E|X| < \infty$, we have

$$E(X - F^{-1}(\alpha))^+ = \int_{F^{-1}(\alpha)}^\infty (x - F^{-1}(\alpha))dF(x) = \int_{F^{-1}(\alpha)}^\infty (1 - F(x))dx$$

by integration by parts.

If $U$ is a random variable, uniformly distributed on $(0, 1)$, then $X$ is equal in distribution to $F^{-1}(U)$, since $F^{-1}(u) \leq x$ if and only if $u \leq F(x)$. Thus,

$$
\begin{aligned}
E(X - F^{-1}(\alpha))^+ &= E(F^{-1}(U) - F^{-1}(\alpha))^+ \\
&= \int_0^1 (F^{-1}(t) - F^{-1}(\alpha))^+ dt \\
&= \int_\alpha^1 (F^{-1}(t) - F^{-1}(\alpha)) dt,
\end{aligned}
$$

since

$$
(F^{-1}(t) - F^{-1}(\alpha))^+ = F^{-1}(t) - F^{-1}(\alpha)
$$

when

$$
F^{-1}(t) \geq F^{-1}(\alpha),
$$

which happens when $t \geq \alpha$. Thus,

$$
E(X - F^{-1}(\alpha))^+ = \int_\alpha^1 F^{-1}(t) dt - (1 - \alpha) F^{-1}(\alpha),
$$

so that

$$
\begin{aligned}
TVaR_\alpha(X) &= (1/(1 - \alpha)) \int_\alpha^1 F^{-1}(t) dt \\
&= F^{-1}(\alpha) + E(X - F^{-1}(\alpha))^+ / (1 - \alpha).
\end{aligned}
$$

It turns out that $TVaR_\alpha(.)$ also admits a Choquet integral representation. Indeed, from the above, we have

$$
\begin{aligned}
TVaR_\alpha(X) &= VaR_\alpha(X) + (1/(1 - \alpha)) E(X - F^{-1}(\alpha))^+ \\
&= \int_0^{F^{-1}(\alpha)} dt + \int_{F^{-1}(\alpha)}^\infty [(1 - F(t))/(1 - \alpha)] dt \\
&= \int_0^\infty \min\{1, (1 - F(t))/(1 - \alpha)\} dt.
\end{aligned}
$$

Thus, if we let

$$
g_\alpha(x) = \min\{1, x/(1 - \alpha)\},
$$

then

$$
g_\alpha(0) = 0, g_\alpha(1) = 1
$$

and $g_\alpha(.)$ is nondecreasing. Therefore, $v(.) = g_\alpha \circ P(.)$ is a capacity, and

$$
TVaR_\alpha(X) = C_v(X) = \int_0^\infty g_\alpha(1 - F(t)) dt.
$$

Unlike $VaR_\alpha(.)$, the risk measure $TVaR_\alpha(.)$ is subadditive. Again, this fact will be explained by using Choquet integral theory. In fact, after having the Choquet integral at our disposal, we will investigate properties of all the risk measures discussed so far, as well as new ones.

**Remark 4.3** $TVaR_\alpha(X)$ can also be written as

$$TVaR_\alpha(X) = \int_0^1 \Psi_\alpha(t)F^{-1}(t)dt$$

where

$$\Psi_\alpha(t) = (1/(1-\alpha))1_{(\alpha,1]}(t).$$

In this form, $TVaR_\alpha(X)$ looks like a weighted average of $F^{-1}$ with the weighting function $\Psi_\alpha(.)$. Risk measures of this type are called spectral risk measures.

A related risk measure to $TVaR_\alpha(.)$ is the Conditional Value-at-Risk, denoted as $CVaR_\alpha(.)$, which is defined as

$$CVaR_\alpha(X) = E(X \mid X > F^{-1}(\alpha)),$$

where its meaning is clear from its definition.

Like $TVaR_\alpha(.)$, $CVaR_\alpha(X) \geq VaR_\alpha(X)$. In fact we have

$$CVaR_\alpha(X) = VaR_\alpha(X) + \frac{E(X - F^{-1}(\alpha))^+}{(1 - F(F^{-1}(\alpha))}.$$

This can be seen as follows:

$$
\begin{aligned}
E(X \mid X > F^{-1}(\alpha)) &= F^{-1}(\alpha) + E((X - F^{-1}(\alpha))|X > F^{-1}(\alpha)) \\
&= F^{-1}(\alpha) + E(X - F^{-1}(\alpha))^+)/P(X > F^{-1}(\alpha)) \\
&= F^{-1}(\alpha) + E(X - F^{-1}(\alpha))^+/(1 - F(F^{-1}(\alpha))).
\end{aligned}
$$

From this, we see that if $F$ is continuous, then $CVaR_\alpha(X) = TVaR_\alpha(X)$, since then $F \circ F^{-1}$ is identity function, resulting in $F(F^{-1}(\alpha)) = \alpha$, in the above equality.

**Remark 4.4** Note that $CVaR_\alpha(X) = TVaR_\beta(X)$ where $\beta = F(F^{-1}(\alpha))$ which depends, not only on $\alpha$, but also on the distribution function $F$ of the variable $X$ for noncontinuous distribution function $F$. Of course, $TVaR_\beta(X)$ has a Choquet integral representation with "distortion function" $g_\beta(x) = \min\{1, x/(1-\beta)\}$, so that $CVaR_\alpha(X)$ is written as

$$CVaR_\alpha(X) = \int_0^\infty g_\beta(1 - F(t))dt,$$

*but in which $g_\beta$ depends on $X$ (and $\alpha$). As such, the operator $CVaR_\alpha(.)$, for each given $\alpha$, is not a Choquet integral with respect to a capacity $v$ of the form $g_\alpha \circ P$.*

## Notes on truncated distributions

Let $X$ be a real-valued random variable with distribution function $F$ where
$$F : \mathbb{R} \to [0, 1], F(x) = P(X \le x).$$

Truncated distributions of $X$ are distributions of $X$ conditioned on various events of the form $(X \in A)$ where $A$ is a Borel set of $\mathbb{R}$.

For example, for $a \in \mathbb{R}$, $(X > a) = (X \in (a, \infty))$ is an event, here $A = (a, \infty)$. Conditioning $X$ on $(X > a)$ is considering only the values of $X$ beyond the level $a$. The distribution function of $X$ conditioned upon $(X > a)$ is called the truncated distribution of $X$ at $a$. This new random variable, denoted *symbolically* as $(E|X > a)$ ("symbolically" since even a "*conditional event*" of the form "$A$ given $B$," written as $(A|B)$ does *not* exist as a formal mathematical entity, but only its conditional probability $P(A|B)$).

This distribution, denoted as $F_{X>a}(.)$, of the truncated random variable, is defined as $F_{X>a}(.) = P(X \le x | X > a)$. Clearly,

$$F_{X>a}(x) - P(a < X \le x)/P(X > a)$$
$$= \begin{cases} [F(x) - F(a)]/P(X > a) & \text{if } x > a \\ 0 & \text{if } x \le a. \end{cases}$$

If $F$ is absolutely continuous so that its (almost everywhere) derivative exists $f(x) = dF(x)/dx$ (its associated probability density function), then the density of the truncated random variable is

$$f_{X>a}(x) = f(x)/P(X > a) = f(x)/[1 - F(a)],$$

and hence the mean of $(E \mid X > a)$ is

$$E(X \mid X > a) = \int_{\mathbb{R}} x f_{X>a}(x)dx$$
$$= [1/(1 - F(a))] \int_a^\infty x f(x)dx.$$

For example, if $X$ is a standard normal random variable with distribution function

$$\Phi(x) = (1/\sqrt{2\pi}) \int_{-\infty}^x e^{-y^2/2}dy,$$

then

$$
\begin{aligned}
E(X \mid X > a) &= [1/(1 - \Phi(a))](1/\sqrt{2\pi}) \int_a^\infty y e^{-y^2/2} dy \\
&= [1/(1 - \Phi(a))](1/\sqrt{2\pi}) \int_a^\infty d(-e^{-y^2/2}) \\
&= [1/(1 - \Phi(a))](1/\sqrt{2\pi})(e^{-a^2/2}) \\
&= \phi(a)/(1 - \Phi(a)).
\end{aligned}
$$

Note that $\phi(a)/(1 - \Phi(a))$ is the *hazard rate function* of the normal distribution (evaluated at $a$) which is also called the *inverse Mills ratio* in econometrics.

**Example 4.5** *Here are some examples of CVaR.*

*(a) CVaR of normal losses. With the notation in examples above,*

$$
CVaR_\alpha(X) = \mu + \sigma[\phi(\Phi^{-1}(\alpha))/(1 - \alpha)].
$$

*(b) CVaR of lognormal losses.*

$$
CVaR_\alpha(Y) = \Phi(\sigma - \Phi^{-1}(\alpha))/(1 - \alpha) \exp(\mu + \sigma^2/2).
$$

### *Note on the computations of CVaR.*

*CVaR* is an expectation, and as such it can be computed by various methods, e.g., using moment generating functions.

Recall that the *moment generating function* of a random variable $X$ with distribution function $F$ is defined when $X$ is such that $E(e^{tX})$ is finite in some interval containing zero, as

$$
M_X : \mathbb{R} \to \mathbb{R} \quad M_X(t) = E(e^{tX}) = \int_{-\infty}^\infty e^{tx} dF(x)
$$

so that $EX = dM_X(t)/dt|_{t=0} = M_X'(0)$.

Now $X \mid X > a$ has distribution function

$$
\begin{aligned}
F_{X>a}(x) &= P(a < X \le x)/P(X > a) \\
&= \begin{cases} [F(x) - F(a)]/[1 - F(a)] & \text{if } x > a \\ 0 & \text{if } x \le a \end{cases}
\end{aligned}
$$

and its moment generating function is

$$
M_{X>a}(t) = E(e^{tX} \mid X > a) = \int_{\mathbb{R}} e^{tx} dF_{X>a}(x).
$$

For example, if $X$ is normal with mean $\mu$ and variance $\sigma^2$, then

$$
\begin{aligned}
M_{X>a}(t) &= [1 - \Phi((a-\mu)/\sigma)] \int_a^\infty e^{tx}\phi(x)dx \\
&= [e^{\mu t + \sigma^2 t^2/2}][1 - \Phi((a-\mu)/\sigma - \sigma t)]/[1 - \Phi((a-\mu)/\sigma)].
\end{aligned}
$$

Thus,

$$
M'_{X>a}(0) = \mu + \sigma\phi(a)/[1 - \Phi(a)].
$$

## 4.3 Desirable properties of risk measures

We consider now a systematic framework for risk modeling. It consists essentially of listing basic desirable properties that a risk measure should possess in order to be qualified as a realistic quantification of risk.

The current literature does not reach a consensus on which risk measures should be used in practice, rather the focus is on studying properties that a risk measure must satisfy to avoid, e.g., inadequate portfolio selections.

As stated in previous sections, we wish to assign a numerical value to each random variable $X$ to describe its risk, where $X$ could stand for the *profit-or-loss* of an investment portfolio, the *loss* of a financial position, or the capital needed to hold for an insurance company to avoid insolvency. Such a "risk measure" $\rho(X)$ would be of course a function of $X$, i.e., only depends on $X$. But the complete information about $X$ is its distribution function $F_X$ (we consider the setting of real-valued random variables for our discussions) so that $\rho(X)$ is of the form $\theta(F_X)$, some population parameter.

Just like probability measures, we first need to specify the *domain* and *range* for risk measures. Note that here we talk only about *risk in financial economics*!

### Range of risk measures

In the financial context, random variables of interest are either total future returns of investment portfolios or total possible claims for insurance companies. The risk of a risky position is thus the possible loss of money. Thus, for investment, the value of a risk could be the maximum possible loss, and for insurance, it could be the minimum amount of money the company should hold to avoid insolvency, i.e., sufficient to meet its obligations. In either case, the numerical value assigned to the qualitative notion of risk could be any real number, with negative numbers representing losses. Thus the range for risk measures will be taken to be the real line $\mathbb{R}$.

## Domain of risk measures

Clearly risk measures operate on real-valued random variables. With the range taken to be $\mathbb{R}$, we could consider the domain of risk measures to be the vector space $\mathcal{L}$ of all possible real-valued random variables. However, for practical settings or to be rigorous, proper subsets of $\mathcal{L}$ should be considered.

As we will see, any proposed risk measure should be consistent with appealing economic principles, such as reducing risk by *diversification*. To investigate such common sense principles, we need to include elements such as $X = \sum_{i=1}^{k} \lambda_i Y_i, \lambda_i > 0, \sum_{i=1}^{k} \lambda_i = 1$, into the domain of risk measures. Note that a total return of an investment portfolio is of the form $X = \sum_{i=1}^{k} \lambda_i Y_i$, where $\lambda_i > 0$, $\sum_{i=1}^{k} \lambda_i = 1$, $Y_i$ being the rate of return of the asset $i$ in the portfolio. Thus if the $Y_i$ are in a domain $\mathcal{X} \subseteq \mathcal{L}$, we want to talk also about risk of $\sum_{i=1}^{k} \lambda_i Y_i$ as well.

A *cone* $\mathcal{X}$ in $\mathcal{L}$ is a subset of $\mathcal{L}$ such that $\lambda X \in \mathcal{X}$ whenever $X \in \mathcal{X}$ and $\lambda > 0$. If a cone is also a convex subset of $\mathcal{L}$, then it is called a *convex cone*. Thus, in the following, the domain for risk measures will be taken as a convex cone of real-valued random variables. Note that, if $\mathcal{X}$ is a convex cone, then $\sum_{i=1}^{k} \lambda_i Y_i \in \mathcal{X}$, whenever $Y_i \in \mathcal{X}$ and $\lambda_i > 0$, so that $X + Y \in \mathcal{X}$, when $X, Y \in \mathcal{X}$. In fact $\mathcal{X}$ is a convex cone if and only if it is a cone and $X + Y \in \mathcal{X}$, when $X, Y \in \mathcal{X}$.

Examples of convex cones of $\mathcal{L}$ are the class of almost surely finite random variables, and the class of essentially bounded random variables. Such a structure for the domain $\mathcal{X}$ of risk measures is sufficient for stating desirable properties of risk measures as we will see next.

Basically, assigning a number to a random variable is defining a map from a class $\mathcal{X}$ of random variables to the real line, i.e., a *functional* $\rho : \mathcal{X} \to \mathbb{R}$. Such functionals are meant to quantify the concept of risk of random variables. When appropriate, they are called *risk measures*. This explains a host of risk measures proposed in the literature! For example, if we are interested in defining quantitative risks for future net worths of financial positions, we could want to consider the risk $\rho(X)$ of the future net worth $X$ of a financial position to be the amount of money which needs to be added to $X$ to make the position acceptable. In this case, risk is interpreted as a *capital requirement*. On the other hand, we could consider $\mathcal{X}$ as the class of *losses* of financial positions. In this case, $\rho(X)$

is interpreted as the risk of a possible loss.

The problem of quantitative modeling of risks is still an art. Any "reasonable" $\rho : \mathcal{X} \to \mathbb{R}$ can be considered as a risk measure. Are there ways to judge the appropriateness of a proposed risk measure? Perhaps one obvious way is to see whether a given risk measure satisfies some common sense properties in a given economic context. For such a purpose, we need to have a list of desirable properties of risk. The following is such a list from current economic literature. Of course, the list can be modified or added to. It is used to set some standard for risk measurement.

In the following, $X$ and $Y$ are any elements of $\mathcal{X}$ which is a convex cone of *loss* random variables of financial positions, containing constants. We follow Artzner [8].

Consider the following properties for a risk measure $\rho(.)$.

**Axiom 1. (Monotonicity)** If $X \leq Y$, almost surely (a.s.), i.e., $P(X \leq Y) = 1$, then $\rho(X) \leq \rho(Y)$.

**Axiom 2. (Positive homogeneity)** $\rho(\lambda X) = \lambda \rho(X)$ for $\lambda \geq 0$.

**Axiom 3. (Translation invariance)** $\rho(X + a) = \rho(X) + a$ for $a \in \mathbb{R}$.

**Axiom 4. (Subadditivity)** $\rho(X + Y) \leq \rho(X) + \rho(Y)$.

**Definition 4.6** *A functional $\rho : \mathcal{X} \to \mathbb{R}$ satisfying the above axioms is called a coherent risk measure.*

While, formally, *any* such functional could be considered as a reasonable candidate for measuring risks of a given context, we do not use the above axioms as an "axiomatic approach" for constructing risk measures. For example, if we search for functionals satisfying the above axioms, then the mean operator $\rho(X) = E(X)$, on the class of random variables with finite means, is obviously a candidate, but we never use it as a risk measure! The reason is this. As stated earlier, in financial economics, we are dealing with random variables which are themselves considered as risks. We wish to assess quantitatively these risks by what we call risk measures for decision-making. Given the context, we could *propose* some adequate risk measure. Then, we use the above list of axioms, for example, to examine whether our proposed risk measure has these desirable properties.

Here are the reasons leading to the above axioms. In the case of loss variables, small losses should have smaller risks. That is the natural motivation for Axiom 1. If we take the viewpoint of the capital requirement for future net worths of financial positions, then Axiom 1 will be written in reverse order, i.e., $X \leq Y$ a.s. $\implies \rho(X) \geq \rho(Y)$, since the risk is reduced if payoff profile is increased.

The risk of a loss of a financial position should be proportional to the size of the loss of the position. This leads to Axioms 2 and 3.

If we take the viewpoint of the capital requirement for future net worths of financial positions, then Axiom 3 takes the form $\rho(X + a) = \rho(X) - a$ for $a \in \mathbb{R}$, since here, $\rho(X)$ is interpreted as the (minimum) amount of money, which, if added to $X$, and invested in a risk-free manner, makes the position acceptable.

All the above can be seen clearly when the risk measure $\rho(.)$ is a Choquet integral with respect to some capacity $v$ (see next section). The coherence axioms given above are for $\rho(X) = C_v(X)$. Other forms of axioms correspond to defining $\rho(X) = C_v(-X)$. For example, if the capacity $v(.)$ is of the form $g \circ P(.)$ for some $g : [0,1] \to [0,1]$, we seek conditions on the function $g$ for $\rho_g(X) = C_{v_g}(X)$ to be a coherent risk measure. Similarly, we seek conditions on the weighting function $\Psi$ to make the spectral risk measure $\rho_\Psi(X) = \int_0^1 \Psi(t) F^{-1}(t) dt$ coherent. In either case, the choice of distortion function $g$ or of the weighting function $\Psi$ remains to be determined.

A self-evidence in investment science is this. *Diversification* should reduce risk. Specifically, one way to reduce risk in, say, portfolio selection is *diversification*, according to the old saying "don't put all your eggs in one basket," meaning that we should allocate our resources to different activities whose outcomes are not closely related.

In terms of risk measures, this economic principle is translated into the *convexity of* $\rho$, namely

$$\rho(\lambda X + (1 - \lambda)Y) \leq \lambda \rho(X) + (1 - \lambda)\rho(Y), \ \lambda \in [0,1].$$

This convexity of $\rho$ is a consequence of Axioms 2 and 4. In fact, Axiom 2 and convexity imply Axiom 4, since for $\lambda = 1/2$, we have

$$(1/2)\rho(X + Y) = \rho((X + Y)/2) \leq (1/2)(\rho(X) + \rho(Y)).$$

Convexity could be also taken as a *basic* desirable property for risk measures, independently with others.

While the problem of risk modeling seems to be an art, the above concept of coherent risk measures provides a reasonable class of risk measures. Before discussing existing risk measures in economics, let's take a closer look at the above axioms.

First of all, unlike probability measures, the above four axioms do not specify the definition of a risk measure. They are only "reasonable" properties for risks. With such a list, every time we have a risk measure at our disposition, we should check whether it is coherent or not. This constitutes an advance in the art of risk modeling.

Clearly $E_F(X) = \int_\Omega X(\omega)dP(\omega) = \int_\mathbb{R} xdF(x)$ satisfies all the above axioms, but we do not use the Lebesgue integral operator $E(.)$ as a risk measure, for obvious reasons. Examples of coherent risk measures will be given in the next chapter.

It turns out that another integral operator, called a *Choquet inte-gral*, generalizing the Lebesgue integral slightly, is appropriate for risk modeling, since it also satisfies the above axioms and hence provides constructions of coherent risk measures in a fairly general setting. To make the presentation smooth, we consider in the next chapter a tutorial on this type of integral.

## 4.4 Exercises

1. Let $X$ be a random variable with distribution function $F$. Let

$$F^{-1} : (0,1) \to \mathbb{R}, \ F^{-1}(t) = \inf\{x \in \mathbb{R} : F(x) \geq t\}.$$

(a) Verify that, for any $t \in (0,1)$,

$$F(F^{-1}(t)-) \leq t \leq F(F^{-1}(t)),$$

where $F(x-)$ denotes the left limit of $F(.)$ at $x$.

(b) Let $g : (0,1) \to \mathbb{R}$ is such that, for any $t \in (0,1)$

$$\sup\{x \in \mathbb{R} : F(x) < t\} \leq g(t) \leq \inf\{x \in \mathbb{R} : F(x) > t\}.$$

Show that $g(.)$ is nondecreasing with

$$g(t-) = \sup\{x \in \mathbb{R} : F(x) < t\}, \ g(t+) = \inf\{x \in \mathbb{R} : F(x) > t\}.$$

(c) Show that if $f : (0,1) \to \mathbb{R}$ is another function satisfying

$$\sup\{x \in \mathbb{R} : F(x) < t\} \leq f(t) \leq \inf\{x \in \mathbb{R} : F(x) > t\},$$

then $f(.) = g(.)$ almost everywhere.

2. Let $\varphi : [0,1] \rightarrow \mathbb{R}^+$, nondecreasing with $\int_0^1 \varphi(x)dx = 1$. The spectral risk measure with weighting function $\varphi$ is defined as

$$\rho_\varphi(X) = \int_0^1 \varphi(x) F_X^{-1}(x)dx,$$

where $F_X$ denotes the distribution function of $X$.

(a) Show that $X \leq Y$ (a.s.) implies $\rho_\varphi(X) \leq \rho_\varphi(Y)$.

(b) Show that $\rho_\varphi(.)$ is subadditive.

(c) Is $VaR_\alpha(.)$ a spectral risk measure?

3. For $\alpha \in (0,1)$, the $\alpha$-quantile set of a random variable $X$, with distribution function $F$, is defined to be the set

$$Q_\alpha(X) = \{x \in \mathbb{R} : P(X < x) \leq \alpha \leq P(X \leq x)\}.$$

Verify that

(a) $Q_\alpha(X) = [F^{-1}(\alpha), F^+(\alpha)]$ where

$$F^{-1}(t) = \inf\{x \in \mathbb{R} : F(x) \geq t\}, \ F^+(t) = \inf\{x \in \mathbb{R} : F(x) > t\}.$$

(b) $F^{-1}(t) = \sup\{x \in \mathbb{R} : P(X < x) < t\}, \ F^+(t) = \sup\{x \in \mathbb{R} : P(X < x) \geq t\}$.

4. With the above notation, show that if $g : \mathbb{R} \rightarrow \mathbb{R}$ is a measurable, nondecreasing and left-continuous function, then

$$F_{g(X)}^{-1}(.) = g(F_X^{-1}(.)).$$

5. Let $X$ be normally distributed with mean $\mu$ and variance $\sigma^2$. Compute

(a) $E(X - d)^+$, for $d \in \mathbb{R}$,

(b) $E(X - F_X^{-1}(\alpha))^+$, and

(c) $TVaR_\alpha(X)$.

6. We write $X \preceq_{sl} Y$ when $E(X - d)^+ \leq E(Y - d)^+$ for all $d \in \mathbb{R}$. Show that $X \preceq_{sl} Y$ if and only if $TVaR_\alpha(X) \leq TVaR_\alpha(X)$ for all $\alpha \in (0,1)$.

7. Let $(X, Y)$ have joint distribution function

$$F(x, y) = P(X \le x, Y \le y) = (2 - x)^{-1}(2 - y)^{-1} \text{ for } x, y \le 1.$$

   Let $\rho_\alpha(X) = F^{-1}_{(-X)}(\alpha)$, for some $\alpha \in (0, 1)$. Verify that $\rho_\alpha(.)$ is not subadditive.

8. Let $X$ be a random variable with distribution function $F$. Compute $VaR_\alpha(X) = F^{-1}(\alpha)$ when

   (a) $X$ is uniformly distributed on $[a, b]$,

   (b) $X$ is exponentially distributed with mean $\lambda$,

   (c) $X$ is normally distributed with mean $\mu$ and variance $\sigma^2$, and

   (d) $X$ is lognormal, i.e., $\log X$ is normal with mean $\mu$ and variance $\sigma^2$.

9. Let $\mathcal{X}$ be the class of profit or loss random variables with finite variances.

   (a) Verify that $\mathcal{X}$ is a convex cone.

   (b) Let $\rho : \mathcal{X} \to \mathbb{R}^+$, $\rho(X) =$ variance of $X$. What are the axioms of coherent risk measures that $\rho(.)$ satisfies?

10. Let $X$ be normal $N(\mu, \sigma^2)$. Show that

    (a) For each $\alpha \in (0, 1)$, $F^{-1}_X(\alpha)$ is proportional to $\sigma$,

    (b) If $X \le Y$, a.s. then $F^{-1}_X(\alpha) \le F^{-1}_Y(\alpha)$,

    (c) $F^{-1}_{\lambda X}(\alpha) = \lambda F^{-1}_X(\alpha)$ for any $\lambda > 0$, and

    (d) $F^{-1}_{X+a}(\alpha) = F^{-1}_X(\alpha) + a$ for any $a \in \mathbb{R}$.

11. Let $X$ be a real-valued random variable with distribution function $F$. The *quantile function* of $F$ is $F^{-1} : (0, 1) \to \mathbb{R}$, where

$$F^{-1}(\alpha) = \inf\{x \in \mathbb{R} : F(x) \ge \alpha\}.$$

    (a) Let $Y = g(X)$ where $g$ is strictly increasing and continuous. Show that $F^{-1}_Y(\alpha) = g(F^{-1}_X(\alpha))$.

    (b) Let $X > 0$, and $Y = \log X$. Compute the quantile function of $Y$ in terms of that of $X$.

12. Let $X$ be a random variable with distribution function $F$, and $u$ be a utility function. The *certainty equivalent* of $F$ is the real number $c(F)$ that satisfies

$$u(c(F)) = Eu(X) = \int u(x)dF(x).$$

The *risk premium* (i.e., the maximum amount of money that a risk averse decision-maker would pay to avoid taking a risk) in $F$ is the real number $q(F)$ that satisfies

$$q(F) = EX - c(F).$$

Show that a person with utility function $u$ is risk averse, neutral or seeking according to $q(F)$ is $>$, $=$ or $< 0$.

13. Let $X$ taking values $a$ and $b$ with $P(X = a) = p(a) = 1 - P(X = b) = 1 - p(b)$. Show that $u(EX) = u[ap(a) + bp(b)]$ is $>, =$ or $<$ $p(a)u(a) + p(b)u(b)$ according to whether the decision-maker with utility function $u$ is risk averse, neutral or seeking. What is the shape of $u$ corresponding to each of the above risk attitudes?

14. Let $W_o$ be an initial capital to invest in a prospect, and let $X$ be its future value at the end of some given time horizon. Let $F$ be the distribution function of $X$. For $\alpha \in (0, 1]$, let

$$VaR_\alpha(X) = \sup\{x \in \mathbb{R} : P(W_o - X \geq x) \geq \alpha\}.$$

(a) Verify that if $a > VaR_\alpha(X)$ then $P(W_o - X \geq a) < \alpha$.

(b) Show that $VaR_\alpha(X) = W_o - F^{-1}(\alpha)$.

15. Let $X$ be a normal random variable with mean $\mu$ and variance $\sigma^2$.

(a) Compute $E(X \mid X > a)$.

(b) Find the density function of the random variable $X$ truncated by $X \in [a, b]$.

(c) Find the moment generating function of $X \mid X \in [a, b]$.

(d) Find $E(X \mid X < b)$.

16. For $\alpha \in (0, 1)$, show that

$$\begin{aligned} \sup\{x \in \ & \mathbb{R} : P(X \geq x) > 1 - \alpha) = \inf\{x : P(X > x) \leq 1 - \alpha) \\ = \ & \inf\{x : F(x) \geq \alpha\} = F^{-1}(\alpha). \end{aligned}$$

17. Let $X, Y$ be independent normal random variables $N(\mu_1, \sigma_1^2), N(\mu_2, \sigma_2^2)$, respectively. For $VaR_\alpha(X) = F_X^{-1}(\alpha)$, show that for $\alpha \geq 0.5$,

$$VaR_\alpha(X + Y) \leq VaR_\alpha(X) + VaR_\alpha(Y).$$

18. Compute $CVaR_\alpha(X)$

    (a) when $X$ is normal $N(\mu, \sigma^2)$, and
    (b) when $X$ is lognormal, i.e., $\log g$ is $N(\mu, \sigma^2)$.

19. Let $F$ be a distribution function on $\mathbb{R}$. Show that if we define $F^{-1} : [0, 1] \to \mathbb{R}$ by

$$F^{-1}(\alpha) = \begin{cases} \inf\{x : F(x) > \alpha\} & \text{for } 0 \leq \alpha < 1 \\ \inf\{x : F(x) = 1\} & \text{for } \alpha = 1 \end{cases}$$

then $F^{-1}$ is right continuous.

# Chapter 5

# Choquet Integrals as Risk Measures

*This chapter provides a tutorial on the concept of the Choquet integral for risk modeling. Within the context of the Choquet integral, we discuss coherent risk measures as well as their consistency with respect to stochastic dominance.*

## 5.1   Extended theory of measures

It could be amazing for students that it sounds like "integrals are everything!" A measure theory is in fact a theory of integration. Abstract Lebesgue measure and integrals are the cornerstones of probabilistic uncertainty analysis. Stieltjes measure and integrals cannot handle irregular observations in stochastic finance and hence we need another type of integral, namely the *Ito integral*. When observation processes go beyond semimartingales, we need some new types of integrals, such as for fractional Brownian motion. And here, in the context of financial risk, we need the Choquet integral, named after Gustave Choquet for his pioneering work on the theory of capacities in potential theory.

This extended theory of measures has entered various areas of applications, such as engineering, decision theory and statistics. And now it enters financial risk analysis! In this theory, measures are replaced by more general set functions called capacities, and Lebesgue abstract integral is extended to nonadditive functionals. The main reasons that the Choquet integral surfaces in risk modeling and analysis are the fact that popular risk measures proposed so far are Choquet integrals on one hand, and on the other hand, desirable properties for risk measures are mostly basic properties of the Choquet integral.

As compared to the Ito integral and other newly developed integrals for mathematical finance, the Choquet integral is much simpler. It enters financial risk analysis since most of popular risk measures are in fact a Choquet integral, and as such they possess desirable properties for coherence. In the form of a Choquet integral representation, a proposed risk measure can be adjusted to be consistent with stochastic dominance rules.

The most important and new property coming from the Choquet integral is *comonotonicity of random variables* (risks). Comonotonic risks in financial and actuarial sciences have become a major concern of economists. As far as risk modeling is concerned, the Choquet integral provides a neat model for desirable risk measures. This is so because of its basic and characteristic properties which we will detail in this section.

This section is a complete *tutorial* on the Choquet integral. It consists of basic properties with focus on *comonotonic additivity*, a concept due to Claude Dellacherie, the related concept of *copulas* of A. Sklar, and the *fundamental characterization theorem* of Schmeidler. When comonotonic risks are of concern, the Schmeidler theorem provides the rationale for modeling risks as a Choquet integral.

The Choquet integral, as a functional, is nonlinear and is even not subadditive in general. When using Choquet integrals (with respect to different capacities) as risk measures we wish to consider those that are *subadditive* in view of the diversification problem. Conditions (on capacities) for the Choquet integral to be subadditive will be investigated leading to the *subadditivity theorem* (essential for coherent risk modeling) with specialization to the case where capacities are taken as distorted probabilities, results and concepts due to Dieter Denneberg [34].

## 5.2   Capacities

We have seen that popular risk measures such as value-at-risk and tail value-at-risk are of the form

$$\int_0^\infty g(1 - F(t))dt + \int_{-\infty}^0 [g(1 - F(t)) - 1]dt$$

where $F$ is the distribution function of the "risk" $X$ (a financial position), for some $g : [0,1] \to [0,1]$, nondecreasing, and $g(0) = 0, g(1) = 1$. Any such function is called a *distortion function*.

Note that we always have our probability space $(\Omega, \mathcal{A}, P)$, on which all random variables are defined, in the back of our mind!

For $g(x) = x$, this integral reduces to

$$\int_0^\infty (1 - F(t))dt - \int_{-\infty}^0 F(t)dt,$$

which is the expectation of $X$

$$EX = \int_{-\infty}^\infty x dF(x).$$

Indeed, using Fubini's theorem, we have

$$
\begin{aligned}
\int_{-\infty}^\infty x dF(x) &= \int_0^\infty \left[\int_0^x dy\right] dF(x) - \int_{-\infty}^0 \left[\int_x^0 dy\right] dF(x) \\
&= \int_0^\infty \left[\int_y^\infty dF(x)\right] dy - \int_{-\infty}^0 \left[\int_{-\infty}^y dF(x)\right] dy \\
&= \int_0^\infty (1 - F(y))dy - \int_{-\infty}^0 F(y)dy.
\end{aligned}
$$

Note that

$$\int_0^\infty (1 - F(t))\, dt - \int_{-\infty}^0 F(t)dt$$

can be written as

$$EX = \int_0^\infty P(X > t)dt + \int_{-\infty}^0 [P(X > t) - 1]dt.$$

For general $g$, observe that

$$g(1 - F(t)) = g(P(X > t)) = (g \circ P)(X > t),$$

so that if we let $\nu = g \circ P$, then the extension of $EX$ takes the form

$$\int_0^\infty \nu(X > t)dt + \int_{-\infty}^0 [\nu(X > t) - 1]dt$$

in which the set function $\nu$, defined on $\mathcal{A}$, is no longer additive (let alone a probability measure). However $\nu$ is nondecreasing, i.e., $A \subseteq B \implies \nu(A) \leq \nu(B)$, and $\nu(\varnothing) = 0, \nu(\Omega) = 1$. Any such set function is called a *capacity*.

Capacities and the Choquet integral can be formulated in more general settings. However, for our purpose here in risk modeling and analysis, the following set-up is sufficient.

**Definition 5.1** *Let $(\Omega, \mathcal{A})$ be a measurable space. A map $v : \mathcal{A} \to [0,1]$ is called a **capacity** if $v$ is increasing, i.e., $A \subseteq B \implies v(A) \leq v(B)$, $\nu(\varnothing) = 0$ and $\nu(\Omega) = 1$.*

**Example 5.2** *Let $(\Omega, \mathcal{A})$ be a measurable space. For a fixed $A$ such that $\varnothing \neq A \in \mathcal{A}$, define $u_A : \mathcal{A} \to [0,1]$ by*

$$u_A(B) = \begin{cases} 1 & \text{if } A \subseteq B \\ 0 & \text{if not.} \end{cases}$$

**Example 5.3** *Let $(\Omega, \mathcal{A}, P)$ be a probability space. Consider the $\sigma$-field $2^\Omega$ (the power set of $\Omega$, i.e., the class of all subsets of $\Omega$). Define $v : 2^\Omega \to [0,1]$ by*

$$v(A) = \sup\{P(B) : B \in \mathcal{A}, B \subseteq A\},$$

*then $v(.)$ is a capacity.*

**Example 5.4** *Let $(\Omega, \mathcal{A})$ be a measurable space. Let $\mathbb{P}$ denote the set of all probability measures defined on $\mathcal{A}$. The following lower and upper probabilities $L, U : \mathcal{A} \to [0,1]$ are capacities:*

$$\begin{aligned} L(A) &= \inf\{P(A) : P \in \mathbb{P}\}, \\ U(A) &= \sup\{P(A) : P \in \mathbb{P}\}. \end{aligned}$$

**Example 5.5** *Let $(\Omega, \mathcal{A})$ be a measurable space. Let $f : \Omega \to [0,1]$ such that $\sup\{f(x) : x \in \Omega\} = 1$. Then $v(A) = \sup_{x \in A} f(x)$ is a capacity of a special kind, namely, it satisfies the property, for any $A_1, A_2, ..., A_n$, $n \geq 2$,*

$$v(\cap_{i=1}^n A_i) \leq \sum_{\varnothing \neq I \subseteq \{1,2,...,n\}} (-1)^{|I|+1} v(\cup_{j \in I} A_j)$$

*where $|I|$ denotes the number of elements in the set $I$ (cardinality of $I$).*

Such a capacity $v$ is referred to as an *alternating of infinite order* capacity. A capacity $v$ where the above inequality holds only for $n = 2$, is called a *2-alternating* capacity:

$$v(A \cap B) \leq v(A) + v(B) - v(A \cup B)$$

or

$$v(A \cup B) \leq v(A) + v(B) - v(A \cap B).$$

Alternating of infinite order can be obtained in general as follows: Let $v : \mathcal{A} \to [0, 1]$ be such that $v(\varnothing) = 0, \nu(\Omega) = 1$ and $v$ is *maxitive*, i.e., for any $A, B$ in $\mathcal{A}$,

$$v(A \cup B) = \max\{v(A), v(B)\}.$$

Then $v$ is alternating of infinite order.

The dual concept to alternating of infinite order is *monotone of infinite order*, namely

$$v(\cup_{i=1}^{n} A_i) \geq \sum_{\varnothing \neq I \subseteq \{1,2,...,n\}} (-1)^{|I|+1} v(\cap_{j \in I} A_j).$$

For example, the capacity in Example 5.6 below is monotone of infinite order.

**Example 5.6** *In risk modeling, capacities are in general of the form $v = g \circ P$, where $g$ is a distortion function, i.e., $g : [0, 1] \to [0, 1]$, nondecreasing and $g(0) = 0, g(1) = 1$. Depending upon additional properties of $g$, these capacities, called distorted probabilities, could have more special properties. Distorted probabilities are special cases of capacities which are functions of probability measures. For example, for $\Omega$ finite, and $\mathcal{A} = 2^\Omega$, take $g(x) = x^2$. Then clearly $v_2(A) = (P(A))^2$ is no longer a probability, but is a capacity monotone of infinite order.*

To prove the above assertion, it suffices to observe that $v_2$ is of the form

$$v_2(A) = \sum_{B \subseteq A} f(A),$$

where $f : 2^\Omega \to [0, 1]$ is a probability density function on $2^\Omega$ (i.e., a nonnegative function such that $\sum_{A \subseteq \Omega} f(A) = 1$).

Indeed, if we define $f : 2^\Omega \to [0, 1]$ by

$$f(A) = \begin{cases} [P(\{a\})]^2 & \text{if } A = \{a\} \\ 2P(\{a\})P(\{b\} & \text{if } A = \{a, b\} \\ 0 & \text{if } A = \varnothing \text{ or } |A| > 2, \end{cases}$$

then $v_2(A) = \sum_{B \subseteq A} f(B)$, since $[P(A)]^2 = [\sum_{a \in A} P(\{a\})]^2$.

The same result holds for any integer $k \geq 2$, $v_k = P^k$, by using the multinomial theorem

$$(x_1 + x_2 + \cdots + x_n)^k = \sum \binom{k}{k_1 k_2 ... k_n} x_1^{k_1} x_2^{k_2} ... x_n^{k_n},$$

where the summation extends over all nonnegative integral solutions $k_1, k_2, ..., k_n$ of $k_1 + k_2 + ... + k_n = k$, and

$$\binom{k}{k_1 k_2 ... k_n} = \frac{k!}{k_1! k_2! ... k_n!}$$

by noting that $v_k(A) = [\sum_{a \in A} P(a)]^k$.

In a topological context, such as the Euclidian space $\mathbb{R}^n$, a very special type of capacities which are not only alternating of infinite order but also go down on compact sets plays a crucial role in statistical analysis of *coarse data* modeled as *random sets* (random elements taking values in the space of closed sets of $\mathbb{R}^n$). They are called *Choquet capacities* and play the role of distribution functions for random sets.

**Example 5.7** *Extending the Lebesgue-Stieltjes characterization theorem for probability laws of random vectors to this setting of random sets, a remarkable result due to Choquet is the cornerstone of random set theory. An example of such a capacity is this:*

*Let $\psi : \mathbb{R}^n \to [0, 1]$ be an upper semicontinuous function (i.e., the sets $\{x \in \mathbb{R}^n : \psi(x) \geq \alpha\}$ are closed). Let $\mathcal{K}$ be the class of compact sets in $\mathbb{R}^n$. Define $v_\psi : \mathcal{K} \to \mathbb{R}$*

$$v_\psi(K) = \sup_{x \in K} \psi(x)$$

*then $v_\psi$ can be extended to the Borel $\sigma$-field $\mathcal{B}(\mathbb{R}^n)$ to be a Choquet capacity. See a text like Nguyen [121].*

**Example 5.8 (*Robust Bayesian statistics*)** *Capacities surface in robust Bayesian statistics as follows. Let $\{F(.|\theta) : \theta \in \Theta\}$ be a statistical model for some random vector $X$ under consideration. The Bayesian approach to statistical inference consists of assuming that there is a prior probability measure $\pi$ on the measurable space $(\Theta, \mathcal{B}(\Theta))$. It is realistic (or robust) to only specify the priors as members of a set $\mathcal{P}$ of probability measures on $\mathcal{B}(\Theta)$. Knowing only $\mathcal{P}$, we are forced to conduct statistical inference using the lower or upper bounds*

$$L(.) = \inf\{\pi(.) : \pi \in \mathcal{P}\}, \ U(.) = \sup\{\pi(.) : \pi \in \mathcal{P}\}.$$

*Clearly, $L(.), U(.)$ are not probability measures, but capacities.*

## 5.3 The Choquet integral

For a capacity $\nu$ and a random variable $X$ (measurable), the expression

$$\int_0^\infty \nu(X > t)dt + \int_{-\infty}^0 [\nu(X > t) - 1]dt$$

makes sense since the function $t \to \nu(X > t)$ is monotone (decreasing) and hence measurable.

**Definition 5.9** *The **Choquet integral** of $X$ with respect to the capacity $\nu$, denoted as $C_\nu(X)$, is defined to be*

$$C_\nu(X) = \int_0^\infty \nu(X > t)dt + \int_{-\infty}^0 [\nu(X > t) - 1]dt.$$

This type of integral is termed a Choquet integral in honor of Gustave Choquet who considered it in his work on capacity theory [26].

**Example 5.10** *With respect to the capacity $u_A$ in Example 5.2, we have*

$$C_{u_A}(X) = \int_0^\infty u_A(X > t)dt = \inf\{X(\omega) : \omega \in A\}$$

*since*

$$A \subseteq (\omega : X(\omega) \geq t) \Longleftrightarrow t \leq \inf\{X(\omega); \omega \in A\},$$

*so that*

$$u_A(X > t) = \begin{cases} 1 & \text{if } t \leq \inf\{X(\omega); \omega \in A\} \\ 0 & \text{otherwise.} \end{cases}$$

**Example 5.11** *For finite $\Omega$, the computation of Choquet integrals is very simple. Let $\Omega = \{\omega_1, \omega_2, ..., \omega_n\} \subseteq \mathbb{R}$, $X : \Omega \to \mathbb{R}$, and $v$ be a capacity, defined on the power set of $\Omega$. Rearrange $\Omega$ so that*

$$X(\omega_1) \leq X(\omega_2) \leq \cdots \leq X(\omega_n).$$

*Then $C_v(X) = \sum_{i=1}^n X(\omega_i)[v(\{\omega_i, \omega_{i+1}, ..., \omega_n\}) - v(\{\omega_{i+1}, \omega_{i+2}, ..., \omega_n\})]$.*

Observe that $g(\omega_i) = v(\{\omega_i, \omega_{i+1}, ..., \omega_n\}) - v(\{\omega_{i+1}, \omega_{i+2}..., \omega_n\})$ is a probability density function on $\Omega$. Thus $C_v(X)$ is just an ordinary probabilistic expectation, but the density used for this ordinary expectation depends not only on $v$ but also on the ordering of $\Omega$ via $X$. As we will see later, this fact is essential for investigating subadditivity of the Choquet integral.

While the Choquet integral is just the Lebesgue integral of some other function, it presents in fact an unconventional type of integration. If we view it as an integral of $X$ with respect to a "nonadditive measure" $\nu$, then the functional $C_\nu(.)$ has different properties than an ordinary integral operator (details later).

Thus, the above popular risk measures are Choquet integrals of the risky position $X$ with respect to different and special capacities, namely the $g \circ P$. These functions $g$, when composed with $P$, distort the probability measure $P$ (destroying its "measure" properties), so we call them *distortion functions*, and the special capacities $g \circ P$ are called *distorted probabilities*.

What good is such a representation for risks in term of the Choquet integral?

(i) If a risk measure is a Choquet integral, then it is easy to check its coherence.

(ii) In particular, if a risk measure is a Choquet integral with respect to a distorted probability, then consistency with respect to stochastic dominance can be also easily checked.

(iii) The Choquet integral risk measures have some desirable properties in actuarial science.

(iv) Together with the above, the class of risk measures constructed from distortion functions seems to be a plausible class of "good" risk measures to investigate. In fact, this seems to be the current trend in risk modeling.

Of course, there are risk measures which are not Choquet integrals or not Choquet integrals with respect to distorted probabilities. However, if the property of "comonotonicity" is part of a list of desirable properties for risk measures, then risk measures as a Choquet integral can be justified by Schmeidler's theorem [138].

**Some motivating examples**

*Capacities* and the *Choquet integral* appear earlier in statistics and decision theory. We mention below some typical situations. The following example illustrates a situation in the so-called *imprecise probabilities*.

**Example 5.12** *Let $X$ be a random variable with values in*

$$\Theta = \{\theta_1, \theta_2, \theta_3, \theta_4\} \subseteq \mathbb{R}^+.$$

*Suppose the true density function $f_o$ of $X$ is only partially known as*

$$f_o(\theta_1) \geq 0.4, \ f_o(\theta_2) \geq 0.2, \ f_o(\theta_3) \geq 0.2, \ f_o(\theta_4) \geq 0.1$$

*and we wish to compute its mean. An approximation to this uncomputable quantity*

$$E_{f_o}(X) = \sum_{i=1}^{4} \theta_i f_o(\theta_i)$$

*is*

$$\inf\{E_f(X) : f \in \mathcal{F}\},$$

*where $\mathcal{F}$ denotes the class of all densities on $\Theta$ satisfying the above inequalities. It turns out that $\inf\{E_f(X) : f \in \mathcal{F}\}$ can be computed by using the following non-additive set-function $v$.*

*Let $v : 2^\Theta \to [0, 1]$ be defined by*

$$v(A) = \inf\{P_f(A) : f \in \mathcal{F}\},$$

*where $2^\Theta$ is the power set of $\Theta$ and $P_f(A) = \sum_{\theta \in A} f(\theta)$. The set-function $v$ is nondecreasing, $v(\varnothing) = 0, v(\Theta) = 1$. Suppose $\theta_1 < \theta_2 < \theta_3 < \theta_4$ and let*

$$
\begin{aligned}
g(\theta_1) &= v(\{\theta_1, \theta_2, \theta_3, \theta_4\}) - v(\{\theta_2, \theta_3, \theta_4\}) g(\theta_2) \\
&= v(\{\theta_2, \theta_3, \theta_4\}) - v(\{\theta_3, \theta_4\}) g(\theta_3) \\
&= v(\{\theta_3, \theta_4\}) - v(\{\theta_4\}) g(\theta_4) - v(\{\theta_4\}).
\end{aligned}
$$

*Then $g \in \mathcal{F}$ and*

$$E_g(X) = \int_\Theta \theta dP_g(\theta) = \int_0^\infty P_g(\theta_i : \theta_i > t) dt = \int_0^\infty v(\theta_i : \theta_i > t) dt.$$

*This last integral is called the Choquet integral of the function $X(\theta) = \theta$ with respect to the nondecreasing set-function $v$.*

*Of course, it remains to show that $\inf\{E_f(X) : f \in \mathcal{F}\}$ is attained at $g$. Since*

$$E_f(X) = \int_0^\infty P_f(\theta_i : \theta_i > t) dt,$$

*it suffices to show that, for $t \in \mathbb{R}$,*

$$P_g(\theta_i : \theta_i > t) \leq P_f(\theta_i : \theta_i > t)$$

for all $f \in \mathcal{F}$, but that follows from the construction of $g$.

In summary, the Choquet integral of $X$ with respect to $v$ is the infimum for expected values.

**Example 5.13** *Here is another use of the Choquet integral in imprecise information. We are interested in computing the expectation of some random quantity $g(X)$ where $g : \mathbb{R} \to \mathbb{R}^+$ but the observations on $X$ can be only located in some intervals of $\mathbb{R}$. This is typically the situation of coarse data where precise values of $X$ are not observable. Instead, the observations are things which are related to $X$ in some way.*

*Well-known situations are measurement-error models, hidden Markov models in biology, censored data in survival analysis, missing data in econometrics, and indirect observations in auctions. To model the coarse data in our actual example, we use the theory of random sets, namely viewing the observables as sets (here intervals) containing the unobservable $X$. Formally, the random element*

$$S : \Omega \to \{nonempty\ closed\ sets\ of\ \mathbb{R}\}$$

*is such that $X \in S$ almost surely.*

*The computation of $Eg(X)$ in this situation can be carried out as follows: For each $\omega \in \Omega$,*

$$g(S(\omega)) = \{g(x) : x \in S(\omega)\}.$$

*Define $g^*, g_* : \Omega \to \mathbb{R}^+$ by*

$$g^*(\omega) = \sup\{g(x) : x \in S(\omega)\}$$
$$g_*(\omega) = \inf\{g(x) : x \in S(\omega)\}.$$

*Since $g(X(\omega)) \in g(S(\omega))$, we have that*

$$g_*(\omega) \leq g(X(\omega)) \leq g^*(\omega),$$

*resulting in*

$$Eg_* \leq Eg(X) \leq Eg^*.$$

*Thus, we are led to compute bounds on $Eg(X)$.*

*It turns out that $Eg_*, Eg^*$ are Choquet integrals of $g(.)$ with respect to some capacity. To make sure that these Choquet integrals are well-defined, we need to check that the functions $g^*$ and $g_*$ are measurable.*

*The following condition on the random set $S$ will suffice. Extending measurability of single-valued random elements, we say that $S$ is strongly measurable if for each $A \in \mathcal{B}(\mathbb{R})$, we have $A_*$ and $A^* \in \mathcal{A}$, where*

$$A_* = \{\omega \in \Omega : S(\omega) \subseteq A\},$$
$$A^* = \{\omega \in \Omega : S(\omega) \cap A \neq \varnothing\}.$$

*Then the following are equivalent.*

*(a) $S$ is strongly measurable.*

*(b) If $g$ is measurable, then $g^*$ and $g_*$ are measurable.*

*Indeed, suppose $S$ is strongly measurable and $g$ measurable. For the measurability of $g_*$, it suffices to show that $g_*^{-1}([c, \infty)) \in \mathcal{A}$ for all $c \in \mathbb{R}$. Now $\omega \in g_*^{-1}([c, \infty))$ means that $\inf\{g(x) : x \in S(\omega)\} \geq c$. Thus,*

$$S(\omega) \subseteq \{x \in \mathbb{R} : g(x) \geq c\} = g^{-1}([c, \infty)),$$

*and hence $\omega \in [g^{-1}([c, \infty))]_*$.*

*If $\omega \in [g^{-1}([c, \infty))]_*$, then $S(\omega) \subseteq g^{-1}([c, \infty))$. That is, for all $x \in S(\omega)$,*

$$g(x) \geq c \implies \inf\{g(x) : x \in S(\omega)\} \geq c,$$

*and therefore*

$$g_*(\omega) \in [c, \infty) \text{ or } \omega \in g_*^{-1}([c, \infty)).$$

*Thus,*

$$g_*^{-1}([c, \infty)) = [g^{-1}([c, \infty))]_*.$$

*By assumption, $g^{-1}([c, \infty)) \in \mathcal{A}$, and by (a), the measurability of $g_*$ follows. The measurability of $g^*$ is proved similarly.*

*For the converse, let $A \in \mathcal{A}$, then $f(.) = 1_A(.)$ is measurable and*

$$f_*(\omega) = \begin{cases} 1 & \text{if } S(\omega) \subset A \\ 0 & \text{otherwise.} \end{cases}$$

*Hence $f_*^{-1}(\{1\}) = A_*$, and by hypothesis, $A_* \in \mathcal{A}$. Similarly, $A^* \in \mathcal{A}$.*

*Let's return to the computation of $E(g_*)$ and $E(g^*)$. Define the capacity which is well defined since $S$ is assumed to be strongly measurable:*

$$v_* : \mathcal{B}(\mathbb{R}) \to [0, 1]$$
$$v_*(A) = P(\omega : S(\omega) \subseteq A) = P(A_*).$$

*Then,*

$$
\begin{aligned}
v_*(g^{-1}(t,\infty))dtE(g_*) &= \int_\Omega g_*(\omega)dP(\omega) = \int_0^\infty P(\omega : g_*(\omega) > t)dt \\
&= \int_0^\infty P(g_*^{-1}((t,\infty)))dt = \int_0^\infty P[g^{-1}((t,\infty)]_*dt \\
&= \int_0^\infty P\{\omega : S(\omega) \subseteq g^{-1}(t,\infty)\}dt \\
&= \int_0^\infty v_*(g^{-1}(t,\infty))dt = \int_0^\infty v_*(g > t)dt.
\end{aligned}
$$

*Similarly, for the capacity*

$$
\begin{aligned}
v^* &: \quad \mathcal{B}(\mathbb{R}) \to [0,1] \\
v^*(A) &= \quad P(\omega : S(\omega) \cap A \neq \varnothing) = P(A^*),
\end{aligned}
$$

*we have*

$$
E(g^*) = \int_0^\infty v^*(g > t)dt.
$$

*Note that while the distribution function of $X$ cannot be estimated since $X$ is not observable, the capacities $v_*$ and $v^*$ can be estimated from the observable random set $S$.*

**Example 5.14 (*Multicriteria decision-making*)** *One other motivation for using the Choquet integral is in the theory of multicriteria decision-making. The problem of ranking alternatives according to a set of criteria is this. We wish to evaluate alternatives $X : \mathcal{N} \to \mathbb{R}^+$, where $\mathcal{N} = \{1, 2, ..., n\}$ is a set of criteria in some decision problem, with $X(i)$ being the value of $X$ according to the criterion $i$. For this purpose, we first assign degrees of importance $\alpha_i$ to criteria, with $\alpha_i \geq 0$, $\sum_{i=1}^n \alpha_i = 1$. We write $\alpha = (\alpha_1, \alpha_2, ..., \alpha_n)$.*

*For noninteractive criteria, it is reasonable to use a linear weighted average aggregation operator such as*

$$
E_\alpha(X) = \sum_{i=1}^n \alpha_i X(i).
$$

*Viewing $\alpha$ as the probability measure $P_\alpha$ on the power set of $\mathcal{N}$,*

$$
P_\alpha(A) = \sum_{i \in A} \alpha_i,
$$

we see that $E_\alpha(X)$ is the expectation of $X$ with respect to the probability measure $P_\alpha$.

If we write $X(i) = x_i$ and order them so that $x_{(1)} \leq x_{(2)} \leq \cdots \leq x_{(n)}$, and let

$$A_{(i)} = \{(i), (i+1), ..., (n)\}.$$

Then

$$E_\alpha(X) = \sum_{i=1}^n \alpha_{(i)} x_{(i)} = \sum_{i=1}^n (x_{(i)} - x_{(i-1)}) P_\alpha(A_{(i)})$$

with $x_{(0)} = 0$.

When the criteria are interactive, nonlinear aggregation operators are suggested. A plausible way to produce nonlinear operators is to replace $P_\alpha$ by some set-function $v : 2^{\mathcal{N}} \to [0,1]$ such that $v(\varnothing) = 0$, $v(\mathcal{N}) = 1$ and nondecreasing: $A \subseteq B \implies v(A) \leq v(B)$, leading to the evaluation map

$$C_v(X) = \sum_{i=1}^n (x_{(i)} - x_{(i-1)}) v(A_{(i)}).$$

Now observe that this nonlinear aggregation operator can be written as

$$C_v(X) = \int_0^\infty v(X > t) dt.$$

**Example 5.15 (*Notes on random sets as coarse data*)** We have seen that capacities and the Choquet integral play an essential role in modeling coarse data (low quality data) in a variety of applied areas. In principle, this is similar to how these ingredients appear in risk modeling and analysis. Let $S$ be a random closed set in $\mathbb{R}^d$, i.e., a measurable map, defined on $(\Omega, \mathcal{A}, P)$, with values in the space $\mathcal{F}(\mathbb{R}^d)$ of closed sets of $\mathbb{R}^d$ equipped with the hit-or-miss topology.

Let $\mathcal{B}(\mathbb{R}^d)$ be the Borel $\sigma$-field of $\mathbb{R}^d$. Define the capacity

$$v : \mathcal{B}(\mathbb{R}^d) \to [0,1]$$

by

$$v(A) = P(\omega : A \cap S(\omega) \neq \varnothing).$$

For $f : \mathbb{R}^d \to \mathbb{R}^+$, measurable, we have

$$C_v(f) = E(\sup\{f(x) : x \in S\}).$$

This capacity is alternating of infinite order. Such capacities serve as the counterpart of distribution functions for random sets which are generalizations of random vectors. Note that random sets (i.e., sets obtained at random) can be used as models for coarse data in economics.

## 5.4   Basic properties of the Choquet integral

We establish now the fundamentals of the Choquet integral. The following framework is sufficient for risk modeling and analysis.

Let $(\Omega, \mathcal{A})$ be a measurable space, and $v$ a capacity, defined on $\mathcal{A}$. We are going to investigate basic properties of the functional

$$C_v(.) : \mathcal{X} \to \mathbb{R}$$

where $\mathcal{X}$ denotes the space of real-valued (bounded) measurable functions defined on $\Omega$.

First, for $X = 1_A$ with $A \subseteq \Omega$, $A \in \mathcal{A}$, we have

$$(X > t) = \begin{cases} \Omega & \text{for } t < 0 \\ A & \text{for } 0 \leq t < 1 \\ \varnothing & \text{for } t \geq 1 \end{cases}$$

so that

$$C_v(1_A) = v(A).$$

Clearly, the Choquet integral is *monotone increasing*:

$$\text{if } X \leq Y \text{ then } C_v(X) \leq C_v(Y)$$

and *positively homogeneous of degree one*; that is, for $\lambda > 0$,

$$C_v(\lambda X) = \lambda C_v(X).$$

For a nonnegative simple function of the form

$$X = a1_A + b1_B$$

with $A \cap B = \varnothing$ and $0 < a < b$, we have

$$
\begin{aligned}
C_v(X) &= \int_0^b v(X > t)dt \\
&= \int_0^a v(X > t)dt + \int_a^b v(X > t)dt \\
&= av(A) + (b - a)v(B).
\end{aligned}
$$

More generally, for

$$X(\omega) = \sum_{i=1}^n a_i 1_{A_i}(\omega)$$

with the $A_i$'s pairwise disjoint subsets of $\Omega$ and $a_0 = 0 < a_1 < \cdots < a_n$, we have

$$C_v(X) = \sum_{i=1}^{n}(a_i - a_{i-1})v\left(\bigcup_{j=i}^{n} A_j\right).$$

For an arbitrary simple function of the form $X = \sum_{i=1}^{n} a_i 1_{A_i}$, and with the $A_i$'s forming a measurable partition of $\Omega$ and

$$a_1 < \cdots < a_k < 0 < a_{k+1} < \cdots < a_n,$$

we also have

$$C_v(X) = \sum_{i=1}^{n}(a_i - a_{i-1})v\left(\bigcup_{j=i}^{n} A_j\right) \quad \text{with } a_0 = 0.$$

Indeed, since $a_k < 0 < a_{k+1}$, we have

$$
\begin{aligned}
C_v(X) =\ & \int_{a_1}^{a_2}[v(X > t) - 1]dt + \int_{a_2}^{a_3}[v(X > t) - 1]dt + \cdots \\
& + \int_{a_{k-1}}^{a_k}[v(X > t) - 1]dt + \int_{a_k}^{0}[v(X > t) - 1]dt \\
& + \int_{0}^{a_{k+1}}v(X > t)dt + \int_{a_{k+1}}^{a_{k+2}}v(X > t)dt + \cdots \\
& + \int_{a_{n-1}}^{a_n}v(X > t)dt \\
=\ & (a_2 - a_1)\left[v\left(\textstyle\bigcup_{j=2}^{n} A_j\right) - 1\right] + (a_3 - a_2)\left[v\left(\textstyle\bigcup_{j=3}^{n} A_j\right) - 1\right] \\
& + \cdots + (a_k - a_{k-1})\left[v\left(\textstyle\bigcup_{j=k}^{n} A_j\right) - 1\right] \\
& + (0 - a_k)\left[v\left(\textstyle\bigcup_{j=k+1}^{n} A_j\right) - 1\right] + (a_{k+1} - 0)\,v\left(\textstyle\bigcup_{j=k+1}^{n} A_j\right) \\
& + (a_{k+2} - a_{k+1})v\left(\textstyle\bigcup_{j=k+2}^{n} A_j\right) + \cdots + (a_n - a_{n-1})\,v\left(\textstyle\bigcup_{j}^{n} A_j\right) \\
=\ & a_1 + \sum_{i=2}^{n}(a_i - a_{i-1})\,v\left(\textstyle\bigcup_{j=i}^{n} A_j\right) \\
=\ & \sum_{i=1}^{n}(a_i - a_{i-1})\,v\left(\textstyle\bigcup_{j=i}^{n} A_j\right),
\end{aligned}
$$

noting that $v\left(\bigcup_{j=1}^{n} A_j\right) = v(\Omega) = 1$.

Unless $v$ is a probability measure, the Choquet operator $C_v(.)$ is not additive in general. For example, if $X = (1/4)1_A$ and $Y = (1/2)1_B$, and

$A \cap B = \varnothing$, then

$$\begin{aligned} C_v(X+Y) &= (1/4)(v(A \cup B) + (1/2 - 1/4)v(B) \\ &\neq (1/4)v(A) + (1/2)v(B), \end{aligned}$$

i.e.,

$$C_v(X+Y) \neq C_v(X) + C_v(Y).$$

How about *subadditivity*? This question is very important for risk modeling since if we model risk measures as Choquet integrals we would like to know the conditions for them to be subadditive in order to obtain coherent risk measures.

For a general capacity $v$, $C_v(.)$ might not be subadditive. It suffices to translate the example of $VaR_\alpha$ in terms of its associated capacity.

On $(\Omega, \mathcal{A}, P)$, for $\alpha \in (0,1)$, define the capacity $v_\alpha$ as follows.

$$v_\alpha(A) = \begin{cases} 1 \text{ if } P(A) > 1 - \alpha \\ 0 \text{ if } P(A) \leq 1 - \alpha. \end{cases}$$

Let $X, Y : \Omega \to \{0,1\}$ be independent and identically distributed with $P(X = 1) = 1 - P(X = 0) = \theta$. Choose $\alpha$ such that $(1-\theta)^2 < \alpha < 1 - \theta$. Then

$$C_{v_\alpha}(X) = C_{v_\alpha}(Y) = 0$$

because

$$\begin{aligned} C_{v_\alpha}(X) &= \int_0^\infty v_\alpha(X > t)dt \\ &= \int_0^1 v_\alpha(X > t)dt \\ &= \int_0^1 v_\alpha(X = 1)dt = 0 \end{aligned}$$

since $P(X = 1) = \theta < 1 - \alpha$, and hence $v_\alpha(X = 1) = 0$.

On the other hand,

$$\begin{aligned} C_{v_\alpha}(X+Y) &= \int_0^\infty v_\alpha(X + Y > t)dt \\ &= v_\alpha(X + Y \in \{1,2\}) + v_\alpha(X + Y = 2) \\ &\geq v_\alpha(X + Y \in \{1,2\}) \\ &= 1 \end{aligned}$$

since

$$P(X + Y \in \{1, 2\}) = 1 - P(X + Y = 0)$$
$$= 1 - (1 - \theta)^2 > 1 - \alpha.$$

Thus,

$$C_{v_\alpha}(X + Y) > C_{v_\alpha}(X) + C_{v_\alpha}(Y).$$

However, the Choquet integral is subadditive for special capacities $v$, namely those which are 2-*alternating*, i.e.,

$$v(A \cup B) \le v(A) + v(B) - v(A \cap B), A, B \in \mathcal{A}.$$

Specifically,

**Theorem 5.16** $C_v(.)$ *is subadditive if and only if $v$ is 2-alternating.*

**Proof.** First, we prove that the 2-alternating condition is a necessary condition, then next, we prepare some material for proving that it is also sufficient. The proof of sufficiency is long as it requires that we establish it first for simple random variables and then use approximation procedures. Therefore, we only prove it for the simple case of simple random variables. We follow Denneberg. Interested readers should read [34] for the general case.

*The condition is necessary.* Indeed, if $C_v(.)$ is subadditive, then in particular for $X = 1_A, Y = 1_B$, we have

$$C_v(1_A + 1_B) \le C_v(1_A) + C_v(1_B).$$

But,

$$C_v(1_A + 1_B) = \int_0^1 v(1_A + 1_B > t)dt + \int_1^2 v(1_A + 1_B > t)dt$$
$$= \int_0^1 v(A \cup B)dt + \int_1^2 v(A \cap B)dt$$
$$= v(A \cup B) + v(A \cap B),$$

whereas $C_v(1_A) = v(A)$, $C_v(1_B) = v(B)$.

Several observations are needed before we proceed to show that the Choquet integral with respect to 2-alternating capacities is subadditive. In Example 5.11, we have encountered the following situation. If $\Omega$ is a finite set, say $\Omega = \{\omega_1, \omega_2, ..., \omega_n\} \subseteq R$, and $v$ is a capacity on $2^\Omega$, then for each random variable $X : \Omega \to R$, the computation of $C_v(X)$ is simple if we rearrange the elements of $\Omega$ according to the values of $X$, i.e.,

$$X(\omega_1) \leq X(\omega_2) \leq \cdots \leq X(\omega_n),$$

to obtain

$$C_v(X) = \int_{\mathbb{R}} v(X > t)dt = \sum_{i=1}^{n} X(\omega_i)[v(A_i) - v(A_{i+1})]$$

where

$$A_i = \{\omega_i, \omega_{i+1}, ..., \omega_n\}, \; i = 1, 2, ..., n, \text{ with } A_{n+1} = \varnothing.$$

Clearly, the function $g(.)$ defined on $\Omega$ by

$$g(\omega_i) = v(A_i) - v(A_{i+1}), \; i = 1, 2, ..., n$$

is a probability density function, and thus, by setting $P_{v,X}(A) = \sum_{\omega \in A} g(\omega)$, we get a probability measure on $2^{\Omega}$, and

$$C_v(X) = \int_{\Omega} X(\omega)dP_{v,X}(\omega),$$

noting that $P_{v,X}$ depends not only on $v$, but also on $X$ (via its induced ordering on $\Omega$).

Now if we consider various permutations $\pi$ of $\{1, 2, ..., n\}$, then the above situation is just one permutation for which $X(\omega_{\pi(1)}) \leq X(\omega_{\pi(2)}) \leq \cdots \leq X(\omega_{\pi(n)})$ and, with respect to this permutation,

$$C_v(X) = \int_{\Omega} X(\omega)dP_{\pi}(\omega),$$

where $P_{\pi}(A) = \sum_{\omega \in A} g_{\pi}(\omega)$ with $g_{\pi}(\omega_i) = v(A_{\pi(i)}) - v(A_{\pi(i+1)})$.

We are going to show that if we define probability measures $P_{\pi}$ on $2^{\Omega}$ according to permutations $\pi$ of $\{1, 2, ..., n\}$, in the way described below, then in general,

$$C_v(X) \neq \int_{\Omega} X(\omega)dP_{\pi}(\omega),$$

but, if $v$ is 2-alternating, then

$$C_v(X) \geq \int_{\Omega} X(\omega)dP_{\pi}(\omega).$$

The set-up is as follows.

Let $(\Omega, \mathcal{A}, P)$ be a probability space and $\{A_i, i = 1, 2, ..., n\}$ be a measurable partition of $\Omega$. Let $\upsilon$ be a capacity defined on $\mathcal{A}$. Let $\pi$ be a permutation of $\{1, 2, ..., n\}$. We can associate with $\pi$ a (discrete) probability measure $P_\pi$ as follows.

- Let $B_i^\pi = \cup_{j=i}^n A_{\pi(j)}$, with $B_{n+1}^\pi = \varnothing$.

- Let $P_\pi(A_{\pi(i)}) = \upsilon(B_i^\pi) - \upsilon(B_{i+1}^\pi)$, $i = 1, 2, ..., n$.

- Let $X = \sum_{i=1}^n a_i 1_{A_i}$. Then $E_{P_\pi}(X) = \sum_{i=1}^n a_i P_\pi(A_i)$.

As mentioned above, if the permutation $\pi$ orders the values of $X$, i.e.,

$$a_{\pi(1)} \le a_{\pi(2)} \le \cdots \le a_{\pi(n)}$$

then

$$C_\upsilon(X) = E_{P_\pi}(X).$$

Otherwise, for arbitrary permutation $\pi$,

$$C_\upsilon(X) \ne E_{P_\pi}(X).$$

However, if $\upsilon$ is 2-alternating, then we have

$$C_\upsilon(X) \ge E_{P_\pi}(X),$$

as we set out to show.

Without loss of generality, take $\pi$ to be the identity permutation, i.e., $\pi(i) = i$, for all $i \in \{1, 2, ..., n\}$, and suppose $\pi$ does not order the values of $X$. Since $\pi$ does not order the values of $X$, there is some $k$ such that $a_k > a_{k+1}$.

Consider the permutation $\varsigma$ which leaves $\pi$ unchanged except interchanging $k$ and $k+1$ (at which $a_k > a_{k+1}$), i.e.,

$$\varsigma(i) = i \text{ for } i \ne k, k+1 \text{ and } \varsigma(k) = k+1, \ \varsigma(k+1) = k.$$

Clearly, repeating permutations of *this type* (i.e., a permutation interchanging only two values $i, j$ in a previous permutation, at which $a_i > a_j$) with a finite number of times, we will reach a permutation $\lambda$ which orders the values of $X$ so that

$$C_\upsilon(X) = E_{P_\lambda}(X) \ge E_{P_\pi}(X) \text{ if } E_{P_\varsigma}(X) \ge E_{P_\pi}(X).$$

Now

$$E_{P_\varsigma}(X) - E_{P_\pi}(X) = a_k P_\varsigma(A_k) + a_{k+1} P_\varsigma(A_{k+1}) - a_k P_\pi(A_k) - a_{k+1} P_\pi(A_{k+1}).$$

Thus, it suffices to show that if $v$ is 2-alternating then

$$a_k P_\varsigma(A_k) + a_{k+1} P_\varsigma(A_{k+1}) \geq a_k P_\pi(A_k) + a_{k+1} P_\pi(A_{k+1}).$$

Observe that

$$P_\pi(A_k) = v(\cup_{j=k}^n A_j) - v(\cup_{j=k+1}^n A_j)$$

with

$$\cup_{j=k}^n A_j = B_k^\varsigma = (\cup_{j=k+1}^n A_j) \cup B_{k+1}^\varsigma$$

and

$$B_{k+2}^\varsigma = \cup_{j=k+2}^n A_j = (\cup_{j=k+1}^n A_j) \cap B_{k+1}^\varsigma,$$

so that, by 2-alternating of $v$,

$$v(\cup_{j=k}^n A_j) + v(B_{k+2}^\varsigma) \leq v(\cup_{j=k+1}^n A_j) + v(B_{k+1}^\varsigma)$$

or

$$P_\pi(A_k) \leq v(B_{k+1}^\varsigma) - v(B_{k+2}^\varsigma) = P_\varsigma(A_{\varsigma(k+1)}) = P_\varsigma(A_k).$$

Multiplying both sides by $a_k - a_{k+1} > 0$,

$$(a_k - a_{k+1}) P_\pi(A_k) \leq (a_k - a_{k+1}) P_\varsigma(A_k). \qquad (5.1)$$

On the other hand,

$$
\begin{aligned}
P_\pi(A_k) + P_\pi(A_{k+1}) &= v(\cup_{j=k}^n A_j) - v(\cup_{j=k+2}^n A_j) \\
&= v(B_k^\varsigma) - v(B_{k+2}^\varsigma) \\
&= P_\varsigma(A_{\varsigma(k)}) + P_\varsigma(A_{\varsigma(k+1)}) \\
&= P_\varsigma(A_{k+1}) + P_\varsigma(A_k).
\end{aligned}
$$

Multiplying both sides by $a_{k+1}$,

$$a_{k+1}[P_\pi(A_k) + P_\pi(A_{k+1})] = a_{k+1}[P_\varsigma(A_{k+1}) + P_\varsigma(A_k)],$$

then adding to the inequality in (5.1), we get

$$a_k P_\varsigma(A_k) + a_{k+1} P_\varsigma(A_{k+1}) \geq a_k P_\pi(A_k) + a_{k+1} P_\pi(A_{k+1})$$

as desired. Now we are ready to prove that if a capacity $v$ is 2-alternating, then $C_v(.)$ is subadditive for simple random variables.

*The condition is sufficient.* Let $X, Y$ be two simple random variables. Then $X, Y$ and $X + Y$ are of the forms $\sum_{j=1}^n x_j 1_{A_j}, \sum_{j=1}^n y_j 1_{A_j}$ and $\sum_{j=1}^n z_j 1_{A_j}$, respectively, where the $A_i's$ form a measurable partition of $\Omega$.

As mentioned above, $C_v(X+Y)$ can be written as an ordinary integral with respect to the probability measure $P_\pi$ where the permutation $\pi$ of $\{1, 2, ..., n\}$ is chosen to order the values of $X + Y$. With respect to this permutation $\pi$, we have

$$C_v(X + Y) = E_{P_\pi}(X + Y).$$

In one hand, $E_{P_\pi}(X + Y) = E_{P_\pi}(X) + E_{P_\pi}(Y)$ by additivity of Lebesgue integral, and on the other hand,

$$E_{P_\pi}(X) \leq C_v(X), \ E_{P_\pi}(Y) \leq C_v(Y)$$

by using the previous result. Thus,

$$C_v(X + Y) \leq C_v(X) + C_v(Y).$$

∎

## 5.5 Comonotonicity

Consider two simple functions $X$ and $Y$ of the form

$$X = a1_A + b1_B \quad \text{with} \quad A \cap B = \varnothing, \ 0 \leqslant a \leqslant b$$
$$Y = \alpha 1_A + \beta 1_B \quad \text{with} \quad 0 \leqslant \alpha \leqslant \beta,$$

then

$$
\begin{aligned}
C_v(X) &= av(A \cup B) + (b - a)v(B) \\
C_v(Y) &= \alpha v(A \cup B) + (\beta - \alpha)v(B) \\
C_v(X + Y) &= (a + \alpha)v(A \cup B) + [(b + \beta) - (a + \alpha)]v(A),
\end{aligned}
$$

so that

$$C_v(X + Y) = C_v(X) + C_v(Y),$$

i.e., while the Choquet integral is not additive in general, it is additive for some special class of functions. More generally, this equality holds for $X = \sum_{j=1}^n a_j 1_{A_j}$ and $Y = \sum_{j=1}^n b_j 1_{A_j}$, with the $A_j$ pairwise disjoint and the $a_i$ and $b_i$ increasing and nonnegative. Clearly, such pairs of functions satisfy the inequality

$$(X(\omega) - X(\omega'))(Y(\omega) - Y(\omega')) \geqslant 0$$

for any $\omega, \omega' \in \Omega$.

**Definition 5.17** *Two random variables $X, Y$ are said to be **comonotonic**, or **similarly ordered** if for any $\omega, \omega' \in \Omega$, we have*

$$(X(\omega) - X(\omega'))(Y(\omega) - Y(\omega')) \geqslant 0.$$

*Comonotonic random variables are similarly ordered so that they exhibit a strong dependence between themselves.*

**Example 5.18** *Let $Y = aX + b$ with $a > 0$. For any $\omega, \omega' \in \Omega$,*

$$
\begin{aligned}
(X(\omega) - X(\omega'))(Y(\omega) - Y(\omega')) &= a, \text{ and} \\
(X(\omega) - X(\omega'))^2 &\geq 0,
\end{aligned}
$$

*and hence $X$ and $Y$ are comonotonic. In terms of distributions, the joint distribution of $(X, Y)$ is related to its marginals by the strongest copula (see details later). Specifically,*

$$
\begin{aligned}
P(X \leq x, Y \leq y) &= P(X \leq x, X \leq (y - b)/a) \\
&= P(X \leq x \wedge (y - b)/a) \\
&= P(X \leq x) \wedge P(X \leq (y - b)/a) \\
&= P(X \leq x) \wedge P(Y \leq y),
\end{aligned}
$$

*where $\wedge$ denotes minimum. If we let $H(\cdot, \cdot)$ be the joint distribution function of $(X, Y)$, and $F, G$ the marginals of $X, Y$, respectively, then*

$$H(x, y) = C(F(x), G(y)),$$

*where the copula $C$ is $C(u, v) = u \wedge v$. As we will see, this type of dependence is characteristic for comonotonicity in the sense that $X$ and $Y$ are comonotonic if and only if $H(x, y) = F(x) \wedge G(y)$ where $H$ is the joint distribution function of the random vector $(X, Y)$ and $F, G$ are (marginal) distributions of $X, Y$, respectively.*

Note also that, in general, $H(x, y) \leq F(x) \wedge G(y)$.

**Definition 5.19** *If for $X, Y$ comonotonic, if $C_v(X + Y) = C_v(X) + C_v(Y)$, then we say that $C_v(.)$ is **comonotonic additive**.*

It turns out that the concept of comonotonic additivity is essential for the characterization of Choquet integrals.

The concept of *comonotonic functions* as well as *comonotonic additivity of the Choquet integral* were originated from some comments of C.

Dellacherie [32]. Roughly speaking, comonotonicity of $X$ and $Y$ means that $X$ and $Y$ have the same "tableau of variation." It turns out that comonotonic simple random variables are of the previous form. Specifically, if $X = \sum_{i=1}^{n} a_i 1_{A_i}$, $Y = \sum_{j=1}^{m} b_j 1_{B_j}$ and $X, Y$ are comonotonic, then there exist a (measurable) partition $\{D_i, i = 1, 2, ..., k\}$ of $\Omega$, and two nondecreasing sequences $\alpha_i, \beta_i$, $i = 1, 2, ..., k$, such that $X = \sum_{i=1}^{k} \alpha_i 1_{D_i}$ and $Y = \sum_{i=1}^{k} \beta_j 1_{D_j}$. See details later.

Here are a few elementary facts about comonotonic functions.

1. The comonotonic relation is symmetric and reflexive, but not transitive.

2. Any function $X$ is comonotonic with a constant function, so that $C_v(X + a) = C_v(X) + a$, for any $a \in \mathbb{R}$.

3. If $X$ and $Y$ are comonotonic and $r$ and $s$ are positive numbers, then $rX$ and $sY$ are comonotonic.

4. As we saw above, two functions $X = \sum_{j=1}^{n} a_j 1_{A_j}$ and $Y = \sum_{j=1}^{n} b_j 1_{A_j}$, with the $A_j$ pairwise disjoint and the $a_i$ and $b_i$ increasing and nonnegative are comonotonic. In fact, the converse is also true: Two arbitrary simple random variables $X, Y$ are comonotonic if and only if they are of the above forms with $\{a_i\}, \{b_i\}$ increasing, positive or negative.

If $H(X) = \int_\Omega X d\mu$ with $\mu$ a Lebesgue measure, then $H$ is additive, and in particular, comonotonic additive. Here are some facts about comonotonic additivity.

1. If $H$ is comonotonic additive, then $H(0) = 0$. This follows since 0 is comonotonic with itself, whence $H(0) = H(0 + 0) = H(0) + H(0)$.

2. If $H$ is comonotonic additive, then for positive integers $n$, $H(nX)H(X)$. This is an easy induction. It is clearly true for $n = 1$, and for $n > 1$ and using the induction hypothesis,

$$H(nX) = H(X + (n-1)X) = H(X) + H((n-1)X)$$
$$= H(X) + (n-1)H(X)H(X).$$

3. If $H$ is comonotonic additive, then for positive integers $m$ and $n$,

$$H\left(\frac{m}{n}X\right) = \frac{m}{n}H(X).$$

Indeed,

$$\frac{m}{n}H(X) = \frac{m}{n}H\left(n\frac{X}{n}\right) = mH\left(\frac{X}{n}\right) = H\left(\frac{m}{n}X\right).$$

4. If $H$ is comonotonic additive and monotonic increasing, then $H(rX) = rH(X)$ for positive $r$, i.e., $H$ is positively homogeneous of degree one. Just take an increasing sequence of positive rational numbers $r_i$ converging to $r$. Then $H(r_iX) = r_iH(X)$ converges to $rH(X)$ and $r_iX$ converges to $rX$. Thus $H(r_iX)$ converges also to $H(rX)$.

Let us elaborate a little bit more on comonotonicity of variables as well as the comonotonic additivity of the Choquet integral.

First, for two variables $X, Y$, saying that they are comonotonic is the same as saying that any two pairs $(X(\omega), Y(\omega))$ and $(X(\omega'), Y(\omega'))$, in the range of $(X, Y)$, are ordered, i.e., either $(X(\omega), Y(\omega)) \leq (X(\omega'), Y(\omega'))$ or the other way around $(X(\omega'), Y(\omega')) \leq (X(\omega), Y(\omega))$, where, of course, by $\leq$ on $\mathbb{R}^2$ we mean the usual partial order relation $(x, y) \leq (x', y')$ if and only if $x \leq x'$ and $y \leq y'$. To put it differently, $X$ and $Y$ are comonotonic if the range of $(X, Y)$ is a *totally ordered subset* of $\mathbb{R}^2$.

In the example with $X = (1/4)1_A$, $Y = (1/2)1_B$, and $A \cap B = \varnothing$, the range of $(X, Y)$ is $\{(0,0), (1/4, 0), (0, 1/2)\}$ which is not a totally ordered subset of $\mathbb{R}^2$, whereas for $X = a1_A + b1_B$, $Y = \alpha1_A + \beta1_B$ with $A \cap B = \varnothing$, and $0 \leq a \leq b$, $0 \leq \alpha \leq \beta$, the range of $(X, Y)$ is $\{(0,0), (a, \alpha), (b, \beta)\}$ which is a totally ordered subset of $\mathbb{R}^2$.

This equivalent definition of comonotonicity of two variables is extended to more than two variables as follows.

**Definition 5.20** *The random variables $X_1, X_2, ..., X_n$ are said to be (mutually) comonotonic if the range of the random vector $(X_1, X_2, ..., X_n)$ is a totally ordered subset of $\mathbb{R}^n$.*

It turns out that we can also define the comonotonicity of several random variables from that of two random variables. In other words,

**Theorem 5.21** *Mutual comonotonicity is equivalent to pairwise comonotonicity.*

**Proof.** Suppose that $X_1, X_2, ..., X_n$ are mutually comonotonic. Consider $X_i, X_j$, for arbitrary but fixed $i, j \in \{1, 2, ..., n\}$. For any $\omega, \omega' \in \Omega$, either $(X_1(\omega), X_2(\omega), ..., X_n(\omega)) \leq (X_1(\omega'), X_2(\omega'), ..., X_n(\omega'))$ or the other way around. It follows that $(X_i(\omega), X_j(\omega)) \leq (X_i(\omega'), X_j(\omega'))$ or the other way around.

Conversely, suppose $X_i$ and $X_j$ are comonotonic, for all $i, j \in \{1, 2, ..., n\}$. For $\omega, \omega' \in \Omega$, suppose

$$(X_1(\omega), X_2(\omega)) \leq (X_1(\omega'), X_2(\omega')).$$

Then $X_1(\omega) \leq X_1(\omega')$ and $X_2(\omega) \leq X_2(\omega')$. But then, necessarily,

$$(X_2(\omega), X_3(\omega)) \leq (X_2(\omega'), X_3(\omega'))$$

implying that $X_3(\omega) \leq X_3(\omega')$, and so on, resulting in

$$(X_1(\omega), X_2(\omega), ..., X_n(\omega)) \leq (X_1(\omega'), X_2(\omega'), ..., X_n(\omega')).$$

■

There are several equivalent conditions for comonotonicity. For simplicity, we consider the case of two variables. The general case is similar. We write $F_X$ for the distribution function of $X$.

**Theorem 5.22** *The following are equivalent:*

*(i)* $X$ *and* $Y$ *are comonotonic.*

*(ii)* $F_{(X,Y)}(x, y) = F_X(x) \wedge F_Y(y).$

*(iii)* $(X, Y) = (F_X^{-1}(U), F_Y^{-1}(U))$ *in distribution, where* $U$ *is uniformly distributed on* $(0, 1)$.

*(iv)* *There exist a random variable* $Z$ *and nondecreasing functions* $u, v :$ $\mathbb{R} \to \mathbb{R}$ *such that* $X = u(Z)$ *and* $Y = v(Z)$, *both in distribution.*

**Remark 5.23** *Another equivalent condition is this. There exist two nondecreasing and continuous functions* $g, h : \mathbb{R} \to \mathbb{R}$, *with* $g(x) + h(x) = x$, *for all* $x \in \mathbb{R}$, *such that* $X = g(X + Y)$, *and* $Y = h(X + Y)$. *For a proof, see Denneberg [34].*

**Proof.** $(i) \implies (ii)$: $F_{(X,Y)}(x, y) = P(X \leq x, Y \leq y) = P(A \cap B)$, where

$$A = \{\omega \in \Omega : X(\omega) \leq x\}, B = \{\omega \in \Omega : Y(\omega) \leq y\}.$$

When $X$ and $Y$ are comonotonic, we have "either $A \subseteq B$ or $B \subseteq A$," since if not, its negation should be true, i.e., "$A \not\subseteq B$ and $B \not\subseteq A$," meaning that there are $(X(\omega), Y(\omega))$ and $(X(\omega'), Y(\omega'))$ such that $X(\omega) \leq x$, $Y(\omega) > y$, $X(\omega') > x$, $Y(\omega') \leq y$. But then,

$$X(\omega) \leq x < X(\omega') \text{ and } Y(\omega) > y \geq Y(\omega'),$$

so that $(X(\omega), Y(\omega))$ and $(X(\omega'), Y(\omega'))$ are not comparable in the partial order $\leq$ of $\mathbb{R}^2$, contradicting the hypothesis that $X, Y$ are comonotonic.

$(ii) \implies (iii)$: It is well known that if $U$ is a random variable, uniformly distributed on $(0, 1)$, then $F_X^{-1}(U)$ is distributed as $X$, and $F_Y^{-1}(U)$ is distributed as $Y$, but in general, the random vector $(X, Y)$ is not distributed as the random vector $(F_X^{-1}(U), F_Y^{-1}(U))$. This will be the case when $X, Y$ are comonotonic as we show now.

Suppose $(ii)$. We have

$$\begin{aligned} P(F_X^{-1}(U) &\leq x, F_Y^{-1}(U) \leq y) = P(U \leq F_X(x), U \leq F_Y(y)) \\ &= P(U \leq F_X(x) \wedge F_Y(y)) \\ &= F_X(x) \wedge F_Y(y) = P(X \leq x, Y \leq y). \end{aligned}$$

$(iii) \implies (iv)$: Just take $Z = U$, $u(.) = F_X^{-1}(.)$, $v(.) = F_Y^{-1}$.

$(iv) \implies (i)$: For $X = u(Z)$, $Y = v(Z)$ with $u(.), v(.)$ nondecreasing, we have, for any $\omega, \omega' \in \Omega$, with, say, $Z(\omega) \leq Z(\omega')$,

$$X(\omega) = u(Z(\omega)) \leq u(Z(\omega')) = X(\omega')$$

and

$$Y(\omega) = v(Z(\omega)) \leq v(Z(\omega')) = Y(\omega').$$

∎

**Remark 5.24** *For any random variables $X_1, X_2, ..., X_n$, the associated random variables $F_{X_1}^{-1}(U)$, $F_{X_2}^{-1}(U)$, ..., $F_{X_n}^{-1}(U)$ are comonotonic. They are called the comonotonic components of $(X_1, X_2, ..., X_n)$.*

## 5.6    Notes on copulas

*Copulas* (a Latin word meaning a link) are functions relating the (marginal) distributions of the components of a random vector to its joint distribution.

Let $(X_1, X_2, ..., X_n)$ be a random vector. If the joint distribution function of the vector is known, then we can derive all marginal distribution functions. Specifically, given $H : \mathbb{R}^n \to [0, 1]$ with

$$H(x_1, x_2, ..., x_n) = P(X_1 \leq x_1, X_2 \leq x_2, ..., X_n \leq x_n)$$

we get, for $j = 1, 2, ..., n$,

$$F_j(x_j) = P(X_j \le x_j) = H(\infty, ..., \infty, x_j, \infty, ..., \infty).$$

The inverse problem is this. Given the $F_j$'s, $j = 1, 2, ..., n$, what can be said about the joint $H$? More specifically, can we describe the form of $H$ in terms of its marginals?

The answer was given by A. Sklar [148]. See R. Nelsen [119, 120] for a full exposition on copulas. Sklar's result was left unnoticed for quite sometime, despite the fact that his result should be considered as a breakthrough in the art of building multivariate statistical models as well as modeling dependence relations among random variables. Indeed, early in the 1960's, his concept of copulas only led to his joint work with B. Schweizer in the theory of probabilistic metric spaces [140] in view of the relation between copulas and the concept of *t-norms* ("t" for triangular) in probabilistic metric spaces. In the 1970's, both t-norms and copulas entered the field of artificial intelligence, especially as new logical connectives in several new logics, including *fuzzy logics*. Then, slowly and quietly (!), statisticians started paying some attention to copulas. And now, everybody talks about copulas, especially in economics!

So what is a copula?

**Definition 5.25** *An n-copula is a function* $C : [0, 1]^n \to [0, 1]$ *satisfying*

(i) $C(1, ..., 1, x_j, 1, ..., 1) = x_j$ *for* $x_j \in [0, 1]$, $j \in \{1, 2, ..., n\}$,

(ii) $\int C$ *is n-increasing: for any* $J = J_1 \times J_2 \times \cdots \times J_n$ *of closed intervals* $J_i = [a_i, b_i] \subseteq [0, 1]$, $i = 1, 2, ..., n$, *we have*

$$vol(J) = \sum_{\mathbf{v}} sgn(v)c(v) \ge 0$$

*where the summation is over all vertices* $v$ *of* $J$, *and for* $v = (v_1, v_2, ..., v_n)$, $v_i = a_i$ *or* $b_i$, *and*

$$sgn(v) = \begin{cases} 1 & \text{if } v_i = a_i \text{ for an even number of } i\text{'s} \\ -1 & \text{if } v_i = a_i \text{ for an odd number of } i\text{'s}. \end{cases}$$

(iii) $C$ *is* grounded: $C(x_1, x_2, ..., x_n) = 0$ *for all* $(x_1, x_2, ..., x_n) \in [0, 1]^n$ *such that* $x_i = 0$ *for at least one* $i$, $i = 1, 2, ..., n$.

For $n - 2$, the above definition reduces to the follwing.

**Definition 5.26** *A **bivariate copula** is a function* $C : [0,1] \times [0,1] \rightarrow [0,1]$ *satisfying the following:*

(i) $C(x,0) = C(0,x) = 0, C(x,1) = x, C(1,y) = y,$ *and*

(ii) *If* $x \leq y$ *and* $u \leq v$ *then* $C(y,v) - C(y,u) - C(x,v) + C(x,u) \geq 0.$

The now well-known Sklar theorem is this. We state it for the case of two random variables.

**Theorem 5.27 (Sklar)** *Let* $H$ *be the joint distribution function of* $(X,Y)$ *and* $F,G$ *be the (marginal) distribution functions of* $X,Y$, *respectively. Then there exists a copula* $C$ *such that*

$$H(x,y) = C(F(x), G(y))$$

*for all* $x,y$ *in* R. *If* $F$ *and* $G$ *are continuous, then* $C$ *is unique, otherwise* $C$ *is uniquely determined on* $range(F) \times range(G)$. *Conversely, if* $C$ *is a copula and* $F,G$ *are (one-dimensional) distribution functions, then* $H(\cdot,\cdot)$ *defined by* $H(x,y) = C(F(x), G(y))$ *is a joint distribution with marginals* $F,G$.

**Remark 5.28** *For given joint distribution function* $H$ *such that its marginals* $F,G$ *are continuous, we determine its unique associated copula* $C$ *by*

$$C(x,y) = H(F^{-1}(x), G^{-1}(y))$$

*where, as usual,* $F^{-1}$ *is the left-continuous inverse of* $F$. *Such* $C$ *is called the copula of* $H$. *For example, if* $H$ *is a bivariate Gaussian (normal) with standard normal marginals, then the associated* $C$ *is called a Gaussian copula.*

**Example 5.29** *Here are some examples of copulas:*

(a) $C(x,y) = xy.$

(b) $C(x,y) = x \wedge yh.$

(c) $C(x,y) = \max(x + y - 1, 0).$

We turn now to the proof of *comonotonic additivity* of the Choquet integral. We have seen that the Choquet integral is comonotonic additive for special comonotonic simple random variables. This turns out to be true for general comonotonic random variables.

Note that if $C_v(.)$ is comonotonic additive for the sum of any two comonotonic random variables, then it is comonotonic additive for any finite sum of comonotonic random variables. This is so because if $X_1, X_2, ..., X_n$ are comonotonic, then $X_1$ and $\sum_{i=2}^{n} X_i$ are comonotonic. Thus, it suffices to consider the case of two arbitrary random variables.

Here is the theorem.

**Theorem 5.30** *If $X$ and $Y$ are comonotonic, then*

$$C_v(X+Y) = C_v(X) + C_v(Y).$$

We will present two proofs for comonotonic additivity of the Choquet integral since each method is interesting in its own right. The first one is based upon standard method in measure theory, namely, establishing it first for nonnegative simple functions and then passing through the limits. The second method is based upon an extension of quantile functions of probability distribution functions to distribution functions with respect to capacities (see Denneberg [34] or Föllmer and Schied [48]).

**Direct proof.** The comonotonic additivity for simple random variables can be established simply by observing that if $X, Y$ are simple and comonotonic then they can be written as $X = \sum_{i=1}^{l} a_i 1_{A_i}$ and $Y = \sum_{i=1}^{l} b_i 1_{A_i}$ with the $\{a_i\}, \{b_i\}$ both nondecreasing. Indeed, without loss of generality, we can assume that finite step functions $X$ and $Y$ are of the forms: $X = \sum_{i=1}^{n} a_i 1_{A_i}$ and $Y = \sum_{i=1}^{n} b_i 1_{A_i}$ where $(A_i)_{i=1}^{n}$ is a measurable partition of $\Omega$ and $\{a_i\}_{i=1}^{n}$ is an increasing list of numbers.

Now, to prove the claim, we need to show that $X$ and $Y$ are comonotonic if and only if $\{b_i\}_{i=1}^{n}$ is an increasing list of numbers. Suppose that $X$ and $Y$ are comonotonic, for $i < j$ if we take $\omega \in A_i$ and $\omega' \in A_j$ then by comonotonicity of $X$ and $Y$,

$$(X(\omega) - X(\omega'))(Y(\omega) - Y(\omega')) \geq 0$$

or

$$(a_j - a_i)(b_j - b_i) \geq 0,$$

which implies

$$b_j - b_i \geq 0.$$

Conversely, it is very easy to see that if $\{b_i\}_{i=1}^l$ is increasing then $X$ and $Y$ are comonotonic. From the above, it suffices to recall that

$$C_v(X) = \sum_{i=1}^{n}(a_i - a_{i-1})v(\cup_{j=i}^{n}A_j).$$

For arbitrary comonotonic random variables $X, Y$, we use the following approximations. Define

$$\Psi_n(x) = \begin{cases} (k-1)2^{-n} & \text{if } (k-1)2^{-n} \leq x < k2^{-n}, \ k = 1...n2^n \\ n & \text{if } x \geq n. \end{cases}$$

Let

$$X_n = \begin{cases} \Psi_n(X) & \text{if } X \geq 0 \\ -\Psi_n(-X) & \text{if } X < 0 \end{cases}$$

and

$$Y_n = \begin{cases} \Psi_n(Y) & \text{if } Y \geq 0 \\ -\Psi_n(-Y) & \text{if } Y < 0. \end{cases}$$

Then

$$0 \leq X_n(\omega) \uparrow X(\omega) \text{ if } X(\omega) \geq 0$$
$$0 \geq X_n(\omega) \downarrow X(\omega) \text{ if } X(\omega) < 0.$$

Similarly,

$$0 \leq Y_n(\omega) \uparrow Y(\omega) \text{ if } Y(\omega) \geq 0$$
$$0 \geq Y_n(\omega) \downarrow Y(\omega) \text{ if } Y(\omega) < 0.$$

In addition,

$$X(\omega) - \frac{1}{2^n} \leq X_n(\omega) \leq X(\omega) + \frac{1}{2^n}.$$

Similarly,

$$Y(\omega) - \frac{1}{2^n} \leq Y_n(\omega) \leq Y(\omega) + \frac{1}{2^n}.$$

By definition of $X_n$, $X(\omega) \geq X(\omega')$ implies $X_n(\omega) \geq X_n(\omega')$. It is easy to show that comonotonicity of $X$ and $Y$ implies comonotonicity of $X_n$ and $Y_n$.

Now, there are an integer $l_n$, a partition $(A_i)_{i=1}^{l_n}$ of $\Omega$, and two $l_n$-lists of increasing numbers $\{a_i\}_{i=1}^{l_n}$ and $\{b_i\}_{i=1}^{l_n}$ such that

$$X_n = \sum_{i=1}^{l_n} a_i 1_{A_i} \text{ and } Y_n = \sum_{i=1}^{l_n} b_i 1_{A_i}.$$

Hence,

$$
\begin{aligned}
C_v(X_n + Y_n) &= \sum_{i=1}^{l_n} [(a_i + b_i) - (a_{i-1} - b_{i-1})] v \left( \bigcup_{j=i}^{l_n} A_j \right) \\
&= \sum_{i=1}^{l_n} (a_i - a_{i-1}) v \left( \bigcup_{j=i}^{l_n} A_j \right) + \sum_{i=1}^{l_n} (b_i - b_{i-1}) v \left( \bigcup_{j=i}^{l_n} A_j \right) \\
&= C_v(X_n) + C_v(Y_n).
\end{aligned}
$$

By monotonicity of the Choquet integral and

$$
X(\omega) - \frac{1}{2^n} \le X_n(\omega) \le X(\omega) + \frac{1}{2^n},
$$

we have

$$
C_v(X) - \frac{1}{2^n} \le C_v(X_n) \le C_v(X) + \frac{1}{2^n}.
$$

Hence,

$$
\lim_{n \to \infty} C_v(X_n) = C_v(X),
$$

noting that the limit does exist by the monotone convergence theorem. Similarly,

$$
\lim_{n \to \infty} C_v(Y_n) = C_v(Y) \text{ and } \lim_{n \to \infty} C_v(X_n + Y_n) = C_v(X + Y).
$$

From the equality $C_v(X_n + Y_n) = C_v(X_n) + C_v(Y_n)$, let $n \to \infty$, and we get

$$
C_v(X + Y) = C_v(X) + C_v(Y).
$$

∎

**Indirect proof.** Some preparations are needed. These preparations will bring out the useful role that quantile functions play.

Recall that when $v = P$, the following is well known. The (probability) distribution function $F$ of a (nonnegative) random variable $X$ is $F(x) = P(X \le x) = 1 - P(X > x)$, and a quantile function of $X$ (with respect to $P$) is $F^{-1}(t) = \inf\{x : F(x) \ge t\}$. By the Fubini theorem,

$$
EX = \int_0^\infty P(X > x) dx = \int_0^1 F^{-1}(t) dt.
$$

This equality remains valid when we replace the probability measure $P$ on $\mathcal{A}$ by a capacity $v$ on $\mathcal{A}$ and extend the concept of quantile functions of probability distribution functions (i.e., distributions with respect to $P$) to quantile functions of distribution functions with respect to $v$.

Specifically, let the *distribution function of $X$ with respect to $v$* be

$$G_{X,v}(x) = G(x) = 1 - v(X > x)$$

which is nondecreasing. Define

$$\begin{aligned} G^-(\alpha) &= \inf\{x : G(x) \geq \alpha\} \\ G^+(\alpha) &= \inf\{x : G(x) > \alpha\}, \end{aligned}$$

and call a quantile function of $G$, any function $q(.)$ such that

$$G^-(.) \leq q(.) \leq G^+(.),$$

which is equivalent to

$$G(q(\alpha)-) \leq \alpha \leq G(q(\alpha)+) \text{ for all } \alpha \in (0,1).$$

∎

Just like the case for probability distribution functions, quantile functions are equal almost everywhere on $(0,1)$. For details of the above, see Föllmer and Schied [48].

**Lemma 5.31** *If $X = h(Y)$ with $h(.)$ nondecreasing, and $q_Y(.)$ is a quantile function of $Y$ with respect to a capacity $v$, then $h(q_Y)(.)$ is a quantile function of $X$ with respect to $v$.*

**Proof.** First, recall that $q_Y(.)$ is a quantile function of $Y$ with respect to $v$ means that

$$G_Y(q_Y(\alpha)-) \leq \alpha \leq G_Y(q_Y(\alpha)+)$$

for all $\alpha \in (0,1)$, where $G_Y(t) = 1 - v(Y > t)$.

Now, let $q(\alpha) = h(q_Y(\alpha))$. We have

$$G_X(q(\alpha)-) = 1 - v(X > q(\alpha)-) = 1 - v(h(Y) > h(q_Y(\alpha)-)).$$

Observe that, since $h$ nondecreasing,

$$\{\omega : Y(\omega) > q_Y(\alpha)\} \subseteq \{\omega : h(Y(\omega)) > h(q_Y(\alpha))\}.$$

Thus, since $v$ is increasing,

$$\begin{aligned} 1 - v(h(Y) &> h(q_Y(\alpha)-)) \leq \alpha \leq 1 - v(Y > q_Y(\alpha)+) \\ &\leq 1 - v(X > q(\alpha)+) \leq G_X(q(\alpha)+). \end{aligned}$$

∎

**Lemma 5.32** *Let $q_X$ be a quantile function of $X$ with respect to a capacity $v$. Then*

$$C_v(X) = \int_0^1 q_X(\alpha)d\alpha.$$

**Proof.** Since all quantile functions coincide almost everywhere on $(0,1)$, it suffices to show the above equality for $q_X(\alpha) = \sup\{x \geq 0 : G_X(x) \leq \alpha\}$.

By Fubini,

$$
\begin{aligned}
\int_0^1 q_X(\alpha)d\alpha &= \int_0^1 \int_0^\infty 1_{\{x:G_X(x)\leq\alpha\}}dxd\alpha \\
&= \int_0^\infty \left[\int_{G_X(x)}^1 d\alpha\right]dx \\
&= \int_0^\infty (1 - G_X(x))dx \\
&= \int_0^\infty v(X > x)dx.
\end{aligned}
$$

Now if $X, Y$ are comonotonic, then $X = g(Z), Y = h(Z)$ for some random variable $Z$ and nondecreasing functions $g, h$. Thus, by the lemma above, $g(q_Z)$ and $h(q_Z)$ are quantile functions of $X, Y$, respectively. Since $g+h$ is nondecreasing, and $X+Y = (g+h)(Z)$, $(g+h)(q_Z)$ is a quantile function of $X + Y$, and hence $q_{X+Y} = q_X + q_Y$. Since the quantile functions of a random variable coincide almost everywhere on $(0,1)$, integration over $(0,1)$ yields

$$C_v(X + Y) = C_v(X) + C_v(Y)$$

in view of the preceding lemma. ∎

**Remark 5.33** *We have established the proof for the comonotonic additivity of the Choquet integral for nonnegative comonotonic random variables, but the result in fact holds for arbitrary bounded comonotonic random variables by considering $X - \inf X \geq 0$, by noting that $q_{X+a}(.) = q_X(.) + a$.*

## 5.7 A characterization theorem

The comonotonic additivity is characteristic for the Choquet integral in several aspects. When we model a risk measure as a Choquet integral, such as $VaR_\alpha(.)$, we get comonotonic additivity for free. If we have a risk measure which is not comonotonic additive (for comonotonic risks), then that risk measure is not a Choquet integral. Now suppose that we are in a context where additivity of the risk measure is desirable for comonotonic risks (random variables), i.e., we add this comonotonic additivity to our list of desirable properties for an appropriate risk measure, can we justify our choice of that risk measure as a Choquet integral?

The following theorem does just that.

**Theorem 5.34 (Schmeidler)** *Let $\mathbb{B}$ be the class of real-valued bounded random variables defined on $(\Omega, \mathcal{A})$. Let $H$ be a functional on $\mathbb{B}$ satisfying*

*(i)  $H(1_\Omega) = 1$,*

*(ii)  $H(.)$ is monotone increasing, and*

*(iii)  $H(.)$ is comonotonic additive.*

*Then $H(.)$ is of the form $C_\upsilon(.)$ for the capacity $\upsilon$ defined on $\mathcal{A}$ by $\upsilon(A) = H(1_A)$.*

**Proof.** For bounded random variables, it suffices to consider nonnegative ones. From the properties of $H(.)$, we derive that $H(.)$ is homogeneous with degree one, and if we set $\upsilon(A) = H(1_A)$, then $\upsilon(.)$ is a capacity defined on $\mathcal{A}$.

For a nonnegative simple random variable $X$, we can always represent it as $X = \sum_{i=1}^{n} a_i 1_{A_i}$ with $0 < a_1 < a_2 < \cdots < a_n$ and the $A_i$'s are disjoint, nonempty sets. We have

$$\int_0^\infty \upsilon(X > t)dt = \sum_{i=1}^{n}(a_i - a_{i-1})\upsilon(\cup_{j=i}^{n})$$

with $a_0 = 0$. Thus, we need to verify that

$$H\left(\sum_{i=1}^{n} a_i 1_{A_i}\right) = \sum_{i=1}^{n}(a_i - a_{i-1})\upsilon(\cup_{j=i}^{n}) \tag{5.2}$$

for all $n \geq 1$. Now, for $n = 1$, by homogeneity of $H(.)$, we have

$$H(a_1 1_{A_1}) = a_1 H(1_{A_1}) = a_1 \upsilon(A_1),$$

i.e., Equation (5.2) is satisfied.

Suppose Equation (5.2) is true up to $n-1$. Consider $X = \sum_{i=1}^{n} a_i 1_{A_i}$ as above. Write

$$\sum_{i=1}^{n} a_i 1_{A_i} = \sum_{i=2}^{n} (a_i - a_1) 1_{A_i} + a_1 (1_{\cup_{i=1}^{n}}).$$

Now $\sum_{i=2}^{n} (a_i - a_1) 1_{A_i}$ is comonotonic with $a_1 (1_{\cup_{i=1}^{n}})$, so that

$$
\begin{aligned}
H\left( \sum_{i=1}^{n} a_i 1_{A_i} \right) &= H\left( \sum_{i=2}^{n} (a_i - a_1) 1_{A_i} \right) + H(a_1 1_{\cup_{i=1}^{n}}) \\
&= \sum_{i=2}^{n} [(a_i - a_1) - (a_{i-1} - a_1)] \upsilon(\cup_{j=i}^{n}) + a_1 \upsilon(\cup_{i=1}^{n} A_i) \\
&\qquad \text{(by induction hypothesis)} \\
&= \sum_{i=2}^{n} (a_i - a_{i-1}) \upsilon(\cup_{j=i}^{n}) + a_1 \upsilon(\cup_{i=1}^{n} A_i) \\
&= \sum_{i=1}^{n} (a_i - a_{i-1}) \upsilon(\cup_{j=i}^{n}) = \int_{0}^{\infty} \upsilon(X > t) dt.
\end{aligned}
$$

Thus, $H(X) = C_\upsilon(X)$ for nonnegative simple random variables.

Now let $X$ be a nonnegative random variable bounded by $b$, say. For each $n \geq 1$, we partition $(0, b]$ into $2^n$ disjoint intervals $(b(k-1)/2^n, bk/2^n]$, $k = 1, 2, ..., 2^n$. Let

$$A_n^k = \{\omega \in \Omega : b(k-1)/2^n < X(\omega) \leq bk/2^n\}.$$

Define

$$Y_n(\omega) = b(k-1)/2^n \text{ and } Z_n(\omega) = bk/2^n$$

when $\omega \in A_n^k$. Then $Y_n(.) \leq X(.) \leq Z_n(.)$ for all $n \geq 1$, so that $H(Y_n) \leq H(X) \leq H(Z_n)$ since $H(.)$ is monotone increasing. Note that $Y_n(.)$ and $Z_n(.)$ are nonnegative simple random variables.

By comonotonic additivity of $H(.)$, we have

$$0 \leq H(Z_n) - H(Y_n) = b/2^n \to 0 \text{ as } n \to \infty.$$

Now, for all $n$,

$$H(Y_n) = \int_{0}^{b} \upsilon(Y_n > t) dt$$

and

$$H(Z_n) = \int_{0}^{b} \upsilon(Z_n > t) dt.$$

Thus, it suffices to show that, for all $n$,

$$\int_0^b v(Y_n > t)dt \leq \int_0^b v(X > t)dt \leq \int_0^b v(Z_n > t)dt,$$

which follows from $Y_n(.) \leq X(.) \leq Z_n(.)$. In other words, we have

$$H(X) = \int_0^b v(X > t)dt.$$

■

## 5.8   A class of coherent risk measures

Armed with the theory of the Choquet integral, we can now investigate desirable properties of existing risk measures as well as propose new ones. In the next section, we will use Choquet representation of risk measures to investigate their consistency with stochastic dominance rules.

We have seen that some typical risk measures, such as $VaR_\alpha$ and $TVaR_\alpha$, have Choquet integral representations with respect to distorted probabilities, i.e., they are of the form

$$\rho(X) = C_v(X)$$

with $v = g \circ P$, for some distortion function

$$g : [0,1] \to [0,1], g(0) = 0, g(1) = 1,$$

and nondecreasing. Specifically,

$$\rho_g(X) = \int_0^\infty g(1 - F_X(t))dt + \int_{-\infty}^0 [g(1 - F_X(t)) - 1]dt,$$

a special case of a Choquet integral.

Since each $\rho_g(.)$ is a Choquet integral operator, it satisfies automatically the first three axioms for coherent risk measures. Note that the translation invariance axiom is satisfied by comonotonic additivity since every random variable is comonotonic with any constant.

As Choquet integrals, the $\rho_g$ will be coherent if they satisfy the subadditivity axiom. As we have seen, in general Choquet integrals are not subadditive, and they are subadditive if and only if the corresponding capacities are 2-alternating. We specify a sufficient condition for the special case of distorted probabilities now.

**Theorem 5.35** *Let $g$ be a distortion and $v_g = g \circ P$ be its associated distorted probability. If $g$ is concave, then $v_g$ is 2-alternating, and hence $C_{v_g}(.)$ is subadditive.*

**Proof.** For any $A, B \in \mathcal{A}$, with, say, $a = P(A) \le P(B) = b$, let

$$0 \le P(A \cap B) = c \le a \le b \le d = P(A \cup B) \le 1,$$

with $c + d = a + b$, since

$$P(A \cup B) + P(A \cap B) = P(A) + P(B).$$

Since $c \le a \le d$, and $c \le b \le d$, we have

$$a = \left[\frac{d-a}{d-c}\right]c + \left[\frac{a-c}{d-c}\right]d$$

$$b = \left[\frac{d-b}{d-c}\right]c + \left[\frac{b-c}{d-c}\right]d.$$

Since $g$ is concave on $[0, 1]$, we have

$$g(a) \ge \left[\frac{d-a}{d-c}\right]g(c) + \left[\frac{a-c}{d-c}\right]g(d)$$

$$g(b) \ge \left[\frac{d-b}{d-c}\right]g(c) + \left[\frac{b-c}{d-c}\right]g(d),$$

so that

$$g(a) + g(b) \ge \left[\frac{d-a+d-b}{d-c}\right]g(c) + \left[\frac{a-c+b-c}{d-c}\right]g(d)$$

$$= g(c) + g(d),$$

since $c + d = a + b$ implies that

$$d - a + d - b = a - c + b - c = d - c.$$

Thus,

$$v_g(A \cap B) + v_g(A \cup B) \le v_g(A) + v_g(B).$$

∎

**Remark 5.36** *In fact the converse is also true: If $v_g$ is 2-alternating then $g$ must be concave. This is the same as saying that if $C_{v_g}(.)$ is subadditive then $g$ must be concave, in view of the fact that 2-alternating* of a capacity is a necessary and sufficient condition for its associated Choquet integral to be subadditive.

The proof of the above is as follows (from Wirch and Hardy [168]). Suppose the distortion function g is strictly convex on some $(a, b) \subset [0, 1]$. Let $c = (a + b)/2$. For $z < (b - a)/2$, and some positive constant $\lambda$, consider the random vector $(X, Y)$ with joint discrete distribution

| $X \backslash^Y$ | 0 | $\lambda + z/2$ | $\lambda + z$ |
|---|---|---|---|
| 0 | $1 - c - z$ | $z$ | 0 |
| $\lambda + z$ | $z$ | 0 | $c - z$ |

From their joint distribution, we have

$$
\begin{aligned}
C_{v_g}(X) &= (\lambda + z)g(c)C_{v_g}(Y) \\
&= (\lambda + z/2)g(c) + (z/2)g(c - z)C_{v_g}(X + Y) \\
&= (\lambda + z/2)g(c + z) + (z/2)g(c) + (\lambda + z)g(c - z).
\end{aligned}
$$

Now,

$$
\begin{aligned}
&C_{v_g}(X + Y) - C_{v_g}(X) - C_{v_g}(Y) \\
&= (\lambda + z/2)[g(c + z) - g(c) - (g(c) - g(c - z))] \\
&> 0
\end{aligned}
$$

since $c = [(c - z) + (c + z)]/2$, strict convexity gives

$$
2g(c) < g(c - z) + g(c + z)
$$

or

$$
g(c + z) - g(c) > g(c) - g(c - z).
$$

**Example 5.37** *The distortion function for* $VaR_\alpha(.)$ *is*

$$
g_\alpha(x) = 1_{(1-\alpha, 1]}(x)
$$

which is not a concave function, and hence the Value-at-Risk is not subadditive, so that it is not a coherent risk measure. In contrast, the Tail Value-at-Risk $TVaR_\alpha(.)$ has $g_\alpha(x) = \min\{1, x/(1 - \alpha)\}$ as its distortion function, which is concave, and hence subadditive, resulting in a coherent risk measure.

Here is some useful information about concave functions. Since $f : I \to \mathbb{R}$, where $I$ is an interval in $\mathbb{R}$, is concave if $-f$ is convex, we could just consider convex functions. Recall that $f$ is *convex* if, for any $\lambda \in (0, 1)$, and $x, y \in I$,

$$f(\lambda x + (1 - \lambda)y) \leq \lambda f(x) + (1 - \lambda)f(y).$$

If we take $\lambda = 1/2$, then we get

$$f\left(\frac{x+y}{2}\right) \leq \frac{f(x) + f(y)}{2}.$$

A function satisfying this last condition is called a *midconvex function.*

A midconvex function needs not be convex since if $f$ is convex, then it is continuous on open intervals, but there are discontinuous functions that are midconvex.

In fact, a convex function $f$ is absolutely continuous on any $[a, b]$ contained in the interior of $I$, i.e., for any $\varepsilon > 0$, there is $\delta > 0$ such that for any collection $(a_i, b_i), i = 1, 2, ..., n$, of disjoint open subintervals of $[a, b]$ with $\sum_{i=1}^{n}(b_i - a_i) < \delta$, we have

$$\sum_{i=1}^{n} |f(b_i) - f(a_i)| < \varepsilon.$$

It is well known that $f$ is midconvex if and only if

$$f\left(\sum_{i=1}^{n} r_i x_i\right) \leq \sum_{i=1}^{n} r_i f(x_i)$$

for any rational convex combination of elements of $I$, i.e., for any $x_i \in I$, $r_i \geq 0$, rational, $\sum_{i=1}^{n} r_i = 1$. However, a bounded midconvex function $f$ on $[a, b]$ is continuous on $(a, b)$ and hence convex.

Based upon previous analyses, we find the following.

1. Although $VaR_\alpha(.)$ is a Choquet integral, in fact a distorted probability risk measure, it is not coherent since it is not subadditive (since its associated distortion function $g_\alpha(x) = 1_{(1-\alpha, 1]}(x)$ is not concave). Moreover $g_\alpha$ is not above the diagonal of the unit square, and hence $VaR_\alpha(X)$ is not bounded below by $EX$.

2. $TVaR_\alpha(.)$ is a distorted probability risk measure with a concave distortion function and hence subadditive (coherent). Note, however, that its distortion function is not strictly concave.

3. $CVaR_\alpha(.)$ is not a Choquet integral operator since it is not comonotonic additive. Moreover $CVaR_\alpha(.)$ is not subadditive.

4. The Expected Short Fall $ESF_\alpha(X) = E(X - F_X^{-1}(\alpha))^+$ is not subadditive. In fact, $ESF_\alpha(X)$ is not a distorted risk measure.

5. Clearly not all risk measures are distorted probability risk measures. For an arbitrary 2-alternating capacity, the associated risk measure, as a Choquet integral, is a coherent risk measure.

6. Distorted probability risk measures using Wang's (concave) distortion functions are coherent risk measures (see below).

A useful class of distortion functions, known as the *Wang transform*, was proposed by S. Wang [163] for pricing financial and insurance risks. These are distortion functions of the form

$$g_\lambda(x) = \Phi(\Phi^{-1}(x) + \lambda)$$

where $\lambda \in \mathbb{R}$ and $\Phi$ is the distribution function of the standard normal random variable.

We provide here a justification of Wang's distortion functions. Let $X$ be a nonnegative random variable with distribution function $F$. Since

$$EX = \int_0^\infty (1 - F(t))dt$$

it is more convenient to focus on the "survival function"

$$F^*(t) = 1 - F(t) = P(X > t).$$

For a premium calculation $\rho(X)$, it is "desirable" that $\rho(X) \geq EX = \int_0^\infty F^*(t)dt$. That can be achieved by increasing $F^*(.)$ by transforming it by some function $g$, i.e., $g \circ F^*$ so that $g \circ F^* \geq F^*$. For pricing financial risks, we seek $g$ such that $g \circ F^* \leq F^*$.

Also the distortion function $g$ should be such that $g \circ F^*$ is a survival function, so that $\int_0^\infty (g \circ F^*)(t)dt$ is an expectation. Note that, for appropriate distortion functions $g$, $1 - g(F^*(.))$ is a (probability) distribution function of some random variable.

Motivated by normal and lognormal random variables, the distortion function $g$ could be chosen to preserve these distributions. We derive Wang's distortion functions for the normal random variables. The proof is the same for the case of lognormal and hence left as an exercise.

**Proposition 5.38** *Let* $g : [0,1] \to [0,1]$. *Suppose* $g$ *satisfies the following two conditions:*

*(i) If* $F$ *is the distribution function of a normally distributed random variable, then* $g \circ F^*$ *is the survival function of a normal distribution,*

*(ii) For every distribution function* $F$,

$$\int_0^\infty (g \circ F^*)(t)dt \geq \int_0^\infty F^*(t)dt.$$

*Then* $g(x) = \Phi(\Phi^{-1}(x) + \lambda)$ *with some* $\lambda \geq 0$.

**Remark 5.39** *If (ii) is replaced by* $\int_0^\infty (g \circ F^*)(t)dt \leq \int_0^\infty F^*(t)dt$ *then* $\lambda \leq 0$. *The same results hold when* $X$ *is lognormal, and (i) is replaced by* $g \circ F^*$ *is the "survival function" of a lognormal distribution. Note that the survival function* $F^*$ *represents the tail distribution function of* $X$, $g \circ F^*$ *is a new tail distribution with more weight,* $\int_0^\infty (g \circ F^*)(t)dt$ *is viewed as the expectation of* $X$ *with respect to the new tail distribution.*

The proof goes as follows.

**Proof.** Let $\Phi$ denote again the distribution function of the standard normal random variable $Z$. Then the distribution function $F(.)$ of a normal random variable $X$ with mean $\mu$ and variance $\sigma^2$ is $F(x) = \Phi((x - \mu)/\sigma)$.

The condition *(i)* says that $g(F^*(x))$ is of the form $1 - \Phi((x - \lambda)/b)$ for some $\lambda \in \mathbb{R}$ and $b > 0$. In particular, for $Z$, we have

$$g(1 - \Phi(x)) = 1 - \Phi((x - \lambda)/b).$$

Now, since

$$1 - \Phi(x) = \Phi(-x)$$

for all $x \in \mathbb{R}$, we have

$$g(\Phi(-x)) = \Phi(-(x - \lambda)/b).$$

For given $y$, let $x = -\Phi^{-1}(y)$. Then $y = \Phi(-x)$ and hence

$$g(y) = \Phi(-(-\Phi^{-1}(y) - \lambda)/b) = \Phi((\Phi^{-1}(y) + \lambda)/b).$$

It remains to show that $b = 1$ and $\lambda \geq 0$. For this purpose, we use condition *(ii)*. This condition says that $g(y) \geq y$ for all $y \in [0, 1]$. Thus,

$$\Phi((\Phi^{-1}(y) + \lambda)/b) \geq y.$$

Therefore, for each $x \in \mathbb{R}$, this inequality holds for $y = \Phi(x)$, i.e.,

$$\Phi((x + \lambda)/b) \geq \Phi(x)$$

for all $x$. But since $\Phi$ is increasing, we get

$$x \leq (x + \lambda)/b$$

for all $x$, or

$$(b - 1)x \leq \lambda.$$

If $b > 1$ then this inequality cannot hold as $x \to \infty$. If $b < 1$, this inequality cannot hold as $x \to -\infty$. Thus, $b$ must be equal to 1, but then $x \leq (x + \lambda)/b$ becomes $x \leq x + \lambda$, implying that $\lambda \geq 0$.

Note that the functions

$$g_\lambda(x) = \Phi(\Phi^{-1}(x) + \lambda), \lambda \in \mathbb{R}$$

are such that $g_\lambda(0) = 0$, $g_\lambda(1) = 1$, and they are nondecreasing and hence they are distortions functions.

Moreover, they are concave for $\lambda > 0$, and convex for $\lambda < 0$. Indeed,

$$\begin{aligned}
g_\lambda(0) &= \lim_{x \searrow 0} \Phi(\Phi^{-1}(x) + \lambda) = 0 \\
g_\lambda(1) &= \lim_{x \nearrow 1} \Phi(\Phi^{-1}(x) + \lambda) = 1 \\
\frac{d}{dx}[\Phi(\Phi^{-1}(x) + \lambda)] &= \Phi'(\Phi^{-1}(x) + \lambda)/\Phi'(\Phi^{-1}(x)) \\
&= \exp(-(2\lambda\Phi^{-1}(x) + \lambda^2)/2) > 0 \\
\frac{d^2}{dx^2}[\Phi(\Phi^{-1}(x) + \lambda)] &= -\lambda\Phi'(\Phi^{-1}(x) + \lambda)/(\Phi'(\Phi^{-1}(x)))^2
\end{aligned}$$

is negative for $\lambda > 0$, and positive for $\lambda < 0$. ∎

Risk measures built from Wang's concave distortion functions are subadditive (hence coherent).

**Remark 5.40** *Clearly if a distortion function $g$ is such that $g(x) \geq x$ for all $x \in [0, 1]$, then*

$$\int_0^\infty (g \circ F^*)(t)dt \geq \int_0^\infty F^*(t)dt = EX.$$

*The converse is also true (from Wirch and Hardy [168]): If*

$$\int_0^\infty (g \circ F^*)(t)dt \geq \int_0^\infty F^*(t)dt = EX$$

*for any random variable $X$ with distribution function $F$, then $g(x) \geq x$
for all $x \in [0, 1]$.*

*Indeed, if not, then for $t \in (a, b)$ with $0 \leq a < b \leq 1$, $g(t) < t$. For
any $z \in (a, b)$, let $X$ be a random variable with distribution*

$$P(X = 0) = 1 - z = 1 - P(X = c)$$

*with some $c > 0$. Then the survival function of $X$ is*

$$F^*(x) = 1 - F(x) = \begin{cases} z \ for \ 0 \leq x < c \\ 0 \ for \ x \geq c \end{cases}$$

*so that*

$$\int_0^\infty (g \circ F^*)(t)dt = \int_0^c g(z)dx = cg(z) < cz = EX.$$

**Remark 5.41** *Wang's distortion functions are chosen for risk modeling
with some adjusted parameter $\lambda$, but independently of variables under
consideration. We have seen that $CVaR_\alpha(.)$ has a Choquet integral representation but with respect to distortion functions which depends, not
only on $\alpha$, but also on each variable $X$ that we intend to measure its
risk. That, of course, will complicate a lot of things! Here is another
example of distortion functions which do depend on the variables under
consideration.*

*They arise from Lorenz curves in the study of inequalities of incomes
in economics, in fact they are duals of Lorenz curves,*

$$g_F(\alpha) = 1 - L_F(1 - \alpha) \ for \ \alpha \in [0, 1]$$

*where $L_F(.)$ is the Lorenz curve of $F$.*

Recall that if $X$ has distribution function $F$, then the Lorenz curve
of $F$ is $L_F : [0, 1] \to [0, 1]$, where

$$L_F(\alpha) = \frac{\int_0^\alpha F^{-1}(t)dt}{\int_0^1 F^{-1}(t)dt}.$$

Note that $L_F(.)$ is convex (since $F^{-1}$ is nondecreasing), and

$$L_F(0) = 0, L_F(1) = 1.$$

The distortion function $g_F(\alpha) = 1 - L_F(1 - \alpha)$ is nothing else than the
reflection of $L_F$ with respect to the diagonal of the unit square. Clearly,
each $g_F(.)$ is strictly concave.

If $F$ is the distribution function of $X$ which is lognormal $N(\mu, \sigma^2)$,
then $L_F(\alpha) = \Phi(\Phi^{-1}(\alpha) - \sigma)$, i.e., a Wang's distortion function with
$\lambda = -\sigma < 0$.

## 5.9   Consistency with stochastic dominance

So far the quantification of risks in various contexts is this. Given a financial context and the type of random variables for which we wish to define risk, we propose an appropriate risk measure with a clear semantic. Then we check whether this risk measure has desirable properties or not in order to decide whether to use it for real-world applications. Among possible desirable properties for a risk measure, we have the axioms for coherent risk measures to check.

The examination of a proposed risk measure would not be complete if we do not examine its relation with preferences, since after all risk does relate to preferences. Specifically, we need to see how risk measures are *consistent* with stochastic dominance rules which are not only derived from utility but also based on risk attitudes of decision-makers.

Basically, this is the problem of ordering risks, especially in actuarial science. Random variables involved are *loss* variables with finite means. Here, in the context of comparing risks, $X$ is preferred to $Y$ if $\rho(X) \leq \rho(Y)$. For example, if $\rho_g(.)$ is a risk measure based on the distortion function $g$, then $\rho_g(X) \leq \rho_g(Y)$ when $F_X(.) \geq F_Y(.)$, i.e., when $X \preceq_1 Y$. Thus, *consistency* of $\rho_g(.)$ with respect to first-order stochastic dominance means

$$X \preceq_1 Y \implies \rho_g(X) \leq \rho_g(Y),$$

in other words, the risk measure $\rho_g(.)$ preserves FSD. Note that we prefer $X$ which precedes $Y$ in the stochastic dominance order.

In general, if $\preceq$ is some partial order on the space of distribution functions, we say that $X$ *precedes* $Y$ when $X \preceq Y$ and a risk measure $\rho(.)$ is *consistent* with $\preceq$ if

$$X \preceq Y \implies \rho(X) \leq \rho(Y).$$

As stated above, all distorted probability risk measures are consistent with first stochastic dominance since distortion functions are nondecreasing. For $VaR_\alpha(.)$, in fact, we have

$$X \preceq_1 Y \iff F_X^{-1}(.) \leq F_Y^{-1}(.).$$

Recall from Chapter 2 that $X$ dominates $Y$ in *second-order stochastic dominance* (SSD), $X \succeq_2 Y$, if and only if

$$\int_{-\infty}^{t} (F_X - F_Y)(x)dx \geq 0 \text{ for all } t \in \mathbb{R}$$

and, dually, $X$ dominates $Y$ in *risk seeking stochastic dominance* (RSSD), $X \succeq_{2'} Y$, if and only if

$$\int_t^\infty (F_X - F_Y)(x)dx \geq 0 \text{ for all } t \in \mathbb{R}.$$

This is the same as the so-called *stop-loss order*, $X \preceq_{sl} Y$, defined as

$$E(X - t)^+ \leq E(Y - t)^+ \text{ for all } t \in \mathbb{R}$$

by noting that

$$E(X - t)^+ = \int_t^\infty (x - t)dF(x) = \int_t^\infty (1 - F(x))dx.$$

Recall that an insurance company, when facing a risk $X$ can proceed as follows if it does not want to bear all the risk. It can reinsure with another company with a contract which says it will handle the risk as long as the risk does not exceed a determined amount $t$ (called the retention). A stop-loss contract is a contract with a fixed retention. If $X > t$, the reinsurance company will handle the amount $X - t$. The expected cost for the reinsurance company (net premium) is then $E(X - t)^+$.

Following Müller [116] the function $\pi(.) : \mathbb{R}^+ \to \mathbb{R}^+$

$$\pi_X(t) = E(X - t)^+ = \int_t^\infty (1 - F(x))dx$$

is called the *stop-loss transform* (of the distribution function of $X$).

Clearly, $X \preceq_{sl} Y$ if and only if $\pi_X(.) \leq \pi_Y(.)$ which, in turn, is precisely RSSD, $X \preceq_{2'} Y$, since

$$\pi_X(.) - \pi_Y(.) \leq 0 \Longleftrightarrow \int_t^\infty (F_Y - F_X)(x)dx \leq 0 \text{ for all } t \in \mathbb{R}.$$

Thus, consistency of risk measures with respect to stop-loss order is the same as with respect to RSSD.

Recall that the distorted probability risk measure

$$TVaR_\alpha(X) = (1/1 - \alpha) \int_\alpha^1 F_X^{-1}(t)dt$$

has the Choquet integral representation with the concave distortion function $g_\alpha(x) = \min\{1, x/(1 - \alpha)\}$, namely

$$TVaR_\alpha(X) = C_{v_{g_\alpha}}(X) = \int_0^\infty g_\alpha(1 - F(t))dt.$$

Also, for a distortion function $g$, the corresponding risk measure is

$$\rho_g(X) = C_{v_g}(X),$$

where $v_g(.) = (g \circ P)(.)$. It can be shown that

$$X \preceq_{2'} Y \Longleftrightarrow TVaR_\alpha(X) \leq TVaR_\alpha(Y) \text{ for all } \alpha \in (0,1),$$

meaning that $TVaR_\alpha(.)$ is consistent with RSSD, for any $\alpha$ (see Exercises).

It turns out that concavity of chosen distortion functions is a sufficient condition for corresponding distorted probability risk measures to be consistent with respect to RSSD, or equivalently, with respect to stop-loss order. Specifically,

**Theorem 5.42** *If the distortion function $g$ is concave then*

$$X \preceq_{2'} Y \Longrightarrow \rho_g(X) \leq \rho_g(Y).$$

**Proof.** Suppose $X \preceq_{2'} Y$. First, consider an arbitrary concave piecewise linear distortion function $g$. Suppose $g$ has crack points at $a_i$ with $0 = a_0 < a_1 < ... < a_{n-1} < a_n = 1$. In addition, let the equation of the $i$th segment over the interval $[a_{i-1}, a_i)$ be $y = m_i x + b_i$, then $g$ can be written as

$$g(x) = \sum_{i=1}^{n} a_i (m_i - m_{i+1}) \min\{x/a_i, 1\} \text{ with } m_{n+1} = 0.$$

Indeed, for $x \in [a_j, a_{j+1})$, we have

$$\sum_{i=1}^{n} a_i (m_i - m_{i+1}) \min\{x/a_i, 1\}$$

$$= \sum_{i=1}^{j} a_i (m_i - m_{i+1}) + \sum_{i=j+1}^{n} a_i (m_i - m_{i+1}) x/a_i$$

$$= \sum_{i=1}^{j} (b_{i+1} - b_i) + x \sum_{i=j+1}^{n} (m_i - m_{i+1})$$

since $m_i a_i + b_i = m_{i+1} a_i + b_{i+1}$, and

$$\sum_{i=1}^{j} (b_{i+1} - b_i) + x \sum_{i=j+1}^{n} (m_i - m_{i+1}) = b_{j+1} - b_1 + m_{j+1} x$$

$$= m_{j+1} x + b_{j+1} = g(x)$$

noting that $b_1 = g(0) = 0$. Then,

$$\rho_g(X) = \sum_{i=1}^{n} a_i(m_i - m_{i+1})TVaR_{1-a_i}(X).$$

Now the concavity of $g$ implies that $m_i$ is decreasing in $i$. On the other hand, $X \preceq_{2'} Y$ implies $TVaR_\alpha(X) \le TVaR_\alpha(Y)$ for all $\alpha \in [0,1)$. Therefore,

$$\sum_{i=1}^{n} a_i(m_i - m_{i+1})TVaR_{1-a_i}(X) \le \sum_{i=1}^{n} a_i(m_i - m_{i+1})TVaR_{1-a_i}(Y),$$

i.e., $\rho_g(X) \le \rho_g(Y)$. Thus, $X \preceq_{2'} Y \implies \rho_g(X) \le \rho_g(Y)$ for any concave piecewise linear distortion function $g$.

Next, observe that any concave distortion function $g$ can be approximated from below by concave piecewise linear distortion functions $g_n$ where $g_n$ consists of $2^n$ segments with $i$th segment connecting the point $(\frac{i-1}{2^n}, g(\frac{i-1}{2^n}))$ to the point $(\frac{i}{2^n}, g(\frac{i}{2^n}))$. We have just shown that $X \preceq_{2'} Y$ implies $\rho_{g_n}(X) \le \rho_{g_n}(Y)$ for any $n$. In addition, $g_n \le g$ implies $\rho_{g_n}(Y) \le \rho_g(Y)$. Hence, by the monotone convergence theorem, we get

$$\lim_{n\to\infty} \rho_{g_n}(X) = \rho_g(X) \le \rho_g(Y).$$

∎

**Remark 5.43** *We can in fact state that $X \preceq_{2'} Y \iff \rho_g(X) \le \rho_g(Y)$ for all concave distortion functions $g$. Indeed, if $\rho_g(X) \le \rho_g(Y)$ for all concave distortion function $g$, then it holds in particular for the $g_\alpha(x) = \min\{1, x/(1-\alpha)\}$, $\alpha \in [0,1)$. But then, by the result preceding the above theorem, we have $X \preceq_{2'} Y$. The converse follows from the above theorem.*

A finer result is this (see Wirch and Hardy [168].)

**Theorem 5.44** *A necessary and sufficient condition for*

$$X \preceq_{2'} Y \implies \rho_g(X) < \rho_g(Y)$$

*is that the distortion function $g$ is strictly concave.*

**Remark 5.45** *Recall that $g : [0,1] \to \mathbb{R}$ is strictly concave if for $x \ne y$, $\lambda(0,1)$, we have*

$$g(\lambda x + (1-\lambda)y) > \lambda g(x) + (1-\lambda)g(y).$$

Note that $g$ is strictly concave if and only if $g$ is of the form $g(x) = g(c) + \int_c^x h(t)dt$ for some function $h(.)$ decreasing (strictly decreasing), and some $c \in (0, 1)$. In particular, if $g$ is differentiable, then $g$ is strictly concave if and only if its derivative $g'(.)$ is (strictly) decreasing. Moreover if $g$ is twice differentiable, with $g'' < 0$ on $(0, 1)$, then $g$ is strictly concave.

The above results provide criteria for choosing coherent risk measures which are also consistent with some stochastic dominance rules. For example, within the class of coherent risk measures obtained by using various distortion functions, one should choose strictly concave distortion functions.

## 5.10    Exercises

1. Let $X, Y$ be two random variables. Show that $X$ and $Y$ are comonotonic if and only if there exist $f, g : \mathbb{R} \to \mathbb{R}$ continuous, nondecreasing such that $f(.) + g(.) = $ identity, and

$$X = f(X + Y) \text{ and } Y = g(X + Y).$$

2. Let $X, Y$ be two comonotonic random variables. Suppose $X \succeq_2 Y$ (i.e., in SSD sense). Show that $X \geq Y$ a.s.

3. Let $v$ be a capacity on the measurable space $(\Omega, \mathcal{A})$. Let $X : \Omega \to \mathbb{R}$ be measurable. Show that there exists a probability measure $\pi$ on $\mathcal{A}$, depending on $v$ and $X$, such that

$$C_{v(X)} = E_\pi(X).$$

4. Let $X, Y$ be comonotonic. Is it true that

$$F_{X+Y}^{-1} = F_X^{-1} + F_Y^{-1}?$$

5. Let $(\Omega, \mathcal{A})$ be a measurable space, and $P$ be a probability measure defined on $\mathcal{A}$. The inner measure of $P$ is defined as

$$v : \mathcal{A} \to \mathbb{R}, v(A) = \sup\{P(B) : B \in \mathcal{A}, B \subseteq A\}.$$

Show that $v$ is a capacity satisfying the following property called *monotonicity of infinite order*: for any $n \geq 2$, and $A_1, A_2, ..., A_n$ in $\mathcal{A}$,

$$v(\cup_{i=1}^n A_i) \geq \sum_{\emptyset \neq I \subseteq \{1,2,...,n\}} (-1)^{|I|+1} v(\cap_{i \in I} A_i).$$

6. Let $(\Omega, \mathcal{A})$ be a measurable space and $v : \mathcal{A} \to [0, 1]$ such that $v(\varnothing) = 0$, $v(\Omega) = 1$. Suppose, in addition, $v$ is *maxitive*, i.e.,

$$v(A \cup B) = \max\{v(A), v(B)\}$$

for any $A, B \in \mathcal{A}$.

(a) Verify that $v$ is a capacity.

(b) Show that $v$ is alternating of infinite order.

7. Let $(\Omega, \mathcal{A})$ be a measurable space and $v : \mathcal{A} \to [0, 1]$ a capacity. The distribution function of a random variable $X$ with respect to the capacity $v$ is defined to be

$$G_{v,X}(x) = 1 - v(X > x)$$

and its inverse $G_{v,X}^{-1}(t) = \inf\{x \in \mathbb{R} : G_{v,X}(x) \geq t\}$, $t \in [0, 1]$. Show that

$$C_v(X) = \int_0^1 G_{v,X}^{-1}(t)dt.$$

8. (Continued) Let $X, Y$ be comonotonic, show that

$$G_{v,X+Y}^{-1}(.) = G_{v,X}^{-1}(.) + G_{v,Y}^{-1}(.).$$

9. Let $\rho_\varphi(.)$ be a spectral risk measure. Show that there exists a capacity $v$ such that

$$\rho_\varphi(X) = C_v(X).$$

10. Let $TVaR_\alpha(X) = \int_0^1 (1/(1 - \alpha))1_{(\alpha,1]}(t)F_X^{-1}(t)dt$. What is the distortion function $g_\alpha$ such that $TVaR_\alpha(X) = Cv_\alpha(X)$ where $v_\alpha(.) = g_\alpha \circ P(.)$?

11. Let $g : [0, 1] \to [0, 1]$ satisfying

(a) For any *normal random variable* with distribution function $F$, $g(1 - F(t)) = 1 - G(t)$ where $G$ is the distribution function of some normally distributed random variable, and

(b) For every distribution function $F$, $\int_0^\infty g(1 - F(t))dt \leq \int_0^\infty (1 - F(t))dt$.

Show that $g(.) = \Phi(\Phi^{-1}(x) + \lambda)$ with $\lambda \leq 0$, where $\Phi$ is the distribution function of the standard normal random variable.

12. Let $X$ be normal $N(\mu, \sigma^2)$. Compute its Lorenz curve.

13. The generalized Lorenz curve of a random variable $X$ (with mean $EX = \mu$ such that $0 < \mu < \infty$) is

$$GL_X(t) = (EX) \int_0^t F_X^{-1}(s)ds, \text{ for } t \in [0, 1].$$

Show that $GL_X(.) \geq GL_Y(.)$ if and only if $X \succeq_2 Y$.

14. For $\rho(X) = -F_X^{-1}(\alpha)$, show directly (i.e., using the definition of quantile, without a Choquet integral representation) that

(a) $X \leq Y$, a.s. $\Longrightarrow \rho(X) \geq \rho(Y)$,

(b) $\rho(X + a) = \rho(X) - a$, for any $a \in \mathbb{R}$, and

(c) $\rho(\lambda X) = \lambda \rho(X)$ for any $\lambda > 0$.

15. Recall that $X \succeq_{2'} Y$ if and only if $\int_x^\infty [F_Y(t) - F_X(t)]dt \geq 0$ for all $x \in \mathbb{R}$. Show that

$$X \succeq_{2'} Y \Longleftrightarrow TVaR_\alpha(X) \leq TVaR_\alpha(Y) \text{ for all } \alpha \in (0, 1).$$

# Chapter 6

# Foundational Statistics for Stochastic Dominance

*This chapter provides basic ingredients to study the fundamental problem of how to use empirical data to implement stochastic dominance rules and to assess financial risks. A tutorial on necessary background for statistical inference is presented in great details. It forms not only the basic statistical techniques for investigating stochastic dominance rules but also for related problems in risk estimation as well as in econometrics in general.*

## 6.1   From theory to applications

The applications of statistics to economics in general, and to finance in particular, dominate research activities in these fields. This seems to be obvious since like in almost all human activities, the discovery of knowledge relies essentially on empirical data. The scientific approach to knowledge discovery starts with investigating principles in each domain of study, leading to a conceptional framework in which "formulae" and the like surface. For example, in the quest to understand uncertainty, a theory of probability first appears, where for example, uncertain quantities are postulated as random variables, i.e., bona fide mathematical entities whose random evolution is governed by probability laws.

To apply this theory to real-world problems, we need to find ways to supply unknown quantities which appear in theoretical formulae, such as probability laws and characteristics such as moments. The theory of statistics is invented precisely for this crucial purpose.

We are concerned here with decisions based upon stochastic dominance rules in financial economics. For example, to choose between two risky prospects $X, Y$, we rely on their respective distribution functions $F, G$. If $F(.) \leq G(.)$, then risk neutral decision-makers tend to prefer $X$ to $Y$. But in general, $F$ and $G$ are unknown. How do we supply $F, G$ for decision-making? We need data to estimate $F$ and $G$ as well as to test whether empirical evidence supports a hypothesis like $F(.) \leq G(.)$.

Several issues arise naturally in any problems where we use empirical data to validate theoretical hypotheses.

**(i)**   Suppose the data provide sufficient evidence (through some selected testing procedures) to accept a hypothesis that, say, $F(.) \leq G(.)$, at some chosen level of significance $\alpha$. Then we conclude that $X$ dominates $Y$ in FSD. We might be wrong but that is the "statistically logical" conclusion that we could draw based upon our data. Keep in mind, however, our conclusion depends not only on our chosen $\alpha$, but also on the test we used! As a result, caution should be exercised on the follow-up decisions that we wish to make based on our testing results.

**(ii)**   Suppose the test for FSD fails. Then, we could try SSD, RSSD or TSD. Suppose we are interested in ranking $X, Y$ according to FSD and FSD test fails. Just like in ranking Lorenz curves in the study of income inequality, often the distribution functions $F, G$ may intersect, say, as evidence from their empirical estimates. Then what if we badly need a ranking for decisions? A standard practice in a situation such as this is to apply some "summary" measures, such as risk measures which need to be estimated from data. However, it may be difficult to find a single numerical measure that could gain a wide degree of support. Of course, we just proceed with our recommendations, but leave room for the users to decide, by reporting exactly what we have done. This is similar to the use of the $p$-value in statistical testing.

**(iii)**   Now data! We use data to draw conclusions. The question is "how reliable is our data?" In experimental observed situations where we could design carefully our data collection procedure to obtain, say, a random sample from a population $X$, then in principle, our data are measures with accuracy and do come from $X$, so that our inferential procedures are justified by statistical theory. For observational data where we cannot design the random experiments, but only observe the outcomes of random phenomena, several issues should be examined with care.

First, the correlation among observations should be understood, i.e., the dependence structure of our observed process should be modeled correctly since our inferential procedures depend on it. This is particularly important for predictions where time is an important factor! Clearly we are talking about observed data as time series, so that *statistics of stochastic processes* will enter the picture. Modeling observed processes as well as regressions on variables of interest (such as financial volatility) is essential.

Secondly, how accurate are our data measurements? We might face the problem of having *coarse data*, i.e., data with low quality, in a variety of ways. The typical situation is in collecting data on investment returns. Historical data means data from the past. Now time is an issue. The distribution of the variable $X$, say, of the future return on a risky prospect, while "similar" but not exactly the same, in view of unseen economic variations in the future. Of course, as far as predictions are concerned, the process of extrapolation does take into account of time-varying aspects, as we view our observed process as part of an on-going time evolution process. But when we fix a random variable $X$ of concern and are interested in, say, estimating its distribution function $F$ from historical data, we tacitly assume $F$ does not change with time. And clearly, that is not true. Of course, a historical data set $X_1, X_2, ..., X_n$ has some "link" with the distribution function $F$ of our future $X$, but it is not really a sample "drawn" from $F$. How serious is it to ignore this fact? or should we treat the historical data as coarse data of some sort and develop appropriate statistical procedures to take their low quality into account?

Here it seems to be a good place for readers to recall the two most important aspects of using statistics for real-world applications: *models and data*. Statistical conclusions for decision-making depend heavily on these two factors (and of course, also on our chosen inferential procedures). Unlike physical sciences where models are dictated by dynamic laws of motion, models in uncertainty analysis are proposed based on "empirical grounds." This is exemplified in current research in mathematical finance where various models for *stochastic volatility* of stock markets are proposed. Once a model is put down for study, the next thing to look at is the *type of data* available for making inferences, because statistical procedures are not universal, but are rather tailored made for each given type of data.

We will elaborate on the above issues in subsequent sections of this chapter. But first, we need to review some appropriate background of statistics! While we have stochastic dominance in mind, we will present the material in the general context of using statistics in economics.

## 6.2    Structure of statistical inference

Let $(\Omega, \mathcal{A}, P)$ be our background probability space on which random elements are defined. Random elements are measurable mappings, defined on $(\Omega, \mathcal{A}, P)$ and taking values in some measurable space $(\mathcal{X}, \mathcal{B}(\mathcal{X}))$. They can be *random vectors* when $\mathcal{X}$ is some finite dimensional Euclidean space $\mathbb{R}^d$, the Borel $\sigma$-field $\mathcal{B}(\mathcal{X})$ being the Borel $\sigma$-field generated by the usual topology of $\mathbb{R}^d$; *random processes* when $\mathcal{X}$ is some function space like $C([0,1])$, the space of continuous real-valued functions defined on the interval $[0,1]$ (Brownian motion sample paths), equipped with the sup-norm; or *random sets* when $\mathcal{X}$ is the space of closed subsets of some $\mathbb{R}^d$ (such a space can be topologized by the hit-or-miss topology, or equivalently by the Lawson's topology of the continuous lattice of upper semicontinuous functions, leading to the Borel $\sigma$-field $\mathcal{B}(\mathcal{X})$. See details later in the context of coarse data).

Such a setting is general enough to cover all situations in applications. The random element of interest could be the one having a conditional distribution, based on other covariates, in some regression models. The appearance of infinitely dimensional random elements like stochastic processes cover time series and statistical inference with continuous time-dependent observations.

When $\mathcal{X} = \mathbb{R}^d$, the Euclidean space of dimension $d$, $X$ is called a *random vector*. The Lebesgue-Stieltjes theorem is valid here where the (multivariate) distribution function $F$ of a random vector $X$ is $F : \mathbb{R}^d \to [0,1]$,

$$F(\mathbf{x}) = P\left(\prod_{j=1}^{d}(-\infty, x_j]\right) \text{ for } \mathbf{x} = (x_1, x_2, ..., x_d)$$

and $\prod$ denotes Cartesian product of sets. The probability law of a random vector $X$ is a probability measure $dF$ on $\mathcal{B}(\mathbb{R}^d)$. More generally, sample spaces for, say, infinitely uncountable spaces, such as observations recorded in time (time series), are of the form $\mathbb{R}^T = \{\varphi : T \to \mathbb{R}\}$, where $T$ is a (time) set, such as $\mathbb{Z}$ (the integers), $\mathbb{N}$ (nonnegative integers), or $\mathbb{R}^+$, which extend finite dimensional spaces $\mathbb{R}^d$. The canonical $\sigma$-field on $\mathbb{R}^T$ is constructed as follows. Let $\mathcal{I}$ denote the set of all nonempty finite subsets of $T$. A cylinder is a subset of $\mathbb{R}^T$ of the form $(\prod_{i \in I} B_i) \times \mathbb{R}^T \times \mathbb{R}^T \times \cdots$, for $I \in \mathcal{I}$, $B_i \in \mathcal{B}(\mathbb{R})$, $i \in I$.

By $\mathcal{B}(\mathbb{R}^T)$ we mean the $\sigma$-field of $\mathbb{R}^T$ generated by the class of all cylinders (i.e., the smallest $\sigma$-field containing all cylinders).

A map $X : \Omega \to \mathbb{R}^T$, $\mathcal{A}$-$\mathcal{B}(\mathbb{R}^T)$-measurable, is called a *stochastic process*. Note that, for each $\omega$, the value $X(\omega)$ is a function from $T$ to $\mathbb{R}$, so that, $X$ is in fact a random function, $X(\omega, t)$. Thus, we write also a stochastic process as $X = (X_t, t \in T)$ meaning that a stochastic process is a collection of random variables $X_t$ indexed by a "time" set $T$. The probability law of a stochastic process $X$ is a probability measure on $\mathcal{B}(\mathbb{R}^T)$. Its construction goes as follows.

Since in practice, it is possible to specify probability measures on finite dimensional spaces, such as $\mathbb{R}^d$ above, we ask: Given a family of probability measures $P_I, I \in \mathcal{I}$, where each $P_I$ is a probability measure on $\mathcal{B}(\mathbb{R}^I)$, does there exist a probability measure $Q$ on $\mathcal{B}(\mathbb{R}^T)$ such that $P_I = Q\pi_I^{-1}$, for all $I \in \mathcal{I}$, where $\pi_I$ denotes the canonical projection from $\mathbb{R}^T$ to $\mathbb{R}^I$? The answer is known as the *Kolmogorov consistency theorem*: If the family $P_I, I \in \mathcal{I}$ is consistent, in the sense that

$$P_I = P_J \pi_{JI}^{-1}, \text{ for any } I \subseteq J, \ I, J \in \mathcal{I},$$

where $\pi_{JI}$ is the projection from $\mathbb{R}^J$ to $\mathbb{R}^I$, then there exists a unique probability measure $Q$ on $\mathcal{B}(\mathbb{R}^T)$ such that

$$P_I = Q\pi_I^{-1}, \text{ for all } I \in \mathcal{I}.$$

Note again that on infinite dimensional spaces, there are no counterparts of distribution functions, so that probability laws of random elements have to be specified as probability measures. This theorem provides a rigorous foundation for modeling stochastic processes.

In practice, we specify the *finite dimensional distributions*, i.e., the probability measures $P_I, I \in \mathcal{I}$ and check the consistency condition. For example, if each finite dimensional distribution is the probability measure of a *Gaussian random vector*, i.e., a random vector having a multivariate normal distribution, then the stochastic process is called a *Gaussian process*. Similar to normal random variables or vectors, a Gaussian process is characterized by a mean function $m : T \to \mathbb{R}$, $m(t) = EX_t$ (expectation of $X_t$), and a covariance function $\Gamma : T \times T \to \mathbb{R}$, $\Gamma(s, t) = cov(X_s, X_t)$.

A *Brownian motion* (or *Wiener process*) is a Gaussian process with zero mean function and covariance function of the form $\Gamma(s, t) = \sigma^2 \min(s, t)$, where $\sigma^2$ is called the diffusion coefficient. If $W = (W_t, t \geq 0)$ is a standard Brownian motion (i.e., $\sigma^2 = 1$), then the stochastic process defined by $X_t = W_t - tW_1$, for $t \in [0, 1]$, is a zero mean Gaussian process with covariance function $\Gamma(s, t) = s(1-t)$ for $s \le t$. Such a Gaussian process is called a Brownian bridge. If $X$ is a Brownian motion, then the Gaussian process $Y_t = X_t + t\mu$, for $t \ge 0$, is called a Brownian motion with drift parameter $\mu$, and the the process $Z_t = e^{X_t}$ is called a geometric Brownian motion. A zero mean Gaussian process is called a fractional Brownian motion *of Hurst parameter* $H \in (0, 1)$, if its covariance function is

$$\Gamma(s, t) = (1/2)[s^{2H} + t^{2H} - |t - s|^{2H}].$$

Discovering the probability law of a random element $X$ of interest, from empirical observations on $X$, is the main goal of statistics. In other words, it is for clear reasons that we wish to discover the random process generating our observed data! The probability law of $X$ is the probability measure $P_X = PX^{-1}$ on $\mathcal{B}(\mathcal{X})$. Depending upon our additional information about $P_X$, we can narrow down the search for it by considering various appropriate models, where a *model* refers to a specification of a set $\mathcal{F}$ of possible probability measures containing $P_X$.

We say that the model is *parametric* when $P_X$ is known to lie in a parameter space $\mathcal{F}$ consisting of known probability measures, except some finite dimensional parameter $\theta$ which is known to lie in some subset $\Theta$ of some $\mathbb{R}^d$, i.e., $P_X \in \mathcal{F} = \{P_\theta : \theta \in \Theta\}$. For example, $P_X = dF_{\lambda_0}$ but we only know that $\lambda_0 \in \Phi = (0, \infty)$, and $F_\lambda(x) = (1 - e^{-\lambda x})1_{(0,\infty)}(x)$, $\lambda > 0$. In this case, it suffices to estimate the unknown true $\theta_0 \in \Theta$, corresponding to $P_X$. Thus $X$ is parametric means its distribution is specified to belong to a class of distributions $\mathcal{F} = \{F(.|\theta) : \theta \in \Theta \subseteq \mathbb{R}^d\}$.

The model is called *semiparametric* if the parameter space $\mathcal{F}$ consists of probability measures indexed by parameters with two components $(\theta, \gamma)$ where $\theta \in \Theta \subseteq \mathbb{R}^d$, and $\gamma \in \Gamma$ of infinite dimensions. Location model is a typical semiparametric model. Here the true and unknown distribution of the random variable of interest $X$ is known only to lie in the class of distributions $\mathcal{F}$ whose elements are of the form $F(x|\theta) = F(x - \theta)$, $F$ belongs to some specified class of distributions (e.g., logconcave) and $\theta$ a finite dimensional parameter.

The model is called (fully) *nonparametric* when $\mathcal{F}$ does not have any index component of finite dimension, for example, $\mathcal{F}$ consists of all probability measures on $\mathcal{B}(\mathbb{R})$, absolutely continuous with respect to the Lebesgue measure $dx$ on $\mathcal{B}(\mathbb{R})$; equivalently, the distribution function $F_X$ of $X$ is only known to be some distribution function on $\mathbb{R}$ which is absolutely continuous.

Depending on models, methods for estimating $P_X$ can be developed appropriately. It is clear that model specifications are very important since a misspecification of $\mathcal{F}$ leads to incorrect identification of $P_X$. Usually the specification of a model is guided by empirical data under various forms. Also, models can be proposed as approximations, for example, the probability law underlying a relationship between two random variables $X$ and $Y$ could be modeled as a linear relationship between them. Thus, a linear model of the form $Y = aX + \varepsilon$ is a specification that the conditional distribution of $Y$ given $X$ is related to the distribution of the noise $\varepsilon$ in some specific form.

A model for the volatility in some financial market (viewed as a random variable) is a specification of its possible random dynamics, or a specification of how the current volatility is depending upon previous volatilities. A familiar model specification is the well-known *logit model* in regression in many fields, including econometrics. Other model specifications are models for *stochastic volatility* in financial economics. Note that each such model describes a specific form of the probability law of the random element of interest. While we will not discuss the problem of how to propose a model, we will address the problem of how to justify proposed models with illustrations from some existing proposed models, namely logistic regression models and asymmetric heteroskedasticity models. In a related context of risk modeling, we will also illustrate the justification of some financial risk measures.

In stochastic dominance analysis, often it is hard to propose parametric models for distribution functions of random variables involved. Thus, we will face nonparametric models, and hence nonparametric statistics.

Suppose we have a model $\mathcal{F}$ for $P_X$. In order to infer $P_X$ or some function of it, $\varphi(P_X)$, such as the mean of $X$, we need data observed on $X$. Data can be obtained in discrete or continuous (time) forms. They can be random samples from $X$ (observational or derived from controlled experiments). Depending upon the type of observed data, e.g., independent and identically distributed (i.i.d.) samples or realizations of a stochastic process (a time series), appropriate statistical inference procedures are called for. Besides the dependence structures of data, we have to examine the quality of our observed data, since as stated earlier, it is the data which dictate our inference procedures. Often we face *coarse data*, i.e., data of "low" quality in a variety of forms, such as measurement error data, censored data, hidden data, missing data, indirect observed data, approximate data, imprecise data.

The point is this. As we are going to discuss statistical problems in a simple setting to bring out the basics of statistical inference theory, we should keep in mind that statistical procedures depend heavily on proposed models and types of available data.

Basically, statistical theory consists of using available data to infer unknown probability laws of random elements of interest. This can be cast into problems of estimation, testing hypotheses and predictions. We will mainly focus on statistical aspects related to stochastic dominance, namely nonparametric statistics. The following gives a small picture of the context.

In using stochastic dominance rules for financial decision-making, the crucial problem is *how to use empirical data to apply the theory.*

Let's consider the simplest setting where the choice between $X$ and $Y$ is based only on the knowledge of their probabilistic distribution functions, $F$ and $G$, respectively. Markowitz [104, 105] proposed a conservative method based only on some characteristics of their distribution functions: the mean-variance (MV) approach where a partial order on random variables is defined as follows: $X$ is preferred to $Y$ if $EX \geq EY$ and $Var(X) \leq Var(Y)$. The meaning is clear. For applications, how do we check this partial order, assuming that we have random samples $X_1, X_2, ..., X_n$ and $Y_1, Y_2, ..., Y_m$ from $X$ and $Y$, respectively? In fact they are only past data on "similar" projects (not exactly the same). We could rely on their empirical counterparts (estimates) to infer the MV partial order.

Specifically, it seems plausible that if the sample mean of the $X$-sample is greater than the sample mean of the $Y$-sample, and the variance of the $X$-sample is smaller than the variance of the $Y$-sample, then it is a good indication that $EX \geq EY$ and $Var(X) \leq Var(Y)$. It is just an *indication*, but not a fact yet! To validate it, we need to carry out two testing problems. As a result, only the MV order can be verified or denied at some level $\alpha$ of significance.

How can we check stochastic dominance orders? As in the case of the MV order, since $F, G$ are unknown, we can only rely on empirical data, and hence face the same inference problem, namely, "how can we justify the transfer of empirical evidence to theoretical fact." Here are some heuristics.

Since $F$ and $G$ are unknown, one of the ways to check whether or not $G - F \geq 0$ is to look at their estimators as empirical counterparts. The application of stochastic dominance orders (up to third order) was sometimes illustrated by examples in which estimators of distribution functions are used as counterparts of (unknown) population distribution functions in the assessment of risks via theory (which is based on unknown true distribution functions of risky payoffs). Given samples $X_1, X_2, ..., X_n$ and $Y_1, Y_2, ..., Y_m$, let $F_n(x)$, $G_m(x)$ be estimators of $F(x), G(x)$, respectively. Suppose $F_n(x) \leq G_m(x)$, i.e., $G_m(x) - F_n(x) \geq 0$, for fixed $n, m$. How do we infer that $G(x) - F(x) \geq 0$? Clearly, we need to justify such a procedure in real applications.

Note that we can estimate *consistently* $F$ and $G$, say, by empirical distributions $F_n$, $G_m$, namely

$$F_n(x, X_2, ..., X_n) = \frac{1}{n} \sum_{i=1}^{n} 1_{(X_i \leq x)}$$

$$G_m(y, Y_1, Y_2, ..., Y_m) = \frac{1}{m} \sum_{i=1}^{m} 1_{(Y_i \leq y)},$$

where $1_A$ denotes the indicator function of the set $A$.

The strong law of large numbers implies that, with probability one, $F_n(.) \to F(.)$, and $G_n(.) \to G(.)$, (i.e., almost surely, a.s.). If in addition, $F_n(.) \leq G_n(.)$, almost surely, for each $n$ (it is unlikely that we can have this), then $F(x) \leq G(x)$ for all $x \in \mathbb{R}$. This follows from

**Lemma 6.1** *If $a_n$ and $b_n$ are sequences of numbers converging to numbers $a$ and $b$, respectively, and $a_n \leq b_n$ for each $n$, then $a \leq b$.*

**Proof.** If $a > b$, we take $\delta < a - b$, then for all $n \geq N(\delta)$, we have $a_n \in (a - \delta, a + \delta)$. But then for all $n \geq N(\delta)$, $b_n > a - \delta > b$ since $b_n \geq a_n$. Hence $b_n$ cannot converge to $b$. ■

However, this is not quite a justification for inferring from empirical evidence to stochastic dominance (SD), since we only use one sample size $n$, i.e., compare $F_n(x)$ and $G_n(x)$ for some fixed $n$. Since it is hard to specify parametric forms for $F$ and $G$, we often face non-parametric testing problems, and SD can be only asserted with some level of significance.

Next, in the process of risk assessment, if the first SD fails (via empirical counterparts), we could try our luck by looking at the second SD criterion. Again, since $\int_a^x [G(t) - F(t)]dt$ is unknown, to "test" whether it is greater than 0, we rely on its empirical counterpart, namely,

$$\int_a^x [G_n(t) - F_n(t)]dt, \text{ for each } x \in \mathbb{R}.$$

Now if $a > -\infty$, then $F_n(t) \to F(t)$, a.s., and $|F_n| \leq 1$ which is integrable on $(a, x)$, so that, by Lebesgue dominated convergence theorem,

$\int_a^x F_n(t)dt \rightarrow \int_a^x F(t)dt$ (with probability one). Hence, our empirical procedure is justified.

This applies also to third SD. Suppose that the third SD fails. What can we do? Of course, SD is a partial order relation on random variables, there are situations that we cannot rank them according to SD ordering. However, the investor needs to decide to invest!

Of course, the investor can do several things to arrive at a decision (when SD theory cannot help), such as ask an expert, or place his decision within the context of, say, a multi-criteria decision-making problem (by listing explicitly factors which could lead to risk. Here, as a note, a technique such as Choquet integral, a non-linear aggregation method for interactive criteria, could be used to produce a complete ranking procedure).

In our context of SD theory where we have only random samples, $X_1, X_2, ..., X_n$ from $X$ and $Y_1, Y_2, ..., Y_m$ from $Y$, we could look more closely at the data to at least advise the investor, even with a weaker order, for example, a weaker (total) order based on some probability quantification (e.g., some risk measure).

Obviously, any such weaker order should be implied by SD order. This is the problem of consistency of risk measures with respect to stochastic dominance that we addressed in Chapter 4.

At the other end, suppose data are coarse, i.e., of low quality, such as each sample value $X_i$ is only known to lie in an interval $S_i$ (i.e., $X_i \in S_i$ with probability one). How do we estimate $F(x)$ from the random sample of sets $S_1, S_2, ..., S_n$?

Another important case of coarse data is hidden (or missing) data, as in bioinformatics (computational biology) and auction theory. Specifically, the $X_i$'s are not directly observable, instead, another related sample $Z_i$'s are observed. To estimate the distribution $F$ of $X$ from the unobserved sample $X_1, X_2, ..., X_n$, we need, of course, to exhibit some sort of relationship (link) between the $X_i$ and the $Z_i$.

In computational biology (such as for identification of DNA sequences), the hidden Markov model is used successfully. In econometrics of auction data, for the first-price sealed-bid auction, for example, the situation is nicer, since, viewing auctions as games of conflict (dynamic or static) with incomplete information, observed data $Z_i$ (bids from buyers in the past) can be related explicitly to the unobserved data $X_i$ (valuations of the buyers) via the Bayes-Nash equilibrium, so that both estimation and identification on the distribution $F$ of $X$ can be carried out.

## 6.3 Generalities on statistical estimation

Let us start out with the estimation problem. While the context of estimation can be formulated in a general setting, we choose the simplest case to bring out the essentials of the art. The following has the flavor of a parametric setting, but the ideas can be extended to general settings.

Suppose our data consists of a random sample $X_1, X_2, ..., X_n$, i.e., an i.i.d. collection of random variables, drawn from the population $X$ with unknown probability law $P_X$. Let $\varphi(P_X)$ be our "parameter" of interest. To estimate $\varphi(P_X)$ we will use the data $X_1, X_2, ..., X_n$ (for the time being, we ignore the Bayesian approach, but will return to it later). Thus, an estimator of $\varphi(P_X)$ will be some function of $X_1, X_2, ..., X_n$, say, $T_n(X_1, X_2, ..., X_n)$, called a statistic. Obviously there are many statistics which we could use to estimate our target parameter $\varphi(P_X)$ provided of course these statistics take values in the space $\varphi(\mathcal{F})$. More precisely, an estimator of $\varphi(P_X)$ is a statistic $T$ which takes values in $\varphi(\mathcal{F})$. So the first questions are "which one?" and "why?"

In order to answer this question we need to specify our choice problem. Why do we prefer one estimator to another? Obviously, we want to use "good" or if possible "best" estimators! For that, what do we mean by "best"? Clearly, performance of estimators comes to mind. Every time we use an estimator $T_n$ to estimate $\varphi(P_X)$, we commit an error, i.e., the value of $T_n$ might be different from $\varphi(P_X)$. We need to specify this error. Unfortunately, there is more than one way to do that. Indeed, the difference between $T_n$ and $\varphi(P_X)$ can be measured by various "metrics" such as $|T_n - \varphi(P_X)|$ or $(T_n - \varphi(P_X))^2$, or more realistically by $E|T_n - \varphi(P_X)|$, $E(T_n - \varphi(P_X))^2$, assuming, for concreteness, of course $T_n$ and $\varphi(P_X)$ are real-valued.

Suppose we choose a criterion for comparing estimators, say, the mean-squared error $E(T_n - \varphi(P_X))^2 = MSE(T_n, P_X)$, noting that $P_X \in \mathcal{F}$. Then we can compare two estimators $T_n$ and $T_n'$ of $\varphi(P_X)$ using this criterion. Specifically we say that $T_n$ is better than $T_n'$ if $MSE(T_n, P_X) \leq MSE(T_n', P_X)$ for all $P_X \in \mathcal{F}$ with strict inequality for at least some $P_X$. Note that the phrase "for all $P_X \in \mathcal{F}$" is necessary since the "true" $P_X$ is unknown.

Unfortunately, whatever estimator we choose, we can always find another one which is better in the above sense at some parameter point $P_X$! Indeed, if $MSE(T_n, P_X) > 0$ at $P_X = P_X^*$, then if we take $T_n' = P_X^*$, identically, then $MSE(T_n', P_X^*) = 0 < MSE(T_n, P_X^*)$. The point is this. If we let all possible estimators into the competition, then the best one may not exist.

However, these "pathological" estimators, such as the above $T'_n$, are such that $E_{P_X}(T'_n) \neq \varphi(P_X)$ for some $P_X \in \mathcal{F}$. Such estimators are called *biased estimators*, where the bias for $\varphi(P_X)$ is $b(T'_n, P_X) = E_{P_X}(T'_n) - \varphi(P_X)$. This suggests that in order to define best estimators, we may want to restrict the class of possible estimators of $\varphi(P_X)$ to a smaller class of estimators $T_n$ such that $E_{P_X}(T_n) = \varphi(P_X)$ for all $P_X \in \mathcal{F}$. Such estimators are called unbiased estimators of $\varphi(P_X)$.

Note that we do not say that unbiased estimators are better than biased estimators, as far as errors are concerned. As we will see, estimators in nonparametric models, such as pointwise estimators of probability density functions are not unbiased, but asymptotically unbiased, i.e., $\lim_{n \to \infty} b(T_n, P_X) = 0$, a necessary condition for consistency. It turns out that within the class of unbiased estimators, often it is possible to identify a best estimator.

**Example 6.2** *Let $X$ be a real-valued random variable with unknown distribution function $F$ and let*

$$\varphi(F) = \int_{-\infty}^{\infty} \int_{-\infty}^{\infty} |x - y| dF(x) dF(y)$$

*be the Gini mean difference.*

The first question is whether there exist unbiased estimators for $\varphi(F)$. Here the answer is yes since, in general, it can be easily shown that a parameter $\varphi(F)$ is estimable (i.e., can be estimated by some unbiased estimator) if and only if it is of the form

$$\varphi(F) = \int_{-\infty}^{\infty} \cdots \int_{-\infty}^{\infty} h(x_1, x_2, ..., x_m) dF(x_1) dF(x_2) ... dF(x_m)$$

for some function (called a kernel) $h(x_1, x_2, ..., x_m)$, for some $m$.

Here, $m = 2$ and $h(x, y) = |x - y|$ (a symmetric function of two variables). It is so, since clearly the statistic

$$U_n(X_1, X_2, ..., X_n) = \binom{n}{2}^{-1} \sum_{1 \leq i < j \leq n} |X_i - X_j|$$

is not only an unbiased estimator for $\varphi(F)$, but also the best unbiased estimator. The construction of the best unbiased estimator in this example falls into the well-known theory of $U$-statistics (see, e.g., Serfling

[141]). The optimality property of $U$-statistics (i.e., being minimum variance unbiased estimators) remains valid for nonparametric models which are sufficiently rich such as when $\mathcal{F}$ consists of all absolutely continuous distribution functions on $\mathbb{R}$. This is so because, first of all, we can replace an arbitrary kernel $h$ in the representation of $\varphi(F)$ by a symmetric kernel; secondly, any symmetric statistic is a function of the order statistics $X_{(1)} \leq X_{(2)} \leq \cdots \leq X_{(n)}$ which is sufficient for any $\mathcal{F}$; thirdly, when $\mathcal{F}$ is sufficiently rich, the order statistic is also complete, and the result follows by Rao-Blackwell theorem in standard statistics, by noting that $U$-statistics are obtained by conditioning upon the order statistics. Note that we have not only answered the question of how to choose a best estimator for $\varphi(F)$, but also answered the obvious follow-up question, "how to obtain a best estimator." The theory of $U$-statistics provides a construction of best unbiased estimators. Interested readers should study also asymptotic behavior of $U$-statistics, such as consistency and limiting distributions, for applications in large samples.

Having a "good" estimator, in terms of error, is just the very first step in the estimation process. For setting up confidence bands or regions as well as testing hypotheses, we need to know the probability law of our estimators (sampling distributions). Unless the population $X$ is "well understood" in the sense that we can postulate a simple parametric form of its distribution, such as $X$ is lognormal, sampling distributions of estimators are hard to find with fixed sample sizes. We need more information from the data to be able to approximate sampling distributions. One obvious type of information needed is more data! *Panel data* in econometrics and *high frequency data* in financial economics are examples of this obvious quest for more information. Increasing the sample sizes leads to more accurate estimates and more powerful tests.

Remember also the obvious fact from sampling from a finite population: if we have enough time, money and energy to conduct a census throughout the population, then we should get accurate values for population parameters of interest! For infinite populations, we could increase the sample size, if we can, to have more information, in view of the information principle which says, "more data, more accuracy." Thus, we need to consider the case of large samples. How large is large is a matter of degree of accuracy of estimators.

The setting for *large sample statistics* is this. Again, to be concrete, we consider discrete "time." Thus our data are a first segment of an infinite sequence of observations $(X_n, n \geq 1)$ from $X$. Now with large samples we can request more appropriate properties of our estimators.

We have seen that asymptotic unbiasedness is one desirable property. From the above information principle, it is natural to consider an estimator $T_n$ (in fact, a sequence of estimators) of $\varphi(P_X)$ to be "good" when it approaches $\varphi(P_X)$ as the sample size $n$ increases. This asymptotic property of estimators is referred to as *consistency*. Since, at the design phase, the $T_n$ are random elements (i.e., functions and not numbers), and as such, the convergence of $T_n$ to $\varphi(P_X)$ could be interpreted in several ways. The pointwise convergence $\lim_{n\to\infty} T_n(\omega) = \varphi(P_X)$, for all $\omega \in \Omega$ is too strong to require, as exemplified by the following phenomenon.

Consider the experiment consisting of tossing a fair coin indefinitely. Let $X_n$ denote the outcome of the $n$th toss. The sample space associated with this random experiment is $\Omega = \{0,1\}^\mathbb{N}$ where $\mathbb{N}$ is the set of positive natural integers, each $\omega = (x_n, n \geq 1)$, $x_n \in \{0,1\}$ (1 for heads, 0 for tails) represents the result of an infinite sequences of tosses. We have $X_n(\omega) = x_n$.

Clearly we expect

$$T_n(\omega) = \frac{1}{n} \sum_{i=1}^{n} X_n(\omega)$$

to converge to $1/2$. But for $\omega$ being a sequence containing only a finite number of 1's,

$$\lim_{n\to\infty} T_n(\omega) = \lim_{n\to\infty} \frac{1}{n} \sum_{i=1}^{n} X_n(\omega) = 0.$$

Thus, we should avoid such situations by requiring only that

$$\lim_{n\to\infty} T_n(\omega) = \varphi(P_X), \text{ for almost all } \omega \in \Omega,$$

i.e., only on some subset $A \subseteq \Omega$ such that $P(A) = 1$. The interpretation is that with probability one, $T_n$ will converge to $\varphi(P_X)$. We refer to this almost sure (a.s.) convergence of the sequence of random elements $T_n$ to the constant $\varphi(P_X)$ (or to another random element) as the *strong consistency* property. Specifically, $T_n$ is strongly consistent for $\varphi(P_X)$ if

$$P(\omega \in \Omega : \lim_{n\to\infty} T_n(\omega) = \varphi(P_X)) = 1.$$

When such a requirement cannot be met, we might like to settle down for less!, i.e., for some weaker type of consistency, called *weak consistency*. This can be formulated as saying: with "high probability" $T_n$

will converge to $\varphi(P_X)$. Clearly this is a weaker requirement than strong consistency since 1 is maximum probability and hence definitely means "high," whereas the linguistic term "high" means anything just short of being 1, say, $1 - \varepsilon$, for any $\varepsilon > 0$, as small as you please. Specifically, the estimator $T_n$ is said to be a weakly consistent estimator for $\varphi(P_X)$ if for any $\varepsilon > 0$,

$$\lim_{n \to \infty} P(|T_n - \varphi(P_X)| > \varepsilon) = 0.$$

How do we check consistency of estimators? Having a desirable property of estimators in mind, we use mathematical language to specify it by a definition so that we can check it when necessary. Sometimes the situations are nicer in the sense that we can recognize consistency of estimators by simple applying a limit theorem in probability theory known as *laws of large numbers*. Roughly speaking, the strong law of large numbers (SLLN) (Kolmogorov) is this. If a population $X$ has a finite mean, then the sample mean of a random sample from it is a strongly consistent estimator of its mean.

**Example 6.3 (Estimation of quantiles)** *The estimation of quantiles is useful for estimating quantile-based risk measures in financial economics. Let $X$ be, say, a loss variable of a financial position, with unknown distribution function $F$. Assume you are given a random sample $X_1, X_2, ..., X_n$ from $X$, and $\alpha \in (0, 1)$. To estimate the $\alpha$-quantile $F^{-1}(\alpha)$, we use the $\alpha$-quantile $F_n^{-1}(\alpha)$ of the empirical distribution function $F_n$. Since $F(F^{-1}(\alpha)-) \leq \alpha \leq F(F^{-1}(\alpha))$, we have that $F_n^{-1}(\alpha)$ is (strongly) consistent for $F^{-1}(\alpha)$ when $F^{-1}(\alpha)$ is the unique solution of $F(x-) \leq \alpha \leq F(x)$.*

Turning to sampling distributions, we need to specify what we mean when we say that when $n$ gets large, the distribution of $T_n$ will be approximately that of some random element $Y$, so that, when $n$ is actually "large" (depending on our demand for accuracy of the approximation), we can use the distribution of $Y$ as that of $T_n$. This type of convergence, while related to the sequence $T_n$, is not about the $T_n$ per se, but about their probability laws (distributions), say, $P_n = PT_n^{-1}$, a sequence of probability measures defined on the $\sigma$-field of the common range of the $T_n$'s.

Since a concept of convergence for sequences of probability measures should generalize that of sequences of distribution functions of random vectors (where probability measures are identified with distribution functions via the Lebesgue-Stieltjes theorem), let's see what happens in the case of random variables. Let $X$, $X_n$ be real-valued random elements, with distribution functions $F$, $F_n$, respectively. The following phenomenon suggests an appropriate definition of the convergence in distribution.

Let $X_n = 1/n$, $n \geq 1$. Then $F_n(x) = 1_{[1/n,\infty)}(x)$ which converges pointwise to 0 for $x \leq 0$ and to 1 for $x > 0$. Now clearly $X_n = 1/n$ should converge "in distribution" to $X = 0$ as $n \to \infty$, where the distribution function $F$ of 0 if $F(x) = 1_{[0,\infty)}(x)$. But $F_n(x) \to F(x)$ only for $x \neq 0$, since $F(0) = 1$, but $F_n(0) = 0$ for any $n \geq 1$, so that $\lim_{n\to\infty} F_n(0) = 0 \neq F(0)$. Now we note that 0 is the only point at which the limiting distribution function $F$ is not continuous. Thus, we say that $X_n$ (with distribution function $F_n$) converges in distribution to $X$ (with distribution function $F$) if $F_n \to F$ on the set $C(F)$ of continuity points of $F$.

To extend this type of convergence to probability measures for general random elements (when there is no concept of distribution functions), we need to explore some equivalent criteria of the above convergence defined in terms of distribution functions suitable for extension. The following will do. We know $F_n \to F$ on the set $C(F)$ of continuity points of $F$ if and only if for any continuous and bounded function $f : \mathbb{R} \to \mathbb{R}$,

$$\int_{\mathbb{R}} f(x)dF_n(x) \to \int_{\mathbb{R}} f(x)dF(x).$$

Thus, we generalize the convergence in distribution of random variables to general random elements as follows. We say that the sequence of random elements $X_n$ (with probability law $P_n$) converge in distribution to the random element $X$ (with probability law $Q$), if $P_n$ converges weakly to $Q$ in the sense that

$$\int_{\mathcal{X}} f(x)dP_n(x) \to \int_{\mathcal{X}} f(x)dQ(x), n \to \infty$$

for any $f : \mathcal{X} \to \mathbb{R}$, continuous and bounded. Of course, we assume $P_n, Q$ are probability measures defined on $\mathcal{B}(\mathcal{X})$ where $\mathcal{X}$ is a metric space. For more details, see [15].

**Remark 6.4** *For probability measures $P_n, Q$ on a metric space $\mathcal{X}$, and $f : \mathcal{X} \to \mathbb{R}$, continuous and bounded with $\|f\| = \sup_{x \in \mathcal{X}} |f(x)|$, we have $\left|\int_{\mathcal{X}} f(x)dQ(x)\right| \leq \int_{\mathcal{X}} \|f\| \, dQ(x) \leq \|f\|$. Let $B = \{f$ continuous and bounded: $\|f\| \leq 1\}$, and $\|Q\| = \sup_{f \in B} \left|\int_{\mathcal{X}} f(x)dQ(x)\right|$. Define $P_n \to Q$, strongly, if $\|P_n - Q\| \to 0$ as $n \to \infty$. Now, let $x_n \to x$ in $\mathcal{X}$ (in the sense of the metric on $\mathcal{X}$), then it is natural to expect that their probability laws*

*(Dirac measures)* $\delta_{x_n}$ *should converge to* $\delta_x$. *But suppose* $x_n \neq x$ *for any* $n \geq 1$, *then for each* $n$, *let* $f_n$ *be continuous and bounded with values in* $[-1, 1]$ *such that* $f_n(x_n) = 1$, $f_n(x) = -1$. *Then* $\|\delta_{x_n} - \delta_x\| = 2$ *so that* $\delta_{x_n}$ *does not converge strongly to* $\delta_x$, *whereas* $\delta_{x_n}$ *converges weakly to* $\delta_x$. *Hence the weak convergence is a more natural concept.*

With this concept of weak convergence of probability measures, we can examine limiting distributions in large samples. Just like the case of consistency, a clear definition of convergence in distribution will allow us to search for limiting distributions of our estimators. Thanks to probability theory again, various limit theorems will help us in this search. Central to this search is the now well-known *central limit theorem* (CLT).

Here is an example of empirical distribution functions for i.i.d. samples, where we only exhibit limiting distributions in the context of random vectors, but if we consider the empirical distribution functions as a collection of random variables indexed by $\mathbb{R}$, i.e., as a stochastic process, called the *empirical process*, then we can use weak convergence of probability measures to discuss its limiting distribution.

**Example 6.5** *Let* $F$ *be the unknown distribution function of the random variable* $X$. *Suppose we wish to estimate* $F(x)$ *for a given* $x \in \mathbb{R}$. *We write*

$$F(x) = \int_{-\infty}^{x} dF(y) = \int 1_{(-\infty, x]}(y) dF(y)$$

*so that* $F(x)$ *is estimable. Its U-statistic, constructed with the kernel* $h(y) = 1_{(-\infty, x]}(y)$, *is*

$$F_n(x; X_1, X_2, ..., X_n) = (1/n) \sum_{i=1}^{n} 1_{(X_i \leq x)},$$

*which we call the empirical distribution function based on the random sample* $X_1, X_2, ..., X_n$. *By SLLN for i.i.d. random variables,* $F_n(x; X_1, X_2, ..., X_n)$ *converges almost surely to* $F(x)$, *and by CLT,* $F_n(x; X_1, X_2, ..., X_n)$ *is asymptotically normal* $N(F(x), F(x)(1 - F(x)/n)$, *i.e.,*

$$\sqrt{n}(F_n - F(x))/\sqrt{F(x)(1 - F(x))}$$

*converges in distribution to* $N(0, 1)$.

**Example 6.6** *For many models, sample quantiles are asymptotically normal. For example, if the distribution function* $F$ *in the model is absolutely continuous with density* $f$ *positive and continuous at* $F^{-1}(\alpha)$, *then* $F_n^{-1}(\alpha)$ *is asymptotically normal* $N(F^{-1}(\alpha), \alpha(1-\alpha)/nf^2(F^{-1}(\alpha)))$.

*Of course, the situation is simpler in parametric models. For example, if $F$ is the distribution function of a normal population $X$ with mean $\mu$ and variance $\sigma^2$, then*

$$F^{-1}(\alpha) = \mu + \sigma \Phi^{-1}(\alpha)$$

*where $\Phi$ is the distribution function of the standard normal random variable, so that an estimator of $F^{-1}(\alpha)$ could be taken to be $X_n^* + S_n \Phi^{-1}(\alpha)$, where $X_n^*$ is the sample mean and $S_n = ((1/n-1)\sum_{i=1}^n (X_i - X_n^*)^2)^{1/2}$ (sample standard deviation).*

**Remark 6.7** *The (nonparametric) estimation of distribution functions can be used to estimate related parameters as well as constructing tests. For example, in estimating or testing about second-order stochastic dominance, the parameters of interest are*

$$\hat{F}(t) = \int_{-\infty}^t F(x)dx, \text{ for each } t \in R.$$

*The plug-in estimators are*

$$\hat{F}_n(t) = \int_{-\infty}^t F_n(x)dx = (1/n)\sum_{i=1}^n (t - X_i)1_{(X_i \le t)}.$$

*Estimation of $\hat{G}(t) - \hat{F}(t)$, and tests for second-order stochastic dominance can be based upon the statistics $\hat{G}_m(t) - \hat{F}_n$. These are statistical functionals and hence fall into the theory of $V$-statistics (see Serfling [141]).*

## 6.4    Nonparametric estimation

We address now the problem of estimation in a nonparametric setting. We start out with a theorem.

### 6.4.1    The Glivenko-Cantelli theorem

The hope to discover the law governing the random evolution of economic quantities when we have sufficient data (large samples) lies in the fact that it is possible to estimate $F$ sufficiently accurate. It forms the very first nonparametric estimation problem.

Let $X_1, X_2, ..., X_n$ be a random sample (i.i.d.) from $X$ with unknown distribution $F \in \mathcal{F}$. We have seen that the empirical distribution function

$$F_n(x; X_1, X_2, ..., X_n) = \frac{1}{n} \sum_{i=1}^{n} 1_{(X_i \leq x)}$$

is a plausible estimator of $F(x)$, for each given $x \in \mathbb{R}$. It is strongly consistent (by SLLN) and asymptotically normal (CLT). Moreover, it is possible to approximate $F$ uniformly. Specifically,

**Theorem 6.8 (*Glivenko-Cantelli*)** $\sup_{x \in \mathbb{R}} |F_n(x) - F(x)| \to 0$, *a.s.,* *as* $n \to \infty$.

This says that for $n$ sufficiently large, $F_n(x)$ is close to $F(x)$ for any $x$, noting that $n$ does not depend on $x$. The (a.s.) uniform convergence of sequence of distribution functions, such as $F_n$, is important for tabulating statistical tables for the limit distribution. For example, to tabulate $F(x)$ at several $x$, we simply need a sample size $n$ *sufficiently large*, in advance, i.e., not depending on which $x$ we want to tabulate.

How large is large? It is a matter of *approximation accuracy*. We need the rate of convergence! For example, if we wish to approximate $F$ by $F_n$ within, say, $\varepsilon = 0.001$, i.e., $|F_n(.) - F(.)| \leq \varepsilon$, then, knowing its rate of convergence, we infer the minimum $n$ which is considered as large.

Note that while $F_n(x)$ is a *pointwise* estimator of $F(x)$, the function $F_n(.)$ is a *global* estimator of $F$, i.e., an estimator of the function $F(.)$. To measure the closeness of $F_n$ to $F$, we will use the Kolmogorov-Smirnov (random) distance.

In order to prove the above Glivenko-Cantelli theorem, we need to spell out the framework of asymptotic statistics.

Let $(\Omega, \mathcal{A}, P)$ be our background probability space on which all random elements are defined. In a fixed sample size setting, the sample space (i.e., the set of data) is $\mathbb{R}^n$ where $(x_1, x_2, ..., x_n)$ is the realization of a random sample $(X_1, X_2, ..., X_n)$, i.e., the $X_i$'s are i.i.d. drawn from the population $X$ with distribution function $F$. Here $X : \Omega \to \mathbb{R}$, measurable with respect to $\mathcal{A}$ and $\mathcal{B}(\mathbb{R})$, the Borel $\sigma$-field of $\mathbb{R}$, and the probability law of $X$ is the probability measure $P_X = PX^{-1} = dF$ on $\mathcal{B}(\mathbb{R})$, by the Lebesgue-Stieltjes theorem. The data are random vectors:

$$(X_1, X_2, ..., X_n) : \Omega \to \mathbb{R}^n.$$

Now, to allow the sample size $n$ to move, including to infinity, we have to view $X^{(n)} = (X_1, X_2, ..., X_n)$ as the initial segment of an infinite sequence $X^\infty = \{X_k : k \geq 1\}$ of i.i.d. random variables. Thus, our sample space will be a set of sequences of real numbers, i.e., a subset of $\mathbb{R}^\infty$. Our random elements of interest map $\Omega$ to $\mathbb{R}^\infty$. Specifically they are measurable maps from $(\Omega, \mathcal{A}, P)$ to $(\mathbb{R}^\infty, \mathcal{B}(\mathbb{R}^\infty), P^*)$, where $P^*$ is the probability measure on $\mathcal{B}(\mathbb{R}^\infty)$ which is determined by Kolmogorov's consistency theorem, from the product measures on each $\mathbb{R}^n$ through $dF$.

Now, from $X^\infty = \{X_k : k \geq 1\}$, we construct a sequence of random distribution functions, called *empirical distribution functions*:

$$F_n : \mathbb{R} \times \Omega \to [0,1] : (t, \omega) \mapsto F_n(t, X^{(n)}(\omega)) = (1/n) \sum_{i=1} 1_{(X_i(\omega) \leq t)},$$

where $1_A$ is the indicator function of a set $A$.

Note that for each $n$, $F_n$ depends on $X^\infty$ only through $X^{(n)}$. Also each $F_n(\cdot, \cdot)$ for fixed $\omega$, is clearly a distribution function as a function of $t$. Now for each given $t \in \mathbb{R}$, and each $n$, the $1_{(X_i \leq t)}$, $i = 1, 2, ..., n$, are i.i.d. from the population $1_{(X \leq t)}$ with mean $P(X \leq t) = F(t)$. It follows from SLLN that $F_n(t, \omega)$ converges $P^*$-almost surely to $F(t)$ as $n \to \infty$. This means that there exists a set $\mathcal{N}_t \in \mathcal{B}(\mathbb{R}^\infty)$ such that $P^*(\mathcal{N}_t) = 0$ and for $\omega$ such that $X^\infty(\omega) \notin \mathcal{N}_t$, we have $F_n(t, X^{(n)}(\omega)) \to F(t)$, as $n \to \infty$.

Note that the above convergence is pointwise in $t$, and each null-set $\mathcal{N}_t$ depends on $t$. Such a result is not enough to define a distance between $F$ and $F_n$ for testing purposes. We need a stronger result, namely a *uniform convergence in t*. Let

$$D_n(X^{(n)}) = \sup_{t \in \mathbb{R}} |F_n(t, X^{(n)}) - F(t)|.$$

Then $D_n(X^{(n)}) \to 0$, with $P^*$-a.s., as $n \to \infty$.

Let $C(F)$ be the set of continuity points of $F$. For $t \notin C(F)$,

$$F_n(t, X^{(n)}) - F_n(t-, X^{(n)}) = (1/n) \sum_{i=1}^{n} 1_{(X_i = t)},$$

which converges, almost surely to $F(t) - F(t-)$, as $n \to \infty$, by SLLN, i.e., there exists $\mathcal{N}_t \in \mathcal{B}(\mathbb{R}^\infty)$ such that $P^*(\mathcal{N}_t) = 0$ and for $\omega$ such that $X^\infty(\omega) \notin \mathcal{N}_t$, we have

$$F_n(t, X^{(n)}) - F_n(t-, X^{(n)}) \to F(t) - F(t-).$$

Since the complement of $C(F)$ is at most countable, $\cup_{t \notin C(F)} \mathcal{N}_t$ is a $P^*$-null set, so that for $X^\infty \notin \cup_{t \notin C(F)} \mathcal{N}_t$,

$$F_n(t, X^{(n)}) - F_n(t-, X^{(n)}) \to F(t) - F(t-) \text{ for all } t \notin C(F).$$

Next, let $\mathbb{Q}$ denote the rationals. For each $t \in \mathbb{Q}$, by (a), there exists a set $\mathcal{N}_t^* \in \mathcal{B}(\mathbb{R}^\infty)$ such that $P^*(\mathcal{N}_t^*) = 0$ and for $\omega$ such that $X^\infty(\omega) \notin \mathcal{N}_t^*$, we have

$$F_n(t, X^{(n)}(\omega)) \to F(t), \text{ as } n \to \infty.$$

But $\mathbb{Q}$ is countable, $\cup_{t \in \mathbb{Q}} \mathcal{N}_t^*$ is a $P^*$-null set, so that for $X^\infty(\omega) \notin \cup_{t \in \mathbb{Q}} \mathcal{N}_t^*$, we have

$$F_n(t, X^{(n)}(\omega)) \to F(t), \text{ for all } t \in \mathbb{Q}.$$

In fact, for $X^\infty(\omega) \notin (\cup_{t \notin C(F)} \mathcal{N}_t) \cup (\cup_{t \in \mathbb{Q}} \mathcal{N}_t^*)$, we have appropriate convergences. Indeed, let $t \in C(F)$ but $t \notin \mathbb{Q}$.

By continuity of $F$ at $t$ and the fact that the rationals are dense in $\mathbb{R}$, for any $\varepsilon > 0$, there are $t_1, t_2 \in \mathbb{Q}$ with $t_1 < t < t_2$ such that

$$F(t_2) - F(t_1) < \varepsilon/2. \tag{6.1}$$

Since $t_1, t_2 \in \mathbb{Q}$, we can find an $N$ such that for $n \geq N$,

$$\begin{aligned} |F_n(t_1) - F(t_1)| &< \varepsilon/2 \\ |F_n(t_2) - F(t_2)| &< \varepsilon/2. \end{aligned} \tag{6.2}$$

Now by monotonicity of $F$, we have

$$F(t_1) \leq F(t) \leq F(t_2).$$

Thus, by (6.1),

$$F(t_2) - \varepsilon/2 < F(t_1) \leq F(t) < F(t_1) + \varepsilon/2.$$

Also, since $F_n$ is monotone increasing, and by (6.2), we have

$$\begin{aligned} F(t) &\leq F(t_2) \leq F(t_1) + \varepsilon/2 \leq [F_n(t_1) + \varepsilon/2] + \varepsilon/2 \leq F_n(t) + \varepsilon \\ F(t) &\geq F(t_2) - \varepsilon/2 \geq [F_n(t_2) - \varepsilon/2] - \varepsilon/2 \geq F_n(t) - \varepsilon \end{aligned}$$

so that

$$|F_n(t) - F(t)| \leq \varepsilon.$$

It remains to show that the convergence is uniform in $t$. Note that from the above analysis, we pick $X^\infty(\omega) \notin (\cup_{t \notin C(F)} \mathcal{N}_t) \cup (\cup_{t \in \mathbb{Q}} \mathcal{N}_t^*)$, and fix it, so that we only need to consider a sequence of deterministic $F_n(t)$, which we know that it converges pointwise to $F(t)$, for all $t \in \mathbb{R}$.

Let's prove it by contradiction. Suppose there are $\varepsilon_0 > 0$, sequences $\{n_k\}, \{t_k\}$ such that $|F_{n_k}(t_k) - F(t_k)| > \varepsilon_0$ for all $k$. Since $F$ and $F_n$ are distribution functions, we cannot have $t_k \to \infty$ or to $-\infty$ (since otherwise $F_{n_k}(t_k) - F(t_k) \to 0$ and hence cannot $> \varepsilon_0 > 0$). Thus, $t_k$ is bounded. Without loss of generality, suppose $t_k \to t^*$.

There exist subsequences $t_k$ converging to $t^*$ monotonically from below or from above. Also, since $|F_{n_k}(t_k) - F(t_k)| > \varepsilon_0$, there exists either a subsequence at which $F_{n_k}(t_k) > F(t_k) + \varepsilon_0$ or a subsequence at which $F_{n_k}(t_k) < F(t_k) - \varepsilon_0$. Restricting to subsequences we have four cases. The first one is

$$t_k \nearrow t^* , t_k < t^* \text{ and } F_{n_k}(t_k) > F(t_k) + \varepsilon_0.$$

Select $t' < t < t''$ with $t', t'' \in \mathbb{Q}$. Using monotonicity of $F, F_{n_k}$, we have, for $k$ sufficiently large,

$$
\begin{aligned}
\varepsilon_0 \quad &< \quad F_{n_k}(t_k) - F(t_k) \le F_{n_k}(t^*-) - F(t') \\
&\le \quad F_{n_k}(t^*-) - F_{n_k}(t^*) + F_{n_k}(t'') - F(t'') + F(t'') - F(t').
\end{aligned}
\tag{6.3}
$$

When $k \to \infty$,

$$F_{n_k}(t^*-) - F_{n_k}(t^*) \to F(t^*-) - F(t^*)$$

and

$$F_{n_k}(t'') - F(t'') \to 0.$$

Next, let $t' \nearrow t^*$ and $t'' \searrow t^*$ along $\mathbb{Q}$, we have

$$F(t'') - F(t') \to F(t^*) - F(t^*-).$$

Thus the right-hand side of (6.3) can be made as small as we please, in particular, strictly smaller than $\varepsilon_0$, yielding a contradiction. The other cases are similar and are left as exercises.

**Remark 6.9 (Empirical processes)** *We simply mention and elaborate a little bit here an asymptotic result concerning empirical processes, as a random element, namely its limiting distribution, which is used in testing for Stochastic Dominance. General results concerning empirical processes are essential in studying large sample statistics. The natural normalization of the (random) distribution function $F_n$ leads to the so-called empirical process*

$$\sqrt{n}[F_n(t) - F(t)], t \in \mathbb{R}.$$

*This is a sequence of stochastic processes (or random functions): for each $n$, $(\sqrt{n}[F_n(t) - F(t)], t \in \mathbb{R})$ is a family of random variables, indexed by* the "time" set $\mathbb{R}$. *Specifically, for each $n$, $(\sqrt{n}[F_n(t) - F(t)],\ t \in [0,1])$ is a random element with values in $D([0,1])$, the space of functions defined on $[0,1]$ that are right continuous and possess left-hand limits at each point. Let $F^{-1}$ denote the left-continuous inverse of $F$, then $F^{-1}(U)$ has distribution $F$, where $U$ is a random variable uniformly distributed on $(0,1)$.*

*This allows us to investigate only $\sqrt{n}[G_n(t) - t]$ where $G_n$ is the empirical distribution function associated to a random sample from $U$, since the sequence $(\sqrt{n}[F_n(t) - F(t)], t \in [0,1])$ has identical probabilistic behavior with the sequence $(\sqrt{n}[G_n(t) - t],\ t \in [0,1])$, i.e., they have the same distribution. Using weak convergence on metric spaces, it can be shown that the "uniform" empirical process $\sqrt{n}[G_n(t) - t]$ converges in distribution to a Brownian bridge (using multivariate CLT). For details, see, e.g., [146].*

## 6.4.2 Estimating probability density functions

Let $F$ be the distribution function of $X$. When $F$ is absolutely continuous, then an equivalent way to represent the random evolution of $X$ is through its probability density function $f$ which is the almost everywhere (a.e.) derivative of $F$. This is so by *the fundamental theorem of calculus,* namely if $F : \mathbb{R} \to \mathbb{R}$ is absolutely continuous, then it is differentiable a.e. and it can be recovered from its a.e. derivative $f$ by the formula

$$F(x) = \int_{-\infty}^{x} f(y)dy, \text{ for any } x \in \mathbb{R}.$$

Now the density function $f(x) = dF(x)/dx$ has the following properties.

(a) $f \geq 0$, since it is the derivative of a nondecreasing function $F$.

(b) $\int_{-\infty}^{\infty} f(y)dy = 1$, since $F(\infty) = 1$.

If you look at the familiar normal density

$$f(x) = (1/\sqrt{2\pi})e^{-x^2/2},$$

you see that, for example, $f(1) \neq 0$ where as $P(X = 1) = 0$, i.e., $f(x) \neq P(X = x)$, since

$$P(X \in [a,b]) = F(b) - F(a) = \int_{a}^{b} f(x)dx.$$

Thus, $f(x)$ does not have a clear probabilistic meaning. It is a device to obtain probabilities via integration. Just like distribution functions, density functions can be also defined axiomatically as any (measurable) function $f$ satisfying (a) and (b) above. This facilitates a lot the modeling of random laws. Specifically, any function $f$ satisfying (a) and (b) can be used as a density for some random variable.

If we know $F$ then of course we can derive its density $f$. But if we do not know $F$ (as in all real-world problems) and only have at our disposition data concerning it, say, a random sample $X_1, X_2, ..., X_n$ from the variable $X$ governing by $F$, then the problem is different. Indeed, in this realistic situation, we only can approximate $F$ by its estimator $F_n$. But clearly $F_n(x)$ is not continuous, let alone absolutely continuous, and hence while it has an a.e. derivative (which is identically zero), this derivative is meaningless as an estimator for $f(.)$. In other words, we cannot derive an estimator for $f$ from the estimator $F_n$ for $F$ by simply taking the derivative of $F_n$.

But why bother about $f$ in the first place, when we have $F$ or its $F_n$? Of course, you can run into many practical situations where $f$ appears in economic analyses, such as in the example on coarse data in auctions where the need to estimate probability density functions is obvious. For completeness in the discussions of that example, let us now say a few words about games.

We present now the most popular method for estimating probability density functions. The model is $\mathcal{F}$, the class of all absolutely continuous distribution functions. Our random variable of interest $X$ has a distribution function $F$ known only to belong to $\mathcal{F}$. There are many different ways to estimate $f(x) = dF(x)/dx$. Here we will only discuss the most popular one, namely, the *kernel method*.

Also, we will only consider pointwise estimation, i.e., estimate the value $f(x)$ of $f(.)$ at each specified point $x$, and not globally, for lack of space. Unlike the pointwise estimation of $F(x)$ (by $F_n(x)$), $f(x)$ (as a parametric parameter $\theta(F) = f(x)$, for a fixed $x$) is *not estimable*, i.e., we cannot find an unbiased estimator for it. However, we should seek asymptotically unbiased estimators for the sake of consistency.

In descriptive statistics, a *histogram* is a graphical display of raw data in the form of a probability density function. It is in fact the very first nonparametric estimator of a probability density function. The construction of a histogram from $X_1, X_2, ..., X_n$ goes as follows. Given an origin $x_0$, and a bin-width $h$, the bins of a histogram are intervals $[x_0 + mh, x_0 + (m + 1)h)$ for positive and negative integers $m$. The *in-*

*terval length* is $h$. The histogram estimator is taken as

$$\hat{f}_n(x) = (1/nh)(\#\{X_i : X_i \in \text{same bin as } x\}),$$

where $\#\{X_i : X_i \in \text{same bin as } x\}$ is the number of the observations $X_i$ which fall into the same interval containing $x$. Clearly we are considering pointwise estimation here.

There are several drawbacks of a histogram: discontinuity of $\hat{f}_n(.)$, the choice of $x_0$ can affect the shape, inefficient usage of data, presentation difficulty in multivariate estimation. A more efficient use of data, as far as density estimation is concerned, is this. Instead of "drawing" a histogram first, then estimating the value $f(x)$ by $\hat{f}_n(x)$, we could first consider an $x$, then look at the observations falling "around" it. This idea turns out to be consistent with the very meaning of $f(x)$ as the derivative of $F(.)$ at $x$.

Specifically, as stated above, while $f(x)$ does not have a probabilistic meaning, it is so "asymptotically." Indeed, we have

$$\begin{aligned} f(x) &= \lim_{h \to 0} (F(x+h) - F(x-h))/2h \\ &= \lim_{h \to 0} P(x - h < X \le x + h)/2h. \end{aligned}$$

The sample counterpart of $P(x - h < X \le x + h)/2h$ is

$$f_n(x) = (1/2nh)\#\{X_i \in (x - h, x + h]\}$$

which is the proportion of the observations falling into the interval $(x - h, x + h]$.

Now, observe that $f_n(x)$ can be also written as

$$(1/nh) \sum_{i=1}^{n} K[(x - X_i)/h]$$

where the *kernel* $K$ is

$$K(x) = \begin{cases} 1/2 & \text{if } x \in [-1, 1) \\ 0 & \text{otherwise.} \end{cases}$$

This kernel estimator is called the *naive estimator* of $f(x)$. The kernel is also referred to as a "window" (open around the point $x$).

The above "naive" kernel is a uniform kernel with the weight (degree of importance) $1/2$ assigned to each observation $X_i$ in the window around $x$. This weight assignment is not sufficiently efficient in the sense that observations closer to $x$ should receive higher weights than those far away. To achieve this, we could modify the kernel accordingly. For example, a kernel of the form

$$K(x) = (3/4)(1 - x^2)1_{(|x| \leq 1)}$$

could reflect an appropriate weight assignment.

In any case, the general form for kernel density estimators should be

$$f_n(x) = (1/nh) \sum_{i=1}^{n} K[(x - X_i)/h]$$

for some choice of kernel $K$, i.e., $K$ is a probability density function, i.e., $K(.) \geq 0$ and $\int_{-\infty}^{\infty} K(x)dx = 1$. Note that the naive kernel satisfies the conditions of a probability density function. Now we have a general form for density estimators, we could proceed ahead to *design* them to obtain "good" estimators. From the above form, we see clearly that estimators' performance depend on the design of the bandwidth $h$ and the kernel $K$. Thus, from a practical viewpoint, the choice of the bandwidth $h$ is crucial since $h$ controls the smoothness of the estimator (just like histogram). Also, as we will see, the smoothness of kernel estimators depends on properties of the kernel $K$.

Here is the analysis leading to optimal design of density estimators.

**Analysis of the bias**    The bias of $f_n(x)$ is $b(f_n(x)) = Ef_n(x) - f(x)$. The mean-squared error (MSE) is

$$E[f_n(x) - f(x)]^2 = Var(f_n) - b^2(f_n(x)).$$

Assuming that $f$ is sufficiently smooth, such as $f''$ exists, we get

$$b(f_n(x)) = (h^2/n)f''(x) \int_{-\infty}^{\infty} y^2 K(y)dy + o(h^2) \text{ as } h \to 0$$

where $o(h^2)$ is a function such that $\lim_{h \to 0} o(h^2)/h^2 = 0$.

Thus we need to choose $K$ such that $\int_{-\infty}^{\infty} y^2 K(y)dy < \infty$. On the other hand, to make $h \to 0$, we choose $h_n$ (a function of $n$) so that $h_n \to 0$ when $n \to \infty$.

Looking at the bias, we see that it is proportional to $h^2$. Thus to reduce the bias, we should choose $h$ small. Also, the bias depends on $f''(x)$, i.e., the curvature of $f(.)$.

**Analysis of the variance** We have

$$Var(f_n(x)) = (1/nh) \int_{-\infty}^{\infty} K^2(y)dy + o(1/nh), \text{ as } nh \to \infty.$$

The variance is proportional to $(nh)^{-1}$. Thus, to reduce the variance, we should choose $h$ large! Also the variance increases with $\int_{-\infty}^{\infty} K^2(y)dy$: flat kernels should reduce the variance. Of course, increasing the sample size $n$ reduces the variance.

How can we balance the choice of $h$ for reducing bias and variance? Note that increasing $h$ will lower the variance but increase the bias and vice versa. A compromise is to minimize the MSE. Now,

$$
\begin{aligned}
MSE(f_n(x)) &= (h^4/4)[f''(x)]^2 \left[ \int_{-\infty}^{\infty} y^2 K(y)dy \right]^2 \\
&+ (1/nh) \left[ \int_{-\infty}^{\infty} K^2(y)dy \right] f(x) + o(1/nh)
\end{aligned}
$$

as $nh \to \infty$. Thus, $MSE(f_n(x)) \to 0$, as $n \to \infty$ (i.e., $f_n(x)$ is MSE-consistent, and hence weakly consistent) when $h_n \to 0$ and $nh_n \to \infty$, as $n \to \infty$. The optimal choice of bandwidth is $h_n = n^{-1/5}$ and the convergence rate is $n^{-4/5}$. All of the above could give you a "flavor" of designing kernel density estimators!

To summarize general results on asymptotic properties of kernel estimators, we list the following: Under suitable choices of $K$ as indicated above, and with $f$ continuous,

(a) $f_n(x)$ is asymptotically unbiased when $h_n \to \infty$, as $n \to \infty$,

(b) $f_n(x)$ is weakly consistent for $f(x)$ provided $h_n \to 0$ and $nh_n \to \infty$.

For nonparametric estimation when data are stochastic processes, see a text like [132].

## 6.4.3 Method of excess mass

We present here another estimation method which seems suitable for estimating density functions where qualitative information, rather than analytic assumptions about the model, is available.

The kernel method [136] for density estimation is only one among a variety of other methods, such as orthogonal functions or excess mass. It is a popular approach. We illustrate briefly the recent method of *excess mass* which seems not to be well-known in econometrics. This method has the advantage that we do not need to assume analytic conditions on $f$ but only some information about its shape.

For each $\alpha \geq 0$, a cross section of $f$ (say, in the multivariate case) at level $\alpha$ is

$$A_\alpha(f) = \{x \in \mathbb{R}^d : f(x) \geq \alpha\}.$$

A piece of information about the shape of $f$ could be $A_\alpha \in \mathcal{C}$, some specified class of geometric objects, such as ellipsoids (e.g., multivariate normal).

Now observe that $f$ can be recovered from the $A_\alpha$'s as

$$f(x) = \int_0^\infty 1_{A_\alpha}(x)d\alpha,$$

so that it suffices to estimate the *sets* $A_\alpha$, $\alpha \geq 0$ by some *set statistics* $A_{\alpha,n}$ (i.e., some *random set statistics*) and use the plug-in estimator

$$f_n(x) = \int_0^\infty 1_{A_{\alpha,n}}(x)d\alpha$$

to obtain an estimator for $f(x)$. But of course, the question is how to obtain $A_{\alpha,n}$ from the sample $X_1, X_2, ..., X_n$. Here is the idea of J. A. Hartigan [69]. Let $\lambda(dx)$ denote the Lebesgue measure on $\mathbb{R}^d$. Then

$$\mathcal{E}_\alpha(A) = (dF - \alpha\lambda)(A)$$

is the *excess mass* of the set $A$ at level $\alpha$. Note that $(dF - \alpha\lambda)$ is a *signed measure*.

**Theorem 6.10** $A_\alpha$ *maximizes* $\mathcal{E}_\alpha(A)$ *over* $A \in \mathcal{B}(\mathbb{R}^d)$.

**Proof.** For each $A \in \mathcal{B}(\mathbb{R}^d)$, write $A = (A \cap A_\alpha) \cup (A \cap A_\alpha^c)$, where $A_\alpha^c$ is the set complement of $A_\alpha$. Then

$$\mathcal{E}_\alpha(A) = \int_{A \cap A_\alpha} (f(x) - \alpha)dx + \int_{A \cap A_\alpha^c} (f(x) - \alpha)dx.$$

Now, on $A \cap A_\alpha$, $f(x) - \alpha \geq 0$, so that

$$\int_{A \cap A_\alpha} (f(x) - \alpha)dx \leq \int_{A_\alpha} (f(x) - \alpha)dx \qquad (A \cap A_\alpha \subseteq A_\alpha).$$

On $A \cap A_\alpha^c$, $f(x) - \alpha \leq 0$. Thus,

$$\int_{A \cap A_\alpha} (f(x) - \alpha)dx + \int_{A \cap A_\alpha^c} (f(x) - \alpha)dx \leq \int_{A \cap A_\alpha} (f(x) - \alpha)dx \leq \mathcal{E}_\alpha(A).$$

∎

Just like MLE, this maximization result suggests a method for estimating $A_\alpha(f)$. The empirical counterpart of $\mathcal{E}_\alpha(A)$ is

$$\mathcal{E}_{\alpha,n}(A) = (dF_n - \alpha\lambda)(A).$$

Thus, a plausible estimator of the $\alpha$-level set $A_\alpha(f)$ is the random set $A_{\alpha,n}$ maximizing $\mathcal{E}_{\alpha,n}(A)$ over $A \in \mathcal{C}$.

While the principle is simple, the rest is not! Some questions of interest are: How do we optimize a *set-function*? That is, is there a *variational calculus for set-functions?* How to formulate the concept of *random set statistics*? Remember: our parameter is a set $(A_\alpha)$, so its estimator also has to be a set. How do we measure the performance of random set estimators? How about *convergence* concepts for random sets for large sample properties of estimators? How do we formulate limiting distributions for random set estimators? Of course, we will not discuss these issues in this text!

### 6.4.4 Nonparametric regression

We discuss now the formulation of nonparametric regression using the kernel method. Noted that the extension of kernel method from one dimension to several dimensions (for multivariate distributions of random vectors) is straightforward.

Recall that a regression takes the form

$$Y = g(X) + \varepsilon$$

where the unknown function $g(.)$ is left unspecified. It is a nonparametric model for the conditional distribution of $Y$ given $X$, and as such, it requires nonparametric estimation. Nonparametric regression analysis traces the dependence of the response variable $Y$ on one or several predictors without specifying in advance the function that relates the response to the predictors. Nonparametric regression distinguishes from linear and traditional nonlinear regression in directly estimating the regression function $E(Y|X)$ rather than estimating the parameters of the models. The nonparametric estimation of regression function is this.

Let $(X_i, Y_i), i = 1, 2, ..., n$ be a random sample from a bivariate distribution with joint density $f(x, y)$. The marginal density of $X$ is

$$f_X(x) = \int_{-\infty}^{\infty} f(x, y)dy$$

and the conditional density of $Y$ given $X = x$ is

$$f_{Y|X}(y|x) = f(x,y)/f_X(x).$$

The conditional mean (or regression of $Y$ on $X$) is

$$\begin{aligned}
r(x) &= E(Y|X = x) \\
&= \int_{-\infty}^{\infty} y f_{Y|X}(y|x) dy \\
&= \int_{-\infty}^{\infty} y f(y,x)/f_X(x) dy.
\end{aligned}$$

Using kernel estimation for both densities $f(x,y)$ and $f(x)$, we arrive at the following. The kernel estimator of $f(x,y)$ is of the form

$$f_n(x,y) = (nh_n^2)^{-1} \sum_{i=1}^{n} K[(x - X_i)/h_n, (y - Y_i)/h_n]$$

and that of $f(x)$ is

$$\hat{f}_n(x) = (nh_n)^{-1} \sum_{i=1}^{n} J\left(\frac{x - X_i}{h_n}\right).$$

Note that $\hat{f}_n(x) = \int_{-\infty}^{\infty} f_n(x,y) dy$.

From the above, a kernel estimator of $r(x)$ is

$$r_n(x) = \int_{-\infty}^{\infty} f_n(x,y)/\hat{f}_n(x) dy.$$

For simplicity, we can take

$$K(x,y) = J(x)J(y)$$

and arrive at

$$r_n(x) = \sum_{i=1}^{n} Y_i J[(x - X_i)/h_n] / \sum_{i=1}^{n} J[(x - X_i)/h_n].$$

**Remark 6.11** *Nonparametric kernel regression can be used to estimate risk measures such as $TVaR_\alpha$ or $CVaR_\alpha$ as they are expectations conditional upon truncated variables.*

## 6.4.5   Risk estimation

The problem of estimating risks of financial positions can be formulated as follows. Let $X$ be the loss variable of interest whose distribution function $F$ is unknown. We might have some additional information about $F$ such as $X$ has a finite mean, but otherwise, $F$ is unspecified, i.e., we face estimation in a nonparametric model. Let $\rho(X)$ be a risk measure. The only unknown quantity in $\rho(X)$ is $F$. To be specific, consider the estimation of a Choquet integral with respect to a capacity generated by a distortion function $g$, namely

$$\rho(X) = \rho(F) = \int_0^\infty g(1 - F(t))dt.$$

Thus, $\rho(X)$ is a parameter, i.e., a function of $F$. As such, a natural way to estimate $\rho(F)$ is to estimate $F(.)$ by some estimator $F_n(.)$ and use the plug-in estimator $\rho(F_n)$.

In applications, the data related to $X$ can be coarse in various forms. Here we consider the simplest situation of random samples to bring out basic aspects of the estimation problem. Thus, we assume that we have at our disposal a sequence of i.i.d. observations $X_n$, $n \geq 1$. Recall that a natural pointwise estimator of $F(x)$ is the empirical distribution function, based on $X_1, X_2, ...X_n$,

$$F_n(x) = (1/n) \sum_{i=1}^n 1_{(X_i \leq x)}.$$

For $\alpha \in (0,1)$, when $g_\alpha(x) = 1_{(1-\alpha,1]}(x)$, $\rho(X) = VaR_\alpha(X) = F^{-1}(\alpha)$, the plug-in estimator of $VaR_\alpha(X)$ is

$$F_n^{-1}(\alpha) = X_{(k)} \text{ if } (k-1)/n < \alpha \leq k/n,$$

where $X_{(k)}$ is the $k$th-order statistic, $k = 1, 2, ..., n$.

**Remark 6.12** *When we can specify that the nonparametric model of $F$ is the class of absolutely continuous distribution functions, we can estimate $F$ with smooth estimators by using nonparametric estimation of density functions, via, say, the kernel method.*

$$\hat{F}_n(x) = \int_{-\infty}^x f_n(y)dy = \int_{-\theta}^x [(1/nh_n) \sum_{i=1}^n K((y - X_i)/h_n)]dy$$

$$= (1/n) \sum_{i=1}^n \hat{K}((x - X_i)/h_n),$$

*where*

$$\hat{K}(x) = \int_{-\infty}^{x} K(y)dy.$$

This is basically the problem of quantile estimation where large sample theory is available. In general, the estimation of statistical functionals $\rho(F)$ can be carried out in the setting of $V$-statistics by looking at $\rho(F_n)$ or in that of $L$-estimates (linear combinations of order statistics), see, e.g., Serfling [141]. This can be achieved by a natural transformation of distorted risk measures to spectral risk measures. Recall that, a spectral risk measure $\rho_\varphi(.)$, with respect to the weighting function (spectrum) $\varphi$ is

$$\rho_\varphi(X) = \int_0^1 \varphi(t)F^{-1}(t)dt,$$

where $\varphi : [0,1] \to \mathbb{R}^+$ is nondecreasing and $\int_0^1 \varphi(t)dt = 1$.

The plug-in estimator of the spectral risk measure $\int_0^1 \varphi(t)F^{-1}(t)dt$ is

$$\int_0^1 \varphi(t)F_n^{-1}(t)dt = \sum_{i=1}^{n} \left[ \int_{(i-1)/n}^{i/n} \varphi(t)dt \right] X_{(i)},$$

which is a linear combination of the order statistics. Thus, it is of interest to transform distorted probability risk measures to spectral risk measure representations (weighted averages of quantile functions).

We are going to show the following fact.

**Theorem 6.13** *If $\rho(X) = C_v(X) = \int_0^\infty v(X > t)dt$ where $v = g \circ P$, with $g$ concave, then*

$$\int_0^\infty g(1 - F_X(t))dt = \int_0^1 g'(1-s)F_X^{-1}(s)ds, \qquad (6.4)$$

*where $g'(.)$ is the derivative of $g$ on $(0,1)$, except possibly on an at-most-countable subset of it.*

Equation (6.4) is a representation of a distorted probability risk measure $\rho(X)$ as a weighted average of the quantile function of its argument $(X)$, so that its estimator is a linear combination of order statistics, namely

$$\rho_n(F) = \sum_{i=1}^{n} X_{(i)} \left[ \int_{(i-1)/n}^{i/n} g'(1-s)ds \right].$$

What is interesting to observe is that in (6.4), the function $\varphi(t) = g'(1-t)$ on $[0, 1]$ is in fact a spectrum so that the right-hand side of (6.4) is a spectral risk measure of $X$. In other words, the theorem says that every distorted probability risk measure (a special form of Choquet integral risk measures) with a concave distortion function is a spectral risk measure.

While it is possible to show the above theorem by integration by parts, we choose to prove it by a more elegant way, and by doing so we obtain the converse, namely that every spectral risk measure is a Choquet integral risk measure (in fact, a distorted probability risk measure with a concave distortion function). Thus, the class of risk measures as Choquet integral is very large.

Thus, let's prove first the following lemma.

**Lemma 6.14** *Let $\rho_\varphi(X) = \int_0^1 \varphi(t) F_X^{-1}(t) dt$ be a spectral risk measure. Then $\rho_\varphi(X) = C_v(X)$, where the capacity $v = g \circ P$ with $g(t) = 1 - \int_0^{1-t} \varphi(s) ds$ (so that $g'(t) = \varphi(1 - t)$).*

**Proof.** To show that spectral risk measures have Choquet integral representations, we use Schmeidler's theorem, noting that the boundedness of random variables can be relaxed. Since $F_{1_\Omega}^{-1}(\alpha) = 1$, for $\alpha \in (0, 1)$, we have $\rho_\varphi(1_\Omega) = \int_0^1 \varphi(\alpha) d\alpha = 1$. Next,

$$X \geq Y \Rightarrow F_X(.) \leq F_Y(.) \Leftrightarrow F_X^{-1}(.) \geq F_Y^{-1}(.),$$

and hence $\rho_\varphi(X) \geq \rho_\varphi(Y)$.

If $X$ and $Y$ are comonotonic then $F_{X+Y}^{-1} = F_X^{-1} + F_Y^{-1}$. It follows that $\rho_\varphi(X + Y) = \rho_\varphi(X) + \rho_\varphi(Y)$. Thus $\rho_\varphi(X) = C_v(X)$ where the capacity $v(.)$ is determined by $v(A) = \rho_\varphi(1_A)$.

Now,

$$F_{1_A}(x) = \begin{cases} 0 & \text{if } x < 0 \\ 1 - P(A) & \text{if } 0 \leq x < 1 \\ 1 & \text{if } x \geq 1, \end{cases}$$

so that

$$F_{1_A}^{-1}(\alpha) = \begin{cases} 0 & \text{if } 0 \leq \alpha \leq 1 - P(A) \\ 1 & \text{if } 1 - P(A) < \alpha < 1. \end{cases}$$

Hence,

$$\begin{aligned} v(A) &= \rho_\varphi(1_A) = \int_0^1 \varphi(\alpha) F_{1_A}^{-1}(\alpha) d\alpha \\ &= \int_{1-P(A)}^1 \varphi(\alpha) d\alpha = 1 - \int_0^{1-P(A)} \varphi(\alpha) d\alpha. \end{aligned}$$

For $g(t) = 1 - \int_0^{1-t} \varphi(\alpha) d\alpha$, we have, $\forall A \in \mathcal{A}$,

$$g(P(A)) = 1 - \int_0^{1-P(A)} \varphi(\alpha) d\alpha = v(A).$$

■

**Remark 6.15** *Since $C_{g \circ P}(.)$, with $g$ concave, is a coherent risk measure, it follows that spectral risk measures are coherent.*

**Proof.** (of the Theorem) Since $g$ is concave, it is differentiable on $(0, 1)$, except possibly on an at most countable subset $D_g$ of $(0, 1)$. Define $\varphi : [0, 1] \to [0, 1]$ as

$$\varphi(s) = \begin{cases} g'(1 - s) & \text{if } 1 - s \notin D_g \\ a_s \geq 0 & \text{if } 1 - s \in D_g. \end{cases}$$

Then $\varphi(.)$ is a spectrum. Indeed, $\varphi(.) \geq 0$ since $g$ is nondecreasing, and

$$\begin{aligned} \int_0^1 \varphi(s) ds &= \int_0^1 g'(1 - s) ds = \int_0^1 g'(u) du \\ &= g(1) - g(0) = 1 - 0 = 1. \end{aligned}$$

Thus taking $\varphi(.)$ as a spectrum, we apply the above lemma:

$$\int_0^1 \varphi(s) F_X^{-1}(s) ds = \int_0^\infty h(1 - F_X(t)) dt,$$

where

$$h(t) = 1 - \int_0^{1-t} \varphi(s) ds = 1 - \int_0^{1-t} g'(1 - s) ds.$$

But

$$\int_0^{1-t} g'(1 - s) ds = \int_t^1 g'(u) du = g(1) - g(t) = 1 - g(t),$$

so that $h(.) = g(.)$, and hence (6.4). ■

## 6.5 Basics of hypothesis testing

Consider the situation where we wish to use empirical evidence to assess whether $F(.) \leq G(.)$, where $F, G$ are unknown distribution functions of $X, Y$, respectively. Our evidence consists of random samples $X_1, X_2, ..., X_n$ ("drawn" from $X$) and $Y_1, Y_2, ..., Y_m$ (from $Y$). Here is a quick tutorial on testing statistical hypotheses. First, let's give a name to what we want to test: $F(.) \leq G(.)$ is called a hypothesis. Since we are going to test this hypothesis against its negation, we are in fact considering two hypotheses: $F(.) \leq G(.)$ and its negation "$F(.)$ is not greater than $G(.)$." We call the hypothesis $F(.) \leq G(.)$ the null hypothesis and denote it as $H_0$ ("null" for "no effect," as statistical testing was traditionally originated for treatment differences), and its negation hypothesis as the alternative hypothesis, denoted as $H_a$.

More formally, in the above example, let $\Theta$ denote the parameter space consisting of all pairs $(F, G)$ of distributions. $\Theta$ is partitioned into $\Theta_0 = \{(F, G) : F(.) \leq G(.)\}$ and its set complement $\Theta_a = \Theta \backslash \Theta_0$. The elements of $\Theta$ are denoted as $\theta$. Thus

$$H_0 : \theta \in \Theta_0, H_a : \theta \in \Theta_a.$$

We will use data to decide between $H_0$ and $H_a$. How? That will be the heart of the theory of statistical testing! Assume for the moment that we have a way to do that. The outcome of our test will be either accepting $H_0$ or rejecting it. Whatever the outcome, we will commit errors of two different kinds: rejecting $H_0$ when in fact it is true (called type I error) and accepting $H_0$ when $H_a$ is true (called type II error). Our testing procedure will be probabilistic in nature since we use random samples to form our testing strategy. As such, the above two types of error are expressed as probabilities as follows:

$$\alpha = P(\text{reject } H_0 | H_0 \text{ is true}) = \text{type I error,}$$
$$\beta = P(\text{accept } H_0 | H_0 \text{ is false}) = \text{type II error.}$$

An ideal test procedure is the one that minimizes these errors. Unfortunately, since these errors behave in opposite directions, no such ideal test procedure is possible. The Neyman-Pearson approach (1933) is to treat these hypotheses asymmetrically. We fix a level $\alpha$, and look for testing procedures which have smallest $\beta$. As we will see, such testing procedures are possible and that is the best that we can do. Thus, testing is a matter of degree! For example, after a test at level $\alpha = 0.05$, we accept that $F(.) \leq G(.)$ since the data support that hypothesis. We proceed to declare that $X \succeq_1 Y$, then make economic decisions based on that, with the "confidence" that we can only be wrong 5% of the time.

Since we use data to make decisions in the above testing problem, any testing procedure (i.e., a way to decide which hypothesis we hold as true) should be a statistic (i.e., a function of the data). Continued with our example, let $\mathbb{R}^{n+m}$ be our sample space. A testing procedure is a way to decide which hypothesis is likely to be true after observing a sample point (our actual data). Thus, a testing procedure will partition the sample space into a rejection region (or critical region) $R \subseteq \mathbb{R}^{n+m}$ (if our data point belongs to $R$, we reject $H_0$, otherwise we do not reject $H_0$) and its set complement. Since a set $A$ is equivalent to its indicator function $1_A$, a *test* is a random variable (in fact a statistic) $T_{n+m}$ based on $X_1, X_2, ..., X_n, Y_1, Y_2, ..., Y_m$, with values in $\{0, 1\}$, where $T_{n+m} = 1$ means we reject $H_0$, in other words $T_{n+m} = 1_R$. Note that we are talking about *nonrandomized* tests here. As we will see, $T_{n+m}$ will be constructed from some *test statistic*, i.e., some statistic with the view to accept or reject $H_0$.

The parametric setting of testing is this. The random variable or vector of interest is $X$ whose distribution is $F_\theta$ where the true parameter $\theta_0$ is only known to lie in a (finite dimensional) parameter space $\Theta$. Let $\Theta = \Theta_0 \cup \Theta_a$ with $\Theta_0 \cap \Theta_a = \varnothing$. A test $H_0$ against $H_a$ at the $\alpha$ level of significance, where

$$H_0 : \theta \in \Theta_0, H_a : \theta \in \Theta_a$$

is a test with rejection region $R$ such that

$$\sup_{\theta \in \Theta_0} P(\tilde{X} \in R | \theta) \leq \alpha$$

where $\tilde{X}$ denotes the observation.

For given $\alpha \in (0, 1)$, a test at that level will have its $\beta$ as its type II error. We seek the one with smallest $\beta$. To give a good name for such a test (when it exists), we observe that $\beta$ is smallest when $1 - \beta$ is largest! The quantity $1 - \beta$ of that test is called the power of the test. Thus, the best test is called the most powerful test (a test with highest power among all tests at level $\alpha$). A test is more powerful than another one if it has a higher power. The *power function* of a test with critical region $R$ is defined to be

$$\pi(\theta) = P_\theta(\tilde{X} \in R)$$

where $\tilde{X}$ denotes our data point

$$(X_1, X_2, ..., X_n, Y_1, Y_2, ..., Y_m).$$

Before discussing other requirements for tests for large samples, like in estimation, let's show that it is possible to carry out the above program.

## 6.5.1 The Neyman-Pearson lemma

Consider now the simplest situation where $H_0$ completely determines the probability law $P_0$ of the observable $X$, i.e., $H_0$ is a *simple hypothesis*. The alternative hypothesis $H_a$ is also assumed to be simple: under $H_a$, the probability measure $P_a$ of $X$ is completely determined. Furthermore, we assumed that these probability measures are absolutely continuous with respect to some common measure $\lambda(dx)$ on $\mathcal{B}(\mathcal{X})$, so that $P_0(dx) = f_0(x)d\lambda(x)$, $P_a(dx) = f_a(x)d\lambda(x)$, i.e., $f_0$ and $f_a$ are probability density functions of $X$ under $H_0$, $H_a$, respectively.

Let's specify the errors $\alpha, \beta$ for a test at level $\alpha$. As stated above, a test is a rule for choosing one of $H_0$ and $H_a$. For a general setting, let $X$ denote our observation data point in the sample space $\mathcal{X}$ and let $R \subseteq \mathcal{X}$ be the rejection region of a test. Then

$$\begin{aligned} \alpha &= P(X \in R|H_0) \\ \beta &= P(X \notin R|H_a) \end{aligned}$$

where $P(X \in R|H_0)$ is the probability that $X \subset R$ when $H_0$ is true.

From the above, it is clear that a decrease in $\alpha$ can be accomplished only by a "decrease" in $R$, accompanied by an "increase" in $R^c$ (the set complement of $R$) and $\beta$. Thus, there is no way to decrease $\alpha$ and $\beta$ simultaneously with a fixed sample space $(\mathcal{X}, \mathcal{B}(\mathcal{X}))$, a measurable space such as $(\mathbb{R}^n, \mathcal{B}(\mathbb{R}^n))$.

For fixed $\alpha$, we seek $R$ (i.e., a test) to minimize the size of $\beta$. The minimization problem is this. Find $R_*$ in

$$\mathcal{C} = \{R \in \mathcal{B}(\mathcal{X}) : \int_R f_0(x)d\lambda(x) \le \alpha\}$$

so that

$$\int_{R_*} f_0(x)d\lambda(x) - \alpha$$

and

$$\int_{R_*^c} f_a(x)d\lambda(x) \le \int_{R^c} f_a(x)d\lambda(x) \text{ for all } R \in \mathcal{C}.$$

Note that the last inequality is equivalent to

$$\int_{R_*} f_a(x)d\lambda(x) \ge \int_R f_a(x)d\lambda(x).$$

If we take $R_*$ as a critical region, then

$$\alpha = \int_{R_*} f_0(x) d\lambda(x)$$

and $1 - \beta = \int_{R_*} f_a(x) d\lambda(x)$ is "as great as can be." In this relative sense, $R_*$ is the best or most powerful critical region.

The above optimization problem, just like the situation for density estimation using the excess mass approach, is somewhat unconventional, in the sense that the variable is a set, i.e., we are facing an *optimization problem of set functions*, where an appropriate variational calculus is not available!

**Lemma 6.16 (Neyman-Pearson)** *Let $c > 0$ be a constant such that*

$$R_* = \{x \in \mathcal{X} : f_a(x) > c f_0(x)\}$$

*has* $\int_{R_*} f_0(x) d\lambda(x) = \alpha$. *Then,*

$$\int_{R_*} f_a(x) d\lambda(x) \geq \int_R f_a(x) d\lambda(x)$$

*for all* $R \in \mathcal{B}(\mathcal{X})$ *such that* $\int_R f_0(x) d\lambda(x) \leq \alpha$.

**Proof.** We have

$$\int_{R_*} f_a(x) d\lambda(x) - \int_R f_a(x) d\lambda(x)$$

$$= \int_{R_* \cap R} f_a(x) d\lambda(x) + \int_{R_* \cap R^c} f_a(x) d\lambda(x)$$

$$- \int_{R_* \cap R} f_a(x) d\lambda(x) - \int_{R_*^c \cap R} f_a(x) d\lambda(x)$$

$$= \int_{R_* \cap R^c} f_a(x) d\lambda(x) - \int_{R_*^c \cap R} f_a(x) d\lambda(x).$$

Now, on $R_*$, $f_a(.) > c f_0(.)$ so that

$$\int_{R_* \cap R^c} f_a(x) d\lambda(x) \geq c \int_{R_* \cap R^c} f_0(x) d\lambda(x).$$

On $R_*^c$, $f_a(.) \leq c f_0(.)$ so that

$$- \int_{R_*^c \cap R} f_a(x) d\lambda(x) \geq -c \int_{R_*^c \cap R} f_0(x) d\lambda(x).$$

Thus,

$$\int_{R_*} f_a(x)d\lambda(x) - \int_{R} f_a(x)d\lambda(x)$$

$$\geq c \left[ \int_{R_* \cap R^c} f_0(x)d\lambda(x) - \int_{R_*^c \cap R} f_0(x)d\lambda(x) \right],$$

but the right-hand side is equal to

$$c \left[ \int_{R_*} f_0(x)d\lambda(x) - \int_{R} f_0(x)d\lambda(x) \right] \geq c(\alpha - \alpha) = 0.$$

It follows that $\int_{R_*} f_a(x)d\lambda(x) \geq \int_{R} f_a(x)d\lambda(x)$, as desired.  ∎

**Remark 6.17** *The lemma says that the proposed $R_*$ is the solution to the set function minimization problem but says nothing about whether such constant $c > 0$ actually exists, the hypothesis of the lemma is a sufficient condition for the integral inequalities to hold.*

Of course, in applications we often face *composite hypotheses*, i.e., not simple, but the Neyman-Pearson lemma can be used in many cases to obtain uniformly most powerful tests.

## 6.5.2  Consistent tests

As in estimation, we expect "good" tests to be consistent with the information principle in the sense that as we get more data (the sample size $n \to \infty$), the error of type II (accepting the wrong hypothesis) should decrease to zero, or equivalently, their powers should approach one.

The set-up for both nonparametric and parametric models is this. Let $X$ be a random element, defined on a probability space $(\Omega, \mathcal{A}, P)$, with values in a measurable space $(\mathcal{X}, \mathcal{B}(\mathcal{X}))$. Its probability law is a probability $Q = PX^{-1}$ on $\mathcal{B}(\mathcal{X})$. The model for $X$ is a class of probability measures $\mathbb{P}$ on $\mathcal{B}(\mathcal{X})$ which we call the model parameter space. For example, $\mathbb{P}$ is an infinitely dimensional space of probability measures such as the space of all absolutely continuous distribution functions, or more generally, some space of probability measures on a measurable space.

Let $\{\mathbb{P}_0, \mathbb{P}_a\}$ be a partition of $\mathbb{P}$. We wish to test

$$H_0 : Q \in \mathbb{P}_0 \text{ against } H_a : Q \in \mathbb{P}_a = \mathbb{P} \backslash \mathbb{P}_0$$

based on a random sample $X_1, X_2, ..., X_n$ observed from $X$. A test is said to be at *level of significance* $\alpha$ if

$$\sup_{Q \in \mathbb{P}_0} P(\text{reject } H_0 | Q \in \mathbb{P}_0) \leq \alpha.$$

The power function of a test $T_n(X_1, X_2, ..., X_n)$ is

$$\pi_{T_n} : \mathbb{P} \to [0, 1]$$

where $\pi_{T_n}(Q) = P(\text{reject } H_0 | Q)$.

As in estimation, to formulate the concept of consistency of tests, we consider a sequence of tests $T_n(X_1, X_2, ..., X_n)$, $n \geq 1$, all at the $\alpha$-level of significance.

**Definition 6.18** *The test $T_n$, based on a sample $(X_1, X_2, ..., X_n)$ from $X$ whose probability law $Q \in \mathbb{P}$ for testing $H_0 : Q \in \mathbb{P}_0$ against $H_a : Q \in \mathbb{P}_a = \mathbb{P} \backslash \mathbb{P}_a$, at level $\alpha$, is said to be* consistent *if $\lim_{n \to \infty} \pi_{T_n}(Q) = 1$ for any $Q \in \mathbb{P}_a$.*

### 6.5.3 The Kolmogorov-Smirnov statistic

The GC theorem suggests that, under the null hypothesis, the statistic $D_n(X^{(n)}) = \sup_{t \in \mathbb{R}} |F_n(X^{(n)}) - F_0(t)|$ could be used to derive a test for $H_0$ versus $H_a$, in the sense that large values of this statistic indicate a departure from $H_0$ and hence $H_0$ should be rejected. Even for fixed sample size $n$, $D_n(X^{(n)})$ is a sort of distance between $H_0$ and the data, and as such, can be used as a test statistic, rejecting $H_0$ for large observed values of it. We need the sampling distribution of $D_n(X^{(n)})$ (under $H_0$) in order to test. For *continuous* $F_0$, this sampling distribution can be found, as follows.

If $X_1, X_2, ..., X_n$ is the initial segment of $X^\infty$, we order it $X_{(1)} \leq X_{(2)} \leq \cdots \leq X_{(n)}$ and, by convention, set $X_{(0)} = -\infty$ and, $X_{(n+1)} = +\infty$, so that

$$F_n(t) = i/n \text{ for } X_{(i)} \leq t < X_{(i+1)}$$

for $i = 0, 1, ..., n$.

Let $D_n(F) = \sup_{t \in \mathbb{R}} |F_n(t) - F(t)|$, then $D_n(F) = \max\{D_n^+(F), D_n^-(F)\}$, where

$$\begin{aligned} D_n^+(F) &= \sup_{t \in \mathbb{R}}(F_n(t) - F(t)) \\ D_n^-(F) &= \sup_{t \in \mathbb{R}}(F(t) - F_n(t)). \end{aligned}$$

Now,

$$
\begin{aligned}
D_n^+(F) &= \sup_{t \in \mathbb{R}}(F_n(t) - F(t)) \\
&= \max_{0 \le i \le n} \sup\{i/n - F(t) : X_{(i)} \le t \le X_{(i+1)}\} \\
&= \max_{0 \le i \le n}[i/n - \inf\{i/n - F(t) : X_{(i)} \le t \le X_{(i+1)}\}] \\
&= \max_{0 \le i \le n}[i/n - F(X_{(i)})].
\end{aligned}
$$

Since the $X_i$'s are i.i.d. drawn from the population $X$ with distribution $F$, the random variable $F(X)$ is uniformly distributed on $(0,1)$ when $F$ is continuous, and hence $F(X_{(i)})$ is nothing else than the $i$th-order statistic from a random sample from the uniform distribution on $(0,1)$. Thus, the distribution of $D_n^+(F)$ does not depend of $F$, i.e., the same for any continuous $F$. Specifically, we have

$$
P(D_n^+(F) \le x) = \begin{cases}
0 & \text{for } x \le 0 \\
n! \displaystyle\prod_{i=1}^{n} \int_{\max\{(n-i+1)/n-x,o\}}^{x_{n-i+2}} dx_1 dx_2 ... dx_n & \text{for } x \in (0,1) \\
1 & \text{for } x \ge 1.
\end{cases}
$$

Similarly,

$$
\begin{aligned}
D_n^-(F) &= \sup_{t \in \mathbb{R}}(F(t) - F_n(t)) \\
&= \max_{0 \le i \le n} \sup\{F(t) - i/n : X_{(i)} \le t < X_{(i+1)}\} \\
&= \max_{0 \le i \le n}[F(X_{(i+1)}) - i/n],
\end{aligned}
$$

noting that $F$ is continuous so that $F(X_{(i+1)}-) = F(X_{(i+1)})$, and its distribution is the same as that of $D_n^+(F)$ in view of symmetry. It follows that the distribution of $D_n(F) = \max\{D_n^+(F), D_n^-(F)\}$ is (with $x_{n+1} = 1$)

$$
P(D_n(F) \le x) = \begin{cases}
0 & \text{for } x \le 1/2n \\
n! \displaystyle\prod_{i=1}^{n} \int_{\max\{(n-i+1)/n-x,o\}}^{\min\{x_{n-i+2},x+(n-i)/n\}} dx_1 dx_2 ... dx_n & \\
& \text{for } x \in (1/2n, 1) \\
1 & \text{for } x \ge 1.
\end{cases}
$$

For larger sample sizes, it is "convenient" to have approximations to the sampling distributions. It turns out that a nondegenerate limiting distribution can be obtained by considering $\sqrt{n}D_n(X^{(n)})$. Specifically, see the following.

**Theorem 6.19 (Kolmogorov-Smirnov)** *Let $F$ be a continuous distribution function, then for*

$$D_n(X^{(n)}) = \sup_{t \in \mathbb{R}} |F_n(X^{(n)}) - F(t)|,$$

*with $X^\infty = (X_1, X_2, \ldots)$ where the $X_i$'s are i.i.d. $F$, we have*

$$\lim_{n \to \infty} P(\sqrt{n} D_n(X^{(n)}) \leq z) = Q(z)$$

*where*

$$Q(z) = 1 - 2 \sum_{k=1}^{\infty} (-1)^{k-1} e^{-2k^2 z^2}.$$

**Remark 6.20** *$Q(.)$ is a continuous distribution function, called the* Kolmogorov distribution. *It is tabulated for applications in the form of tail probabilities*

$$1 - Q(z) = 2 \sum_{k=1}^{\infty} (-1)^{k-1} e^{-2k^2 z^2}.$$

The *Kolmogorov-Smirnov (KS) test* at level $\alpha$ rejects the null-hypothesis when

$$\sqrt{n} D_n(X^{(n)}) = \sqrt{n} \sup_{t \in \mathbb{R}} \left| F_n(X^{(n)}) - F_0(t) \right| > c,$$

where $c$ is determined by $1 - Q(c) = \alpha$. With the table for tail probabilities of $Q$, we make a decision (rejecting or accepting the null hypothesis) by comparing the observed value of $\sqrt{n} \sup_{t \in \mathbb{R}} |F_n(X^{(n)}) - F_0(t)|$ with tabulated quantiles. Note that the sample size $n$ should be "sufficiently large." For small sample sizes, exact sampling distributions can be obtained.

The one-sample KS test rejects $H_0 : F = F_0$ (a continuous, completely specified distribution function), in favor of $H_a : F \neq F_0$, when, for $n$ large, $\sqrt{n} D_n(F_0) > c$, where the critical point $c$ is determined by $Q(c) = 1 - \alpha$.

**Theorem 6.21** *The Kolmogorov-Smirnov test is consistent.*

**Proof.** Recall that, asymptotically, the test statistic is

$$\sqrt{n} D_n(F_0) = \sqrt{n} \sup_t |F_n(t) - F_0(t)| = \sqrt{n} d(F_n, F_0).$$

We reject $H_0$ when $\sqrt{n} D_n(F_0) > q_{n,1-\alpha}$ (the $1 - \alpha$ quantile of the distribution of $\sqrt{n} D_n$ under $H_0$). Thus the power of the test is $P(\sqrt{n} D_n(F_0) >$

$q_{n,1-\alpha}|F \neq F_0$). Now under $F$, $F_n(t) \to F(t)$, uniformly, with probability one (GC theorem), thus, almost surely, as $n \to \infty$,

$$\sup_t |F_n(t) - F_0(t)| \to d(F, F_0) > 0$$

and hence, almost surely,

$$\sqrt{n} D_n(F_0) = \sqrt{n} \sup_t |F_n(t) - F_0(t)| \to \infty.$$

And, by Slutsky's theorem,

$$P(\sqrt{n} D_n(F_0) > q_{n,1-\alpha}|F) \to 1,$$

noting that $q_{n,1-\alpha} \to q_{1-\alpha}$, the $1 - \alpha$ quantile of the Kolmogorov distribution, as $n \to \infty$. ∎

For one-sided tests,

$$H_0 \quad : \quad F = F_0 \text{ versus } H_a : F \geq F_0, F \neq F_0$$
$$\text{use } D_n^+(F_0)$$
$$H_0 \quad : \quad F = F_0 \text{ versus } H_a : F \leq F_0, F \neq F_0$$
$$\text{use } D_n^-(F_0)$$

with rejection regions $D_n^+(F_0) > c$, $D_n^-(F_0) > c$, respectively.

**Remark 6.22** $\lim_{n\to\infty} P(\sqrt{n} D_n^+(F) \leq t) = 1 - e^{-2t^2}$.

### 6.5.4 Two-sample KS tests

Let $X_1, ..., X_n$ and $Y_1, ..., Y_m$ be two independent random samples from $X, Y$ with distribution functions $F, G$, respectively.

One hypothesis of interest is (two-sided test):

$$H_0 : F = G \text{ versus } H_a : F \neq G.$$

Let $F_n$, $G_m$ denote the empirical distribution functions corresponding the above two samples. Let $D_{n,m} = \sup_{t\in\mathbb{R}} |F_n(t) - G_m(t)|$. Then

$$D_{n,m} = \sup_t |F_n(t) - F(t) + F(t) - G(t) + G(t) - G_m(t)|$$
$$\leq \sup_t |F_n(t) - F(t)| + \sup_t |F(t) - G(t)| + \sup_t |G_m(t) - G(t)|$$

and hence, by the GC theorem, the statistic $D_{n,m}$ can be used to test the above hypotheses, with large values of $D_{n,m}$ indicating a departure from the null hypothesis. Since

$$\lim_{\substack{n\to\infty \\ m\to\infty}} P(\sqrt{nm/(n+m)} D_{n,m} \leq t) = Q(t) = 1 - 2\sum_{k=1}^{\infty}(-1)^{k-1}e^{-2k^2t^2},$$

the statistic $\sqrt{nm/(n+m)}D_{n,m}$ can be used as a test in large samples.
For the one-sided test

$$H_0 \quad : \quad F \leq G$$

versus

$$H_a \quad : \quad F > G \quad \text{(at least there is } x \text{ such that } F(x) > G(x)),$$

use

$$D_{n,m}^+ = \sup_t [F_n(t) - G_m(t)].$$

For the one-sided test

$$H_0 \quad : \quad F \geq G$$

versus

$$H_a \quad : \quad F < G \quad \text{(at least there is } x \text{ such that } F(x) < G(x))$$

use

$$D_{n,m}^- = \sup_t [G_m(t) - F_n(t)].$$

## 6.5.5    Chi-squared testing

Testing about distribution functions is basically a "goodness-of-fit" test. The Kolmogorov-Smirnov (KS) test is such a (nonparametric) test. This test can be used when the underlying unknown distribution $F$ of the population $X$, from which we draw a random sample $X_1, X_2, ..., X_n$, is *continuous*, and the continuous distribution $F_0$ in the null hypothesis is *completely specified*. For example, $F_0$ is a normal distribution with *given* mean and variance, and not just a normal distribution with unknown mean and variance. For other situations, another test could be considered.

The name of the test that we are going to discuss is "chi-squared" since, as we will see, for large sample size, the sampling distribution of the test statistic will have approximately a chi-squared distribution. Historically, K. Pearson was considering a goodness-of-fit test for discrete data, as follows. Suppose we perform $n$ independent trials $X_1, X_2, ..., X_n$ (such as tossing a coin), the outcome of each trial is one of $k$ mutually exclusive events $C_j$, $j = 1, 2, ..., k$. Let $p_j = P(C_j)$ with $\sum_{j=1}^k p_j = 1$. The probability vector $\mathbb{P} = (p_1, p_2, \ldots, p_k)$ is unknown, and it is desirable to know whether $\mathbb{P} = \mathbb{P}^{(0)} = (p_1^{(0)}, p_2^{(0)}, \ldots, p_k^{(0)})$ or not. This is nothing else than testing about a multinomial distribution (Karl Pearson, 1900).

Let $\nu_j$ denote the number of outcomes falling in the "cell" $C_j$. This random variable is the (relative) frequency of observations (among $X_1, X_2,$

..., $X_n$) of cell $C_j$. Let $V = (v_1, v_2, ..., v_k)$. Then $V$ is multinomial with parameter $n$ and $\mathbb{P}$, i.e.,

$$P(v_1 = x_1, v_2 = x_2, ..., v_k = x_k) = [n!/(x_1!x_2!...x_k!]p_1^{x_1}, p_2^{x_2}...p_k^{x_k}$$

(of course for $0 \le x_j \le n$ and $\sum_{j=1}^{k} x_j = n$).

The vector of expected frequencies is $EV = (Ev_1, Ev_2, ..., Ev_k)$ where $Ev_j = np_j$. The law of large numbers tells us that, as $n \to \infty$, the statistic $v_j/n$ converges almost surely to $p_j$, for each $j$. Thus, consider the "vector of deviations" $V - n\mathbb{P}$. Let

$$T_n = \left( \frac{v_1 - np_1}{\sqrt{np_1}}, \frac{v_2 - np_2}{\sqrt{np_2}}, ..., \frac{v_k - np_k}{\sqrt{np_k}} \right)$$

where we write column vector as a row for simplicity of notation. The length of this vector is Pearson's statistic:

$$Q_n = \sum_{j=1}^{k}(v_j - np_j)^2/np_j$$

where, of course, we take $p_j = p_j^{(0)}$.

Clearly, $Q_n$ has a discrete distribution. Randomized tests are needed if exact $\alpha$-level is required. Like KS statistics, exact sampling distributions of $Q_n$ can be determined. As $n$ increases, this becomes too complicated to handle. Pearson proceeded by using an approximate value for the critical point of $Q_n$, i.e., $P(Q_n \ge c_\alpha|H_0) \approx P(\mathcal{X}_{k-1}^2 \ge \mathcal{X}_{k-1,1-\alpha}^2)$, where $\mathcal{X}_{k-1,1-\alpha}^2$ is the quantile of level $1 - \alpha$ of a chi-squared distribution with $k - 1$ degrees of freedom, i.e., $\mathcal{X}_{k-1,1-\alpha}^2$ is the root of the equation

$$\alpha = P(\mathcal{X}_{k-1}^2 \ge x)$$
$$= [1/(2^{(k-1)}/\Gamma((k-1)/2)]\int_x^\infty z^{(k-1)/2}e^{-z/2}dz.$$

Using the chi-squared approximation, the test is called a chi-squared test.

The above is extended to a general framework as follows. Again, consider testing $H_0 : F = F_0$ versus $H_a : F \ne F_0$ using a random sample $X_1, X_2, ..., X_n$ drawn from a population $X$ with unknown distribution function $F$. In general, $X$ can be either of continuous type (observations are then quantitative) or discrete (with observations which are counts or qualitative). In either case, to perform a chi-squared test, the sample data must first be grouped according to scheme in order to form a frequency distribution. The categories (or cells) arise naturally in terms of the relevant verbal or numerical classifications for categorical data. For quantitative data, the categories would be numerical classes chosen by the experimenter, so that, in a sense, *chi-squared testing remains an art.*

Assuming that $F_0$ is completely specified, the cell probabilities $C_j$, $j = 1, 2, ..., k$, can be computed, i.e., $P(X \in C_j | F_0)$, from which the "expected" frequency of cell $C_j$ is derived as $e_j = nP(X \in C_j | F_0)$. The idea is to perform a goodness-of-fit, not between $F$ (via $F_n$) and $F_0$, but between frequencies of cells under the actual observations (representing the true $F$) and under $F_0$. One way to measure the departure from $H_0$ is to consider some sort of distance between $F$ and $F_0$ via frequencies. Specifically, let $f_j$ denote the observed frequency of cell $C_j$, i.e., the number of observations falling in $C_j$. The test statistic proposed by K. Pearson in 1900 is

$$Q_n(X_1, X_2, ..., X_n) = \sum_{j=1}^{k} (f_j - e_j)^2 / e_j.$$

The plausible reasoning is this. A large value of $Q_n$ would reflect an incompatibility between the observed and expected (relative) frequencies, and therefore, $H_0$ should be rejected. This leads to one upper-tailed test. To implement this testing idea, of course, we need the distribution of $Q_n$! For even small $n$, the exact distribution of $Q_n$ is quite complicated to determine. Thus, we look for asymptotic distribution.

We provide here a theoretical *justification* for the limiting distribution of $Q_n$. It is based on the result that the likelihood ratio test converges in distribution to a chi-squared distribution. Once the sample points have been classified into cells, the only random variables of concern are the cell frequencies $v_j$ constituting a random vector, multinomial distributed with $k$ possible outcomes, the $j$th outcome being the $i$th category in the classification system. Thus, the likelihood of the $p_j$'s is $L(p_1, p_2, ..., p_k) = \prod_{j=1}^{k} p_j^{f_j}$, and noting that the maximum likelihood estimators are $f_j/n$.

The null hypothesis can be rewritten as

$$H_0 : p_j = p_j^{(0)} = e_j/n, \ j = 1, 2, ..., k.$$

The likelihood-ratio statistics for this hypothesis are

$$\begin{aligned} \lambda_n &= L(p_1^{(0)}, p_2^{(0)}, ..., p_k^{(0)}) / L(f_1/n, f_2/n, ..., f_k/n) \\ &= \prod_{j=1}^{k} [(p_j^{(0)})/(f_j/n)]^{f_j}, \end{aligned}$$

and hence

$$-2\log \lambda_n = -2\sum_{j=1}^{k} f_j(\log p_j^{(0)} - \log f_j/n).$$

Recall that $-2\log \lambda_n$ tends in distribution, as $n \to \infty$, to a chi-squared distribution with $k-1$ degrees of freedom. Thus, to show that

$$Q_n(X_1, X_2, ..., X_n) = \sum_{j=1}^{k} (f_j - e_j)^2/e_j$$

tends in distribution to $\mathcal{X}_{k-1}^2$, it suffices to show that

$$-2\sum_{j=1}^{k} f_j(\log p_j^{(0)} - \log f_j/n)$$

is asymptotically equivalent to

$$\sum_{j=1}^{k} (f_j - e_j)^2/e_j.$$

This can be seen as follows.

We expand $\log p_j^{(0)} - \log f_j/n$ using Taylor series expansion of $\log x$. The Taylor expansion of $\log x$ about $y$ is

$$\log x = \log y + (x-y)/y + [(x-y)^2/2](-1/y^2) + \Delta.$$

Take $x = p_j^{(0)}$ and $y = f_j/n$, we get

$$\begin{aligned}
\log p_j^{(0)} - \log f_j/n &= (p_j^{(0)} - f_j/n)n/f_j - (p_j^{(0)} - f_j/n)^2 n^2/2f_j^2 + \Delta \\
&= (np_j^{(0)} - f_j)/f_j - (np_j^{(0)} - f_j)^2/2f_j^2 + \Delta_j,
\end{aligned}$$

where $\Delta_j$ represents the quantity

$$\sum_{i=3}^{\infty} (-1)^{i+1}(p_j^{(0)} - f_j/n)^i (n^i/i!f_j^i).$$

Putting this into $-2\sum_{j=1}^{k} f_j(\log p_j^{(0)} - \log f_j/n)$, we get

$$\begin{aligned}
-2\log \lambda &= -2\sum_{j=1}^{k}(np_j^{(0)} - f_j) + \sum_{j=1}^{k}(np_j^{(0)} - f_j)^2/f_j + \sum_{j=1}^{k}\Delta_j \\
&= 0 + \sum_{j=1}^{k}(f_j - e_j)^2/f_j + \varepsilon,
\end{aligned}$$

since $\sum_{j=1}^{k}(np_j^{(0)} - f_j) = n\sum_{j=1}^{k}p_j^{(0)} - \sum_{j=1}^{k}f_j = n(1) - n = 0$, and $e_j = np_j^{(0)}$, under $H_0$.

Now writing $f_j = e_j(f_j/e_j)$ and noting that, by the strong law of large numbers, $(f_j/e_j) = (f_j/np_j^{(0)})$, under $H_0$, converges to one, as $n \to \infty$, we see that, by the Slutsky theorem, $Q_n$ has the same limiting distribution as $-2\log\lambda$.

## 6.6    Exercises

1. Let $\{Q_n, n \geq 1\}$ be a sequence of probability measures on the set $\mathbb{N} = \{0, 1, 2, \ldots\}$. Show that $Q_n$ converges weakly to $Q$ (a probability measure on $\mathbb{N}$), as $n \to \infty$, if and only if, for all $k \in \mathbb{N}$, $Q_n(\{k\}) \to Q(\{k\})$ as $n \to \infty$.

2. Let $X_1, X_2, ..., X_n$ be a random sample from $X$ which is uniformly distributed on $(0, \theta)$, $\theta > 0$. Show that

$$T_n(X_1, X_2, ..., X_n) = \left(\prod_{i=1}^{n} X_i\right)^{1/n}$$

is a consistent estimator for $\theta/e$, where $e$ is the base of the natural logarithm.

3. On $\mathbb{R}^d$, we consider the partial order relation: for $x = (x_1, ..., x_k) \in \mathbb{R}^k$, $y = (y_1, ..., y_k) \in \mathbb{R}^k$, $x \leq y$ means $x_i \leq y_i$ for all $i = 1, 2, ..., k$. Note also that $\mathbb{R}^k$ is a topological space which is a complete and separable metric space. Let $\mathcal{B}(\mathbb{R}^k)$ be the Borel $\sigma$-field of $\mathbb{R}^k$. Probability measures on $\mathcal{B}(\mathbb{R}^k)$ are models for laws of random vectors (vector-valued uncertain quantities in applications). If $P$ is a probability measure on $\mathcal{B}(\mathbb{R}^k)$, then its associated distribution function is given by

$$F : \mathbb{R}^k \to [0, 1], \quad F(x) = P\{y \in \mathbb{R}^k : y \leq x\}.$$

   (a) Verify that $F$ is nondecreasing, i.e., $x \leq y \implies F(x) \leq F(y)$.
   (b) Show directly that $F$ is right continuous (or "continuous from above") on $\mathbb{R}^k$.
   (c) Give a condition for $F$ to be continuous on $\mathbb{R}^k$.
   (d) Verify that $F$ is continuous at a point $x \in \mathbb{R}^k$ if and only if the Borel set $A = \{y \in \mathbb{R}^k : y \leq x\}$ is a $P$-*continuity* set, i.e., $P(\delta(A)) = 0$ where $\delta(A)$ denotes the boundary of $A$.

4. This exercise shows that if $T$ is unbiased for $\theta$, in general $\varphi(T)$ is not unbiased for $\varphi(\theta)$.

   (a) Let $X_1, X_2, ..., X_n$ be a random sample from a Bernoulli population $X$. Let $T(X_1, X_2, ..., X_n) = X_1 + \cdots + X_n$. Verify that $T/n$ is unbiased for $\theta$. Is $(T/n)^2$ unbiased for $\theta^2$?

   (b) From the computation of $E(T/n)^2$, find an unbiased estimator for $\theta^2$.

   (c) For a general $X$ with $E|X| < \infty$. Show that $(EX)^2$ is estimable (i.e., can be estimated by some unbiased estimator). From a random sample $X_1, X_2, ..., X_n$ from $X$, find an unbiased estimator for $(EX)^2$.

5. Let $X_1, X_2, ..., X_n$ be a random sample from an exponential population, i.e., $f(x|\theta) = (1/\theta)e^{-x/\theta}1_{(0,\infty)}(x)$, $\theta > 0$. Let $\varphi(\theta) = \theta^2$.

   (a) Verify that $S^2 = (1/(n-1))\sum_{i=1}^{n}(X_i - \bar{X}_n)^2$ is an unbiased estimator for $\theta^2$.

   (b) Is $S^2$ a function of $\bar{X}_n$? If the answer is no, what does it mean?

   (c) Can you find an unbiased estimator of $\theta^2$ which is based (i.e., a function of) $\bar{X}_n$?

6. Let $X_1, X_2, ..., X_n$ be a random sample from $X$ which is uniformly distributed on $[0, \theta]$, $\theta > 0$.

   (a) Let $Z = \max(X_1, X_2, ..., X_n)$. Find the distribution of $Z$.

   (b) If we take $Z$ to be an estimator for $\theta$, is $Z$ unbiased for $\theta$? Why?

   (c) If the answer in (b) is no, then provide an unbiased estimator for $\theta$.

7. Let $X_1, X_2, ..., X_n$ be a random sample from a Poisson population $X$, with density $f(x|\theta) = \theta^x e^{-\theta}/x!$ for $x = 0, 1, 2, ...$.

   (a) Find the mean and the variance of $X$.

   (b) Show that the sample mean $\bar{X}$ and the sample variance $S^2 = (1/(n-1))\sum_{i=1}^{n}(X_i - \bar{X})^2$ are both unbiased estimators of $\theta$.

   (c) Which of the above two unbiased estimators of $\theta$ do you prefer? Why?

8. Let $X_1, X_2, ..., X_n$ be a random sample from $X$ whose density is $f(x|\theta) = \theta x^{\theta-1} 1_{(0,1)}(x)$. Consider estimating $\theta$.

   (a) Compute the Cramér-Rao lower bound.

   (b) Can you find an unbiased estimator for $\theta$? If you cannot, then do the following: Let $Z = -\sum_{i=1}^{n} \log X_i$ and show that $Z$ has a gamma density of parameters $\alpha = n$, $\beta = \theta$. Find $E(1/Z)$. Show that $(n-1)/Z$ is an unbiased estimator of $\theta$.

   (c) Compute the variance of $(n-1)/Z$. Is $(n-1)/Z$ efficient?

9. Let $X_1, X_2, ..., X_n$ be a random sample from $X$ which is $N(\mu, \sigma^2)$, where $\sigma^2$ is known. Consider the testing problem $H_0 : \mu \leq \mu_o$ versus $H_a : \mu > \mu_o$ at level $\alpha$. Let $c_\alpha$ be the critical point of the test determined by $P(\bar{X}_n > c_\alpha) = \alpha$. Show that this test is consistent.

10. Let $X_n$ be a sequence of independent and nonnegative random variables. Let $S_n = X_1 + X_2 + \cdots + X_n$. Show that $S_n$ converges in probability if and only if $S_n$ converges almost surely.

11. Let $X$ be normal $N(0, \sigma^2)$, $\theta = \sigma^2 > 0$.

   (a) Compute the risk of $S^2 = [1/(n-1)] \sum_{i=1}^{n} (X_i - \bar{X})^2$.

   (b) Let $T = (1/(n+1)) \sum_{i=1}^{n} (X_i - \bar{X})^2$. Compute its risk.

   (c) Compare the two above risks. Explain your result.

12. Using your own examples, show that

   (a) $X_n$ converges in distribution to $X$ does not imply that $X_n - X$ converges in distribution to zero.

   (b) If $X, Y, Z$ such that $X$ and $Y$ have the same distribution, this does not imply that $XZ$ and $YZ$ have the same distribution.

13. Let $X_n$, $X$ be random variables, defined on $(\Omega, \mathcal{A}, P)$, with distribution functions $F_n$, $F$, respectively. Let $Q_n = PX_n^{-1}$, $Q = PX^{-1}$ be their probability measures.

   (a) Suppose $F_n(x) = \begin{cases} 0 \text{ if } x \leq -1/n \\ 1 \text{ if } x > -1/n. \end{cases}$ What is the distribution function $F$ of $X$ for which $X_n \to X$ in distribution?

(b) Let $F_n$ be the distribution functions corresponding to the Dirac measure $\delta_n$ (mass 1 at the point $n \in \mathbb{R}$). Is the limit of $F_n$ a distribution function?

(c) Let $X_n = (S_n - np)/\sqrt{np(1sgn)}$ where $S_n$ is binomial $(n, p)$. Let $X$ be normal $(0, 1)$. We know that $X_n \to X$ in distribution. Can you find a Borel set $A$ of $\mathbb{R}$ such that $Q_n(A)$ does not converge to $Q(A)$?

14. An urn contains $N$ marbles numbered 1 through $N$, with $N$ very large but unknown. You are going to draw $n$ marbles, with replacement, from the urn, and note their numbers. Let $X_i$ be the number of the marble you draw at the $i$th draw, $i = 1, 2, ..., n$. You wish to estimate $N$. A reasonable estimator of $N$ is $X_{(n)} = $ maximum of the $X_i$.

(a) How good is this estimator? For example, is this estimator asymptotically unbiased?

(b) Is $X_{(n)}$ consistent for $N$?

15. Let $X_1, X_2, ..., X_n$ be a random sample from $X$ whose distribution function $F$ is unknown. Suppose $\varphi(F)$ is an estimable parameter.

(a) Show that $\varphi(\theta)$ is of the form $E_F[h(X_1, ..., X_m)]$ for some function $h$ and some integer $m \leq n$.

(b) Show that we can assume the function $h$ *symmetric* without loss of generality. (We call $h$ a "kernel" for $\varphi(\theta)$.)

(c) For a given kernel $h$, the corresponding $U$-statistic for estimating $\varphi(\theta)$ is

$$U_n(X_1, X_2, ..., X_n) = \binom{n}{m}^{-1} \sum_c h(X_{i_1}, ..., X_{i_m})$$

where the summation is over the $\binom{n}{m}$ combinations of $m$ distinct elements $\{i_1, ..., i_m\}$ from $\{1, 2, ..., n\}$. Construct the corresponding $U$-statistics for $P_F(X_1 + X_2 \leq 0)$.

16. We are interested in knowing whether or not the popular sample mean is a Bayes' estimator.

(a) Let $\pi$ be a prior distribution of $\theta$ on $\Phi$ and let $f(x|\theta)$, the conditional density of $X$ given $\theta$, be the model. Show that if $T$ is an unbiased Bayes' estimator of $\varphi(\theta)$, then necessarily $E[(T - \varphi(\theta))^2] = 0$.

(b) Consider the normal model $N(\theta, 1)$. Is the sample mean a Bayes' estimator of $\theta$?

17. Let $X$ be a random vector with values in $\mathbb{R}^3$, $X = (X_1, X_2, X_3)$ with independent components, each $X_i$ distributed with mean $\theta_i$ and variance 1. For $i = 1, 2, 3$, let $g_i : \mathbb{R}^3 \to \mathbb{R}$ and $g = (g_1, g_2, g_3)'$. Let $X^1, X^2, ..., X^n$ be a random sample of size $n$, from $X$. The sample mean of the sample is denoted as $\bar{X}$. The James-Stein estimator of $\theta = (\theta_1, \theta_2, \theta_3)'$ is of the form $T = \bar{X} + g(\bar{X})/n$. Let $h : \mathbb{R}^3 \to \mathbb{R}$ be $h(x) = \begin{cases} 1/\|x\| & \text{if } \|x\| \geq 1 \\ e^{(1-\|x\|)/2} & \text{if } \|x\| \leq 1. \end{cases}$ Take $g$ to be the gradient of $\log h$, i.e., $h_i(x) = [\partial h(x)/\partial x_i]/h(x)$.

Verify that with these choices of $h$ and $g$, the sample mean $\bar{X}$ has greater risk (with square loss) than $T$, and hence is inadmissible.

18. Let $X$ be a random variable whose distribution comes, with equal probability, from either $N(0, 1)$ or $N(\mu, \sigma^2)$.

(a) What is the density of $X$?

(b) Let $X_1, X_2, ..., X_n$ be a random sample from $X$. Show that the likelihood of $(\mu, \sigma^2)$ is unbounded.

19. The goal of this exercise is to show that the study of statistical properties of empirical distributions of arbitrary random variables can be reduced to that of uniform random variables on $(0, 1)$. Let $X_1, X_2, ..., X_n$ be i.i.d. as $X$ which has distribution function $F$. Let $F_n$ denote the empirical distribution function. Let $U_1, U_2, ..., U_n$ be i.i.d. as $U$ which is uniformly distributed on $(0, 1)$. Let $G_n$ denote its empirical distribution function.

(a) Let $F^{-1} : (0, 1) \to \mathbb{R}$ be defined as $F^{-1}(t) = \inf\{x : F(x) \geq t\}$. Verify that $F^{-1}(U)$ has $F$ as its distribution function.

(b) Show that $F_n$ and $G_n(F)$ have the same probabilistic behavior, in the following sense: For any $k$, and $x_1 < x_2 < \cdots < x_k$, the random vector $(F_n(x_1), ..., F_n(x_k))$ has the same joint distribution as $(G_n(F(x_1)), ..., G_n(F(x_k)))$.

(c) Compute the mean of $G_n(t)$, and the covariance of $G_n(t)$ and $G_n(s)$.

(d) Compute $G_n^{-1}$ in terms of the order statistics of $U_1, U_2, ..., U_n$.

(e) Let $V_n = \sqrt{n}(G_n - I)$, where $I$ is the distribution function of $U$.

    i. Verify that $V_n(t) = (1/\sqrt{n}) \sum_{i=1}^{n} [1_{(U_i \leq t)} - t]$ for $0 \leq t \leq 1$.

    ii. Compute $EV_n(t)$ and $Cov(V_n(t), V_n(s))$.

20. Let $F_n(x)$ denote the empirical distribution function based on the observed sample. Let $h_n$ be a decreasing sequence of positive numbers such that $h_n \searrow 0$, as $n \to \infty$. Consider

$$f_n(x) = [F_n(x + h_n) - F_n(x - h_n)]/2h_n.$$

(a) Explain why such an estimator of $f(x)$ seems reasonable.

(b) Let $K(x) = 1/2$ for $x \in [-1, 1)$ and zero outside. Verify that $f_n(x) = (1/nh_n) \sum_{i=1}^{n} K((x - X_i)/h_n)$. Show that $E(F_n(x)) = F(x)$.

(c) Compute $E(F_n(x)F_n(y))$ and $Cov(F_n(x), F_n(y))$, in terms of $n$, $F(x)$, $F(y)$ and $h_n$.

(d) Compute $E(f_n(x))$, $Var(f_n(x))$ and $Cov(f_n(x), f_n(y))$. Compute the mean square $E_f[f_n(x) - f(x)]^2$.

(e) Suppose, in addition, that $f$ is continuous, and $nh_n \to \infty$. Show that $f_n(x)$ is weakly consistent.

# Chapter 7

# Models and Data in Econometrics

*We present and discuss in this chapter the two main factors in statistical analysis of economic problems, namely models and data, as well as some related background.*

*Despite the analogy with physical systems leading to the use of engineering inference tools (e.g., Kalman filter, stochastic differential equations), the analysis of economic systems is much more delicate! On one hand, models need to be proposed with great care, and on the other hand, economic data are coarse in a variety of ways as well as rich in structure (e.g., panel and high frequency data). This chapter serves as a complement to statistical analysis of economic problems in the sense it emphasizes an examination of the rationale underlying proposed models and a closer look at the types of available data in order to derive appropriate statistical inference procedures.*

## 7.1 Justifications of models

We illustrate here the justification of some models proposed in econometrics and financial economics. The justifications contribute not only to our understanding of existing models but also to the art of model building.

While we have at our disposal the theory of statistical inference, we still face a challenging task of how to postulate a reasonable model each time we have a real-world application problem. The theory of statistics does not, and cannot, give us a recipe for postulating your models in some systematic way. Unless our problems bear good resemblance with known models, we should proceed with great care.

A lesson to learn could be to look closely at how other great statisticians or economists came up with their successful models. From a pedagogical viewpoint, all we can do here is to provide some ideas underlying the way models are suggested. Note, however, that in the history of science, great models surfaced and most of the time did not need justifications, since the results will justify the models themselves!

### 7.1.1   The logit model

The logistic function was first derived by Pierre F. Verhulst, in 1838, in his *equations logistiques* describing the self-limiting growth of a biological population. In statistics, a *Probit model* (introduced by Chester Ittner Bliss in 1935, in biology, for the calculation of the dosage-mortality curve) is a popular specification of a *generalized linear model*, using the logistic link function. *Logistic regression* is widely used in many fields, including *econometrics* (term first used by Pawel Ciompa in 1910) which is concerned with the tasks of developing and applying quantitative/statistical methods to the study of economic principles.

There are special features of economic data that distinguish econometrics from other branches of statistics. Economic data are generally observational, rather than being derived from controlled experiments. Because the individual units in an economy interact with each other, the observed data tend to reflect complex econometric equilibrium conditions rather than simple behavioral relationships based on preferences or technology.

The *logit model* was introduced by Joseph Berkson in 1944 who coined the term. The term was borrowed by analogy from the very similar probit model developed by Bliss in 1935. In statistics, *logistic regression* is a model used for prediction of the probability of occurrence of an event. It is used extensively in the medical and social sciences as well as marketing applications such as prediction of a customer's propensity to purchase a product or cease a subscription. In *neural networks*, the logistic function is called the *sigmoid function*. A random variable is said to be *logistically distributed* if its distribution function is of the form

$$F(x; \mu, s) = [1 + e^{-(x-\mu)/s}]^{-1}$$

for the location parameter $\mu \in \mathbb{R}$ and the scale parameter $s > 0$.

Its density is

$$f(x; \mu, s) = e^{-(x-\mu)/s}[s(1 + e^{-(x-\mu)/s})^2]^{-1}$$

where $\mu$ = mean (= mode, = median) and the variance is $\pi^2 s^2/3$.

The standard logistic distribution (also called the *sigmoid function*, a name due to the sigmoid shape of its graph) is

$$f(x) = 1/[1 + e^{-x}].$$

It is the solution of the differential equation

$$f'(x) = f(x)[1 - f(x)]$$

with boundary condition $f(0) = 1/2$.

In the context of decision theory, based upon von Neumann's utility theory, it was D. Luce [100] who seems to be the first to propose the *logistic discrete choice model* in his monograph *Individual Choice Behavior*, J. Wiley, 1959. He offered a *theoretical justification* based upon his assumption of "independence of irrelevant alternatives." According to this assumption, the probabilities of selecting two specified alternatives $a, b$ should not change if we add a third alternative $c$. In other words, the probability of selecting $a$ out of two alternatives $a, b$ should be equal to the conditional probability of selecting $a$ from three alternatives $a, b, c$ under the condition that either $a$ or $b$ is selected.

As cited by D. Luce and P. Suppes [101], Marley showed that if the random error term on utilities are i.i.d. with Gumbel distribution, then probabilities of selecting alternatives follow the logit model. Emil Julius Gumbel (1891-1966) was a German statistician. While in exile in France in 1935, he published this extreme-value distribution that bears his name. The parametric family of Gumbel distributions consists of probability density functions of the form

$$f(x) = (1/\beta) \exp[-(x - \alpha)/\beta - \exp(-(x - \alpha)/\beta)] \text{ for } x \in \mathbb{R},$$

in which $\alpha$ is the mode, $\alpha + c\beta$ is the mean ($c = 0.577216...$ being the Euler's constant), and $\beta^2 \pi^2/6$ is the variance. For $\alpha = 0$, $\beta = 1$, the density is $\exp(-x - e^{-x})$ so that its distribution function is $\exp(e^{-x})$.

In 1974, Daniel McFadden showed the converse [110]: If the errors are i.i.d. and the choice probabilities are described by the logit model, then necessarily these errors are distributed according to the Gumbel distribution.

## Limitations of current justifications of the logit model

Luce's assumption of independence of irrelevant alternatives led to the fact that, for two alternatives $a, b$ with corresponding selection probabilities $p_1, p_2$, the ratio $p_1/p_2$ should depend only on $a, b$, and $p_1/p_2$ will be of the form $f(a)/f(b)$ for some function $f$. The requirement that this ratio be shift-invariant then leads to $f(x) = e^{\beta x}$ for some $\beta$.

There are cases where Luce's assumption is counterintuitive as reported e.g., by J. Chipman [25], G. Debreu and R. D. Luce [31] and K. Train [156]. Here is a simple situation! Suppose in some city, all buses are originally yellow. To go from one location to another, an individual could choose either a taxi $(a)$ or a yellow bus $(b)$. Suppose the probability of selecting $a$ is $p_1 = 0.6$ and that for $b$ is $p_2 = 0.4$, resulting in the ratio $p_1/p_2 = 1.5$. Now suppose the city got new buses and painted them red, and the new buses are of the same brand as the old ones (so that, the comfort is the same). While the choice is still between taxis and buses, with $p_1 = 0.6$ unchanged, we are in fact facing three choices: $a, b$ and $c$ where $c$ is the red bus with $p_2 + p_3 = 0.4$ ($p_3$ is the probability of selecting $c$). Thus now $p_2 < 0.4$ and hence $p_1/p_2 \neq 1.5$, violating the above independence assumption. Note also that Luce's assumption only makes sense if we have three or more alternatives to choose from.

The error referred to in Marley's work is between the "true" utility value of choosing an alternative $a$ and its approximated value. It is known that, the logit model works well even in situations where the errors are distributed differently than the Gumbel distribution, e.g., Train [156]. Here, especially when we have only two alternatives to choose from (as in many important binary choice problems), the explanation of why the logit model is plausible seems open.

In view of the above, we ask: Can we justify the logit model without error distributional assumptions?

## A new justification

We consider a binary choice problem. Let the alternatives be $1, 2$ with likelihoods $p_1, p_2$, and let $u_1, u_2$ be the corresponding approximate utility values. Of course $p_1$ should depend only on $u_1 - u_2$, say, of the form $p_1 = F(u_1 - u_2)$. The problem is the determination of the function $F$. Several natural properties of such an $F$ come to mind:

(i) If we can improve 1, i.e., $u_1$ increases, then $p_1$ should increase, so that $F$ should be non-decreasing,

(ii) As long as 1 becomes better and better, i.e., $u_1 \to +\infty$, $F(x) \to 1$ (keeping $u_2$ fixed, $x = u_1 - u_2$),

(iii) Similarly, $F(x) \to 0$ as $x \to -\infty$,

(iv)  Since we have only two alternatives, we should have $F(x)+F(-x) = 1$ for all $x$ (noting that $p_1 = F(u_1 - u_2)$ while $p_2 = F(u_2 - u_1)$).

Thus $F$ is among the functions $F(.)$, non-decreasing, $F(+\infty) = 1$, $F(-\infty) = 0$ and $F(x) + F(-x) = 1$ for all $x \in \mathbb{R}$.

Now let us elaborate on how likelihoods change as a result of changes in utility values. Note that intuitive explanations for improving alternatives can be easily offered. As we improve alternatives, their utility values increase. Since likelihoods are functions of utilities, they change too. The question is how to update probabilities.

Well, as far as we know, the "correct" way to update probabilities in the light of new evidence, is using Bayes' formula!

Specifically, our $\Omega = A_1 \cup A_2$ where $A_i$ is the event of choosing alternative $i = 1, 2$. Let $E$ denote the new evidence, then $p_1 = P(A_1) = F(\Delta u)$, with $\Delta u = u_1 - u_2$ (noting that $p_2 = P(A_2) = 1 - p_1$), is updated to

$$
\begin{aligned}
p_1' &= P(A_1|E) \\
&= [P(E|A_1)P(A_1)]/[P(E|A_1)P(A_1) + P(E|A_2)P(A_2)].
\end{aligned}
$$

Note that $P(E|A_1)$ is the conditional probability with which we can conclude that there was an "improvement" of size $y$ based on the fact that the individual actually selected $A_1$. This conditional probability is in general different for different values of $y$. To take this dependence into account, we denote $P(E|A_1)$ by $\alpha(y)$. Similarly $P(E|A_2) = \beta(y)$.

Writing $\Delta u = x$, the above updating equation becomes:

$$
F(x + y) = [\alpha(y)F(x)]/[\alpha(y)F(x) + \beta(y)(1 - F(x))] \qquad (7.1)
$$

for $x$. Thus, our function $F$ should be such that:

For every $y$, there exist $\alpha(y)$ and $\beta(y)$ such that (7.1) holds.

In other words, the family of functions $\{\alpha F'(x)/[\alpha F'(x) + \beta]\}$ is shift-invariant.

**Theorem 7.1** *A function $F$ satisfying (i),(ii), (iii), (iv) and Equation (7.1) is of the form*

$$
F(x) = 1/[1 + e^{-\gamma x}]
$$

*for some real number $\gamma$.*

Let $t = p/1-p$ (the odds, where we recover $p$ from $t$ via $p = t/(1+t)$). Recall that $F(x) = F(\Delta u) = P(A_1)$, so we write $t(x)$ to denote its odds.

The right-hand side of (7.1) can be written in terms of odds. Upon dividing the numerator and the denominator by $1 - F(x)$, we get

$$F(x + y) = \alpha(y)t(x)/[\alpha(y)t(x) + \beta(y)]. \qquad (7.2)$$

Now

$$1 - F(x + y) = \beta(y)/[\alpha(y)t(x) + \beta(y)]. \qquad (7.3)$$

Divide (7.2) by (7.3), we get

$$t(x + y) = c(y)t(x), \qquad (7.4)$$

where $c(y) = \alpha(y)/\beta(y) > 0$.

Now it is well known that (see any text on functional equations) all monotonic solutions of Equation (7.4) are of the form

$$t(x) = Ce^{\gamma x}.$$

Thus, we can reconstruct the probability $F(x)$ as

$$F(x) = t(x)/[1 + t(x)] = Ce^{\gamma x}/[1 + Ce^{\gamma x}].$$

The condition $F(x) + F(-x) = 1$ leads to $C = 1$. Dividing both the numerator and the denominator by $e^{\gamma x}$, we get

$$F(x) = 1/[1 + e^{-\gamma x}].$$

See [22] for more information.

**Remark 7.2** *The sigmoid function is a popular choice for (non-linear) activation function in neural networks. Using transformation Lie groups, it can be shown that it is one of the three forms suitable for modeling activation functions.*

### 7.1.2   Stochastic volatility

In order to better understand various models of stochastic volatility in financial economics, we provide here their justifications based only on a natural rationale of *scale invariance*. For more details, see [80] and [150], and also [123].

Consider a typical objective of econometrics: using the known values $X_t$, $X_{t-1}$, $X_{t-2}$, $\ldots$, of different economic characteristics $X$ at different moments of (discrete) time $t$, $t-1$, $t-2$, ..., to predict the future values $X_{t+1}$, $X_{t+2}$, ..., of these characteristics. When the process $X_t$ seems to be stationary, we can describe it by an autoregression $(AR(q))$ model

$$X_t = a_0 + \sum_{i=1}^{q} a_i \cdot X_{t-i} + \varepsilon_t, \tag{7.5}$$

in which the random terms $\varepsilon_t$ are independent normally distributed random variables with zero means and standard deviation $\sigma$. For non-stationary (*heteroskedastic*) economic time series, the volatility $\sigma_t$ should be modeled as a stochastic process, i.e., depending not only on time, but also on randomness. To appropriately describe the corresponding time series, we also need to know how this value $\sigma_t$ changes with time. The heteroskedasticity phenomenon was first taken into account by Engle [41] who proposed a linear regression model of the dependence of $\sigma_t$ on the previous deviations:

$$\sigma_t^2 = \alpha_0 + \sum_{i=1}^{q} \alpha_i \cdot \varepsilon_{t-i}^2. \tag{7.6}$$

This model is known as the Autoregressive Conditional Heteroskedasticity model, or $ARCH(q)$, for short. An even more accurate Generalized Autoregressive Conditional Heteroskedasticity model $GARCH(p, q)$ was proposed in [17]. In this model, the new value $\sigma_t^2$ of the variance is determined not only by the previous values of the squared differences, but also by the previous values of the variance:

$$\sigma_t^2 = \alpha_0 + \sum_{i=1}^{q} \alpha_i \cdot \varepsilon_{t-i}^2 + \sum_{i=1}^{p} \beta_i \cdot \sigma_{t-i}^2. \tag{7.7}$$

Several modifications of these models have been proposed. For example, Zakoian [176] proposed to use regression to predict the standard deviation instead of the variance:

$$\sigma_t = \alpha_0 + \sum_{i=1}^{q} \alpha_i \cdot |\varepsilon_{t-i}| + \sum_{i=1}^{p} \beta_i \cdot \sigma_{t-i}. \tag{7.8}$$

D. B. Nelson [118] proposed to take into account that the values of the variance must always be nonnegative—while in most existing autoregression models, it is potentially possible to get negative predictions for $\sigma_t^2$. To avoid negative predictions, Nelson considers the regression for $\log \sigma_t^2$ instead of for $\sigma_t$:

$$\log \sigma_t^2 = \alpha_0 + \sum_{i=1}^{q} \alpha_i \cdot |\varepsilon_{t-i}| + \sum_{i-1}^{p} \beta_i \cdot \log \sigma_{t-i}^2. \tag{7.9}$$

The above models such as $ARCH(q)$ and $GARCH(p,q)$ models are still not always fully adequate in describing the actual econometric time series. One of the main reasons for this fact is that these models do not take into account a clear *asymmetry* between the effects of positive shocks $\varepsilon_t > 0$ and negative shocks $\varepsilon_t < 0$. It is therefore desirable to modify the $ARCH(q)$ and $GARCH(p,q)$ models by taking asymmetry into account. Several modifications of the $ARCH(q)$ and $GARCH(p,q)$ models have been proposed to take asymmetry into account. For example, Glosten et al. [58] proposed the following modification of the $GARCH(p,q)$ model,

$$\sigma_t^2 = \alpha_0 + \sum_{i=1}^{q}(\alpha_i + \gamma_i \cdot I(\varepsilon_{t-i})) \cdot \varepsilon_{t-i}^2 + \sum_{i=1}^{p}\beta_i \cdot \sigma_{t-i}^2, \qquad (7.10)$$

where $I(\varepsilon) = 0$ for $\varepsilon \geq 0$ and $I(\varepsilon) = 1$ for $\varepsilon < 0$.

Similar modifications were proposed by Zakoian [176] and Nelson [118] for their models. The asymmetric version of Zakoian's model has the form

$$\sigma_t = \alpha_0 + \sum_{i=1}^{q}(\alpha_i^+ \cdot \varepsilon_{t-i}^+ + \alpha_i^- \cdot \varepsilon_{t-i}^-) + \sum_{i=1}^{p}\beta_i \cdot \sigma_{t-i}, \qquad (7.11)$$

where

$$\varepsilon^+ = \varepsilon \text{ for } \varepsilon > 0 \text{ and } \varepsilon^+ = 0 \text{ for } \varepsilon \leq 0;$$
$$\varepsilon^- = \varepsilon \text{ for } \varepsilon < 0 \text{ and } \varepsilon^- = 0 \text{ for } \varepsilon \geq 0.$$

The asymmetric version of Nelson's model has the form

$$\log \sigma_t^2 = \alpha_0 + \sum_{i=1}^{q}(\alpha_i \cdot |\varepsilon_{t-i}| + \gamma_i \cdot \varepsilon_{t-i}) + \sum_{i=1}^{p}\beta_i \cdot \log \sigma_{t-i}^2. \qquad (7.12)$$

All the above formulae can be viewed as particular cases of the following general scheme:

$$\sigma_t = f\left(\alpha_0 + \sum_{i=1}^{q}\alpha_i \cdot f_i(\varepsilon_{t-i}) + \sum_{i=1}^{p}\beta_i \cdot g_i(\sigma_{t-i})\right). \qquad (7.13)$$

For example:

- $ARCH(q)$ and $GARCH(p,q)$ correspond to using $f(x) = \sqrt{x}$ and $f_i(x) = g_i(x) = x^2$.

- Glosten's formula corresponds to using $f_i(x) = (1 + (\gamma_i/\alpha_i) \cdot I(x)) \cdot x^2$.

- A symmetric version of the Nelson's formula corresponds to $f(x) = \sqrt{\exp(x)}$, $f_i(x) = |x|$, and $g_i(x) = \log x^2$; the general (asymmetric) version of this formula corresponds to $f_i(x) = |x| + (\gamma_i/\alpha_i) \cdot x$.

The main problem with the existing asymmetric models is that they are very ad hoc. These models are obtained by simply replacing a symmetric expression $\varepsilon^2$ or $|\varepsilon|$ by an asymmetric one, without explaining why these specific asymmetric expressions have been selected. Mathematically speaking, we can envision many other different functions $f(x)$, $f_i(x)$, and $g_i(x)$. Since the above models are specifications of the dependence of the current volatility $\sigma_t^2$ upon previous ones and the errors $\varepsilon_{t-i}$, our justification will consist of justifying this dependence.

The models should not change if we simply change the measuring units of economic variables. When we replace the original values $\varepsilon_{t-i}$ by new numerical values $\varepsilon_{t-i} = \lambda \cdot \varepsilon_{t-i}$ of the same quantity, then each corresponding term $f_i(\varepsilon_i)$ is replaced with a new term $f_i(\varepsilon_i') = f_i(\lambda \cdot \varepsilon_i)$. Thus, the overall contribution of all these terms changes from the original value

$$I = \sum_{i=1}^{q} \alpha_i \cdot f_i(\varepsilon_{t-i})$$

to the new value

$$I' = \sum_{i=1}^{q} \alpha_i \cdot f_i(\lambda \cdot \varepsilon_{t-i}).$$

It is reasonable to require that the relative quantity of different contributions does not change, i.e., that if two different sets $x_i^{(1)} \overset{\text{def}}{=} \varepsilon_{t-i}^{(1)}$ and $x_i^{(2)} \overset{\text{def}}{=} \varepsilon_{t-i}^{(2)}$ lead to the same contributions $I^{(1)} = I^{(2)}$, then after re-scaling, they should also lead to the same contributions $I'$. Thus, we arrive at the following condition:

Let the values $\alpha_1, \ldots, \alpha_q$ be fixed. Then, the functions $f_1(x), \ldots, f_p(x)$ should satisfy the following condition: if for two sets $x_1^{(1)}, \ldots, x_q^{(1)}$ and $x_1^{(2)}, \ldots, x_q^{(2)}$, we have

$$\sum_{i=1}^{q} \alpha_i \cdot f_i(x_i^{(1)}) = \sum_{i=1}^{q} \alpha_i \cdot f_i(x_i^{(2)}), \tag{7.14}$$

then for every $\lambda > 0$, we must have

$$\sum_{i=1}^{q} \alpha_i \cdot f_i(\lambda \cdot x_i^{(1)}) = \sum_{i=1}^{q} \alpha_i \cdot f_i(\lambda \cdot x_i^{(2)}). \tag{7.15}$$

For simplicity, let us start with the case when the values $x_i^{(2)}$ are very close to $x_i^{(1)}$, i.e., when $x_i^{(2)} = x_i^{(1)} + k_i \cdot h$ for some constants $k_i$ and for a very small real number $h$. For small $h$, we have

$$f_i(x_i^{(1)} + k_i \cdot h) = f_i(x_i^{(1)}) + f_i'(x_i^{(1)}) \cdot k_i \cdot h + O(h^2). \tag{7.16}$$

Substituting the expression (7.16) into the formula (7.14), we conclude that

$$\sum_{i=1}^{q} \alpha_i \cdot f_i'(x_i^{(1)}) \cdot k_i \cdot h + O(h^2) = 0. \tag{7.17}$$

Dividing both sides by $h$, we get

$$\sum_{i=1}^{q} \alpha_i \cdot f_i'(x_i^{(1)}) \cdot k_i + O(h) = 0. \tag{7.18}$$

Similarly, the condition (7.15) leads to

$$\sum_{i=1}^{q} \alpha_i \cdot f_i'(\lambda \cdot x_i^{(1)}) \cdot k_i + O(h) = 0. \tag{7.19}$$

In general, the condition (7.18) leads to (7.19). In the limit $h \to 0$, we therefore conclude that for every vector $k = (k_1, \ldots, k_q)$, if

$$\sum_{i=1}^{q} k_i \cdot (\alpha_i \cdot f_i'(x_i^{(1)})) = 0, \tag{7.20}$$

then

$$\sum_{i=1}^{q} k_i \cdot (\alpha_i \cdot f_i'(\lambda \cdot x_i^{(1)})) = 0. \tag{7.21}$$

The sum (7.20) is a scalar (dot) product between the vector $k$ and the vector $a$ with components $\alpha_i \cdot f_i'(x_i^{(1)})$. Similarly, the sum (7.21) is a scalar (dot) product between the vector $k$ and the vector $b$ with components $\alpha_i \cdot f_i'(\lambda \cdot x_i^{(1)})$. Thus, the above implication means that the vector $b$ is orthogonal to every vector $k$ which is orthogonal to $a$, i.e., to all vectors $k$ from the hyperplane consisting of all the vectors orthogonal to $a$.

It is easy to see geometrically that the only vectors which are orthogonal to the hyperplane are vectors collinear with $a$. Thus, we conclude that $b = \delta \cdot a$ for some constant $\delta$, i.e., that

$$\alpha_i \cdot f_i'(\lambda \cdot x_i^{(1)}) = \delta \cdot \alpha_i \cdot f_i'(x_i^{(1)}). \tag{7.22}$$

Dividing both sides by $\alpha_i$, we conclude that

$$f_i'(\lambda \cdot x_i^{(1)}) = \delta \cdot f_i'(x_i^{(1)}). \tag{7.23}$$

In principle, $\delta$ depends on $\lambda$ and on values $x_i^{(1)}$. From Equation (7.23) corresponding to $i = 1$, we see that

$$\delta = \frac{f_1'(\lambda \cdot x_1^{(1)})}{f_1'(x_1^{(1)})}. \tag{7.24}$$

Thus, $\delta$ only depends on $x_1^{(1)}$ and does not depend on any other value $x_i^{(1)}$. Similarly, by considering the case $i = 2$, we conclude that $\delta$ can depend only on $x_2^{(1)}$ and thus, does not depend on $x_1^{(1)}$ either. Thus, $\delta$ only depends on $\lambda$, i.e., the condition (7.23) takes the form

$$f_i'(\lambda \cdot x_i^{(1)}) = \delta(\lambda) \cdot f_i'(x_i^{(1)}). \tag{7.25}$$

It is known that every continuous function $f_i'(x)$ satisfying Equation (7.23) has the following form:

- $f_i'(x) = C_i^+ \cdot x^{a_i}$ for $x > 0$, and
- $f_i'(x) = C_i^- \cdot |x|^{a_i}$ for $x < 0$,

for some values $C_i^\pm$ and $a_i$; see, e.g., [2] Section 3.1.1 or [122]. (This result was first proven in [129].) For differentiable functions, the easiest way to prove this result is to differentiate both sides of (7.23) by $\lambda$, set $\lambda = 1$, and solve the resulting differential equation.

For the corresponding functions, the condition (7.25) is satisfied with $\delta(\lambda) = \lambda^{a_i}$. Since the function $\delta(\lambda)$ is the same for all $i$, the value $a_i$ is therefore also the same for all $i$: $a_1 = \cdots = a_q$. Let us denote the joint value of all these $a_i$ by $a$.

Thus, all the derivatives $f_i'(x)$ are proportional to $x^a$. Hence, the original functions are proportional either to

- $x^{a+1}$ (for $a \neq -1$) or

- $\log(x)$ (when $a = -1$).

The additive integration constant can be absorbed into the additive constant $\alpha_0$, and the multiplicative constants can be absorbed into a factor $\alpha_i$.

Thus, without loss of generality, we can conclude that in the scale invariant case, either $f_i(x) = x_i^a \cdot (1 + b \cdot I(x))$, or $f_i(x) = \log(|x|)$.

We have proven that the natural scale-invariance condition implies that each function $f_i(x)$ has either the form $\log(x)$, or the form $f_i(x) = x_i^a \cdot (1 + b \cdot I(x))$. This conclusion covers all the functions which are efficiently used to describe asymmetric heteroskedasticity:

- The function $f_i(x) = 1 + (\gamma_i/\alpha_i) \cdot I(x)) \cdot x^2$ used in Glosten's model;

- The function $f_i(x) = x^+ + (\alpha^-/\alpha^+) \cdot x^-$ used in Zakoian's model; and

- The function $f_i(x) = (1 + (\gamma_i/\alpha_i) \cdot sgn(x)) \cdot |x|$ used in Nelson's model.

It is worth mentioning that this result also covers the functions $g_i(x) = x^2$ and $g_i(x) = \log(x^2) = 2\log(x)$ used to describe the dependence on $\sigma_{t-i}$.

Thus, the exact form of the dependence on $\varepsilon_{t-i}$ has indeed been justified by the natural scale invariance requirement—as well as the dependence on $\sigma_{t-i}$. Note that scale-invariance of the econometric formulae describing heteroskedasticity was noticed and actively used in [38]. However, our approaches are somewhat different: In [38], the econometric *models* were taken as *given*, and scale invariance was used to analyze heteroskedasticity tests. In contrast, we use scale invariance to *derive* the econometric models.

### 7.1.3  Financial risk models

In Chapter 5 we discussed a class of risk measures known as distorted probability risk measures. They are based upon an approach towards risk from a psychological viewpoint. Specifically, a distorted probability risk measure is a functional $\rho(.)$, defined on some convex cone $\mathcal{X}$ of random variables (all defined on a probability space $(\Omega, \mathcal{A}, P)$), of the form $\rho(X) = \int(g \circ P)(X > t)dt$, where the distortion function $g : [0, 1] \to [0, 1]$ is nondecreasing with $g(0) = 0, g(1) = 1$. The question is: which $g$ to choose?

Most quantile-based risk measures are distorted probability measures with specified $g$. However, these "built-in" distortions are not quite adequate for generating risk measures since they lack basic properties to make the associated risk measures coherent. Now, using Choquet integral representation, it is known that we should choose $g$ to be strictly concave. The question remains: Which strictly concave distortions to choose?

S. S. Wang [165] proposed to use functions $g$ of the form

$$g_\lambda(y) = \Phi(\Phi^{-1}(y) + \lambda), \qquad (7.26)$$

for some $\lambda \in \mathbb{R}$, and $\Phi(y)$ a cumulative distribution function of some appropriate distribution, e.g., of the standard normal distribution.

A partial explanation of Wang's formula comes from the difference between subjective and objective probabilities. Indeed, how can we explain the fact that, in spite of the seeming naturalness of the mean, people's pricing under risk is usually different from the mean $E[X]$? A natural explanation comes from the observation that the mean $E[X]$ is based on the "objective" probabilities (frequencies) of different events, while people use "subjective" estimates of these probabilities when making decisions.

For a long time, researchers thought that subjective probabilities are approximately equal to the objective ones, and often they are equal. However, in 1979, a classical paper by D. Kahneman and A. Tversky ([75], reproduced in [157]) showed that our "subjective" probability of different events is, in general, different from the actual ("objective") probabilities (frequencies) of these events. This difference is especially drastic for events with low probabilities and probabilities close to 1, i.e., for the events that are most important when estimating risk. Specifically, Kahneman and Tversky have shown that there is a one-to-one correspondence $g(y)$ between objective and subjective probabilities, so that once we know the (objective) probability $P(E)$ of an event $E$, the subjective probability $P_{\text{subj}}(E)$ of this event $E$ is (approximately) equal to

$$P_{\text{subj}}(E) = g(P(E)). \qquad (7.27)$$

This idea was further explored by other researchers; see, e.g., [6] and references therein.

Let us apply the Kahneman-Tversky idea to the events $X > x$ corresponding to different values $x$. For such an event $E$, its objective probability is equal to $\bar{F}(x)$. Thus, the corresponding subjective probability of this event is equal to

$$P_{\text{subj}}(X > x) = g(\bar{F}(x)).  \qquad (7.28)$$

If we compute the "subjective" mean $E_{\text{subj}}[X]$ of $X$ based on these subjective probabilities, we get

$$E_{\text{subj}}[X] = \int_0^\infty P_{\text{subj}}(X > x)\, dx = \int_0^\infty g(\bar{F}(x))\, dx.  \qquad (7.29)$$

There exist several justifications of the Wang transform method, justifications that show that this method is uniquely determined by some reasonable properties. The first justification was proposed by Wang himself, in [165]; several other justifications are described in [78]. The main limitations of the existing justifications of the Wang transform are that these justifications are too complicated, too mathematical, and not very intuitive.

Our first observation is that the Wang transform is not a single function, but a family of functions indexed by $\lambda \in \mathbb{R}$. They form a *transformation group* under composition of functions:

- Identity $g_0(.)$

- $g_\alpha^{-1}(.) = g_{-\alpha}(.)$

- $g_\alpha \circ g_\beta(.) = g_{\alpha+\beta}(.)$

It is reasonable to require that we have a family of transformation that forms a one-parameter group. It is also reasonable to require that the transformations are continuous, and that the dependence on the parameter is also continuous, i.e., that this is a *Lie group*; see, e.g., [23]. It is known that every one-dimensional Lie group (i.e., a Lie group described by a single parameter) is (at least locally) isomorphic to the additive group of real numbers. In precise terms, this means that instead of using the original values of the parameter $v$ describing different elements of this group, we can use a rescaled parameter $\alpha = h(v)$ for some non-linear function $h$, and in this new parameter scale:

- The identity element corresponds to $\alpha = 0$;

- The inverse element to an element with parameter $\alpha$ is an element with parameter $-\alpha : g_{-\alpha} = g_\alpha^{-1}$; and

- The composition of elements with parameters $\alpha$ and $\beta$ is an element with the parameter $\alpha + \beta$: $g_\alpha \circ g_\beta = g_{\alpha+\beta}$.

Let us denote $\Phi(x) \stackrel{\text{def}}{=} g_x(0.5)$. Then, for every $y = \Phi(x)$ and for every real number $\alpha$, we get

$$g_\alpha(y) = g_\alpha(\Phi(x)) = g_\alpha(g_x(0.5)) = (g_\alpha \circ g_x)(0.5). \qquad (7.30)$$

Because of our choice of parameters, we have

$$g_\alpha \circ g_x = g_{x+\alpha}$$

and thus, the formula (7.30) takes the form

$$g_\alpha(y) = g_{x+\alpha}(0.5). \qquad (7.31)$$

By definition of the function $\Phi(x)$, this means that

$$g_\alpha(y) = \Phi(x + \alpha). \qquad (7.32)$$

Here, $y = \Phi(x)$, hence $x = \Phi^{-1}(y)$ and thus,

$$g_\alpha(y) = \Phi(\Phi^{-1}(y) + \alpha). \qquad (7.33)$$

This is exactly the Wang transform formula (7.26).

The above derivation of Wang's transforms is done on a one-parameter family of transformations. It remains to justify this structure of transformations. In view of the following proposition, it suffices to argue that a natural invariance principle will lead to a commutative group of transformations.

It is known that often, group decisions are reasonably stable, even when the roles of different individuals within a group slightly change. In other words, whether the participant $e$ collects the original estimates and passes them to the participant $d$, or, vice versa, $d$ collects the original estimates and passes them to $e$, the final decisions are (largely) the same. In terms of composition, the difference between these two situations is that in the first situation, we have the composition $g_d \circ g_e$, while in the second situation, we have a different composition $g_e \circ g_d$. Thus, the above invariance means that the order of the composition does not matter: $g_d \circ g_e = g_e \circ g_d$. Now, the following simple mathematical result shows that commutative groups are indeed one-parametric.

**Proposition 7.3** *Let $X$ be a set, and let $G$ be a commutative group of transformations $g : X \to X$ with the property that for every two values $x \in X$ and $x' \in X$ there exists a transformation $g_{x,x'} \in G$ that transforms $x$ into $x'$: $g_{x,x'}(x) = x'$. Then, for every $x \in X$ and $x' \in X$, there exists only one transformation $g \in G$ for which $g(x) = x'$.*

**Proof.** Let $x_1 \in X$ and $x_2 \in X$, and let $f$ and $g$ be transformations for which $f(x_1) = g(x_1) = x_2$. We need to prove that $f = g$, i.e., that for every $x \in X$, we have $f(x) = g(x)$. Indeed, let us pick any value $x \in X$, and let us prove that $f(x) = g(x)$. Since the group $G$ is transitive, there exists a transformation $h$ for which $h(x_1) = x$. The group $G$ is also commutative, so we have $f \circ h = h \circ f$; in particular, for $x_1$, we have $f(h(x_1)) = h(f(x_1))$. We know that $f(x_1) = x_2$. By our choice of $h$, we have $h(x_1) = x$. Thus, we conclude that $f(x) = h(x_2)$. Similarly, we have $g \circ h = h \circ g$, hence $g(h(x_1)) = h(g(x_1))$. We know that $g(x_1) = x_2$. By our choice of $h$, we have $h(x_1) = x$. Thus, we conclude that $g(x) = h(x_2)$. Therefore, both $f(x)$ and $g(x)$ are equal to the exact same value $h(x_2)$ and are, hence, equal to each other. ∎

## 7.2    Coarse data

As an empirical science, statistics is data driven. As such, it is important to examine carefully the type of available data in order to develop appropriate statistical inference procedures. We illustrate here several types of coarse data together with their associated statistical methods of analysis.

### 7.2.1    Indirect observations in auctions

We elaborate below on an interesting situation in economics where desirable data cannot be obtained directly.

There is a striking analogy (in structure) between physical and social systems. In control engineering, for example, there are two basic equations: the structural equation and the observational equation. The structure equation comes from the theory under study, namely, dynamics of systems, which is a (stochastic) differential equation. To "identify" the system equation, we need to estimate its parameters from observed data about the system. However, data are coarse since they can be corrupted by noise, and thus a second equation expressing how data are gathered is specified.

In many social studies, the situation is similar. Consider the context of auctions in economics. In the simplest auction format, namely the "first-price, sealed-bid auction," there is one seller and $n$ bidders. For each given object, the seller likes to know how the bidders evaluate the object, for obvious reasons! However, bidders' valuations are only known to the bidders themselves, and not to the seller. Each bidder submits a bid in a sealed envelope. At a given time, the seller opens these envelopes and the bidder who tendered the highest bid wins. So observables are only the bids and not the valuations.

Let $V$ denote the random variable representing the bidders' valuation of the object, and $B$ the bid random variable, with distributions $F_V$, $F_B$ and densities $f_V, f_B$, respectively. Assume also that the valuations $V_i$ and the bids $B_i$, $i = 1, 2, ..., n$, are i.i.d. $V$, $B$, respectively. To estimate $F_V$, we need the data $V_i$, but unfortunately they are not observable. Instead, we observe a realization of another variable $B$. Fortunately, there should be a "link" between $V$ and $B$! This situation reminds us of the popular "hidden Markov models" in biology. Rather than postulating some form of links, i.e., proposing a model, we take advantage of the fact that auctions are special "static games of incomplete information," i.e., can be viewed through the lens of Harsanyi's theory of noncooperative games of incomplete information. Without going into details now (see Section 7.2.2 for some background on game theory), we simply say this: At (Bayes-Nash) equilibrium, we have

$$V_i = B_i + [F_B(B_i)/(n-1)f_B(B_i)], \ i = 1, 2, ..., n. \tag{7.34}$$

We need data from our variable of interest, namely $V$, to estimate its distribution function $F_V$, but we don't have it! Instead, we observe another variable $B$ which is linked to $V$ by the above equation. Unfortunately the functions $F_B$ and $f_B$ in that equation are unknown. However, they can be estimated by the random sample $B_1, B_2, ..., B_n$. If we do so and then compute the corresponding $V_i$, $i = 1, 2, ..., n$, then we only obtain an "approximate" random sample rather than the exact random sample. This reminds us of the familiar measurement-error models in linear (and nonlinear) regression models. However, the type of approximate data is different. The "errors" in our available data come from different sources.

Let's pursue our quest for obtaining the approximate sample. A two-stage estimation is needed.

**Step one:**   We need to "estimate" the unobserved $V_i$ from the observed $B_i$ via (7.34). But both $F_B$ and $f_B$ are unknown! We need to estimate both $F_B$ and $f_B$ from the $B_i$'s. This is a *nonparametric estimation problem*. Of course $F_B(x)$ can be consistently estimated by the empirical distribution function

$$F_{B,n}(x) = (1/n) \sum_{i=1}^{n} 1_{(B_i \leq x)},$$

but we also need an estimate for its density function $f_B$. Such an estimate cannot be derived from the estimator of $F_B$ since $F_{B,n}(x)$ is a step function, resulting in identically zero derivative! Thus, we need *nonparametric estimation methods* for estimating probability density functions.

**Step two:**   With the estimates of both $F_B$ and $f_B$ in the link equation (7.34), we can at least recover approximately the unobserved values on the variable $V$. These are "pseudo-values" of the $V_i$'s. They can be viewed as coarse data, in the sense that they are the $V_i$ but recorded with "error." The source of the "measurement error" comes from Equation (7.34) and the estimation error in our nonparametric estimation procedure. With the approximate values, say, $V_i'$, we estimate $F_V$ as usual. How good is our estimator of $F_V$ using the approximate random sample $V_1', ..., V_n'$? How can we assess performance of hypothesis test statistics based on approximate random samples? How can we evaluate our predictors?

## 7.2.2   Game theory

Game theory provides mathematical models for economic behavior. Game theory is a name given to the important problem of understanding how people make decisions in societies. We call the persons involved *the players* since they participate in a game, where we interpret games in a general sense. As we will see, a statistical testing problem about a population parameter is a two-person game (a game with two players), namely "nature" (player I) and you (the statistician) being player II. Player I moves first: she chooses one action among two possible actions $H_0$, $H_a$ (but you don't see it!), then you choose one action among "accept $H_0$" or "reject $H_0$" (of course by collecting data, and use your, say, uniformly most powerful test procedure!), and the game ends.

   An $n$-person game is a game with $n$ players. Game theory was invented by John von Neumann in 1928, and became popular (and essential) in economics in his book with Oskar Morgenstern in 1943. The structure of a game is this. There are $n$ players. Each player $i$ has her set of strategies $S_i$, and a payoff function

$$u_i : S_i \times S_2 \times \cdots \times S_n \to \mathbb{R}.$$

The main problem in game theory is how to predict the players' behavior, i.e., to make an educated guess at how the players will likely play. Each player will act in such a way to maximize her payoff, but her payoff depends essentially on other players' actions. The solution is in the concept of Nash's equilibrium (the proof of its existence is based on fixed point theorems).

The equilibrium of a game is the expected behavior of that game (of course, assuming that all players are "rational"), i.e., we could expect that, in their own interests, players will likely play strategies forming the equilibrium, as predicted by theory. The relevance of game theory to economics seems obvious. Auctions are *static games of complete information,* i.e., game in which the players simultaneously choose actions, then the players receive payoffs that depend on the combination of actions just chosen. "Static" (or "simultaneous-move") refers to just one move. The game is of "incomplete information" since some player is uncertain about another's payoff function. In auctions, each bidder's (player's) willingness to pay for the object is unknown to the other bidders. Thus, auctions are static games of incomplete information.

The concept of Nash's equilibrium in static games of complete information is as follows. The strategies $(s_1^*, ..., s_n^*)$ of the players form a (Nash) equilibrium of the game if, for each player $i$, $s_i^*$ is player $i$'s best response to the strategies $(s_1^*, ..., s_{i-1}^*, s_{i+1}^*, ..., s_n^*)$ specified for the $n-1$ other players, i.e., if the other players choose $(s_1^*, ..., s_{i-1}^*, s_{i+1}^*, ..., s_n^*)$, then player $i$ will make less "profit" (utility) if she chooses any other strategy $s_i$ different than $s_i^*$, specifically, for any $i$, and $s_i \in S_i$,

$$u_i(s_1^*, ..., s_{i-1}^*, s_i, s_{i+1}^*, ..., s_n^*) \le u_i(s_1^*, ..., s_{i-1}^*, s_i^*, s_{i+1}^*, ..., s_n^*).$$

In a static game of incomplete information, each player cannot solve the above optimization problem since $u_i(s_1^*, ..., s_{i-1}^*, s_i, s_{i+1}^*, ..., s_n^*)$ is not available to player $i$. There is a subtle point in games of incomplete information as compared to games of complete information. In games of complete information, we specify the sets $S_i$ of players' strategies. In games of incomplete information, we only specify the sets of players' *actions,* $A_i$. Their strategies are "rules" leading to their actions from their "private knowledge" (about their own payoff functions) and "subjective beliefs" about other players' payoff functions.

Let's specify the uncertainty of players about the payoff functions of other players. The idea that a player $i$ knows her own payoff function but may be uncertain about others' payoff functions is described as follows. Each player $i$ knows the set of possible payoff functions of another player $j$, say $T_j$, while not specifically which one. Thus each player $i$ has set of possible payoff functions indexed by her "type $T_i$," i.e., the payoff function of player $i$, when players choose their actions $a_1, ..., a_n$, is of the form $u_i(a_1, ..., a_n; t_i)$ where $t_i \in T_i$. Player $i$ knows her $t_i$, but does not know the $t_j$ of other players. In such a situation, player $i$ has to guess the types of other players in order to choose her best course of action. Such a guess is possible if player $i$ has some personal (subjective) belief on other players' possible payoff functions. This sounds familiar to statisticians, especially to Bayesian statisticians! That's why static games of incomplete information are also called *Bayesian games*.

The following specific model of static games of incomplete information is due to Harsanyi (1967). First, the uncertainty about players' actual payoff functions is viewed as follows: each $(t_1, ..., t_n)$ is "drawn" by nature with some prior distribution. Let $t_{-i} = (t_1, ..., t_{i-1}, t_{i+1}, ..., t_n) \in T_{-i}$. The player $i$'s belief about other players' types $t_{-i}$ is the conditional probability $p_i(t_{-i}|t_i)$. Thus, a Bayesian game is specified as

$$G = \{A_i, T_i, p_i, u_i, i = 1, 2, ..., n\}.$$

Now, a strategy for player $i$ is a function $s_i : T_i \rightarrow A_i$. The concept of a Bayes-Nash equilibrium in a Bayesian game is this. The strategies $(s_1^*, ..., s_n^*)$ are a Bayes-Nash equilibrium if, for each player $i$, and for each of $i$'s types $t_i \in T_i$, $s_i^*(t_i)$ solves

$$\max_{a_i \in A_i} \sum_{t_{-i} \in T_{-i}} u_i(s_1^*(t_1), ..., s_{i-1}^*(t_{i-1}), a_i, s_{i+1}(t_{i+1}), ..., s_n^*(t_n); t_i) p_i(t_{-i}|t_i).$$

The interpretation is simple: each player's strategy must be a best response to the other players' strategies. At equilibrium, no player wants to change her strategy. Of course, existence of such equilibrium and how to get it are main issues in game theory. It is precisely this concept of equilibrium (and its rationale) that gives economists the idea of how to predict players' behavior. The concept of Bayes-Nash equilibrium provides the link between observed and unobserved auction data.

### 7.2.3   Measurement-error data in linear models

Consider the classical normal linear model

$$Y = \beta X + \varepsilon. \tag{7.35}$$

If the data, say, $(X_i, Y_i)$, $i = 1, 2, ..., n$ are available, then $\beta$ can be estimated consistently by least squares. Now suppose that the observed data $X_i$ are imperfectly measured, say, as

$$Z = X + u \tag{7.36}$$

where $u$ is $N(0, \sigma_u^2)$. Let's see what happens if we simply use the approximate values of $Z_i$ to estimate $\beta$. Insert (7.36) into (7.35), and we get

$$Y_i = \beta(Z_i - u_i) + \varepsilon_i = \beta Z_i + w_i \tag{7.37}$$

where the new error term is $w = \varepsilon - \beta u$. We have, under standard assumption of uncorrelatedness of all errors ($Z, \varepsilon, u$ are mutually independent),

$$Cov(Z, w) = Cov(X + u, \varepsilon - \beta u) = -\beta \sigma_u^2 \neq 0,$$

i.e., $w$ is correlated with $Z$, violating a basic assumption in standard linear model, and as such the least squares estimated of $\beta$ from (7.37), namely

$$\beta_n = \left[ (1/n) \sum_{i=1}^{n} Z_i Y_i \right] \left[ (1/n) \sum_{i=1}^{n} Z_i^2 \right]^{-1}$$

is not consistent. In fact, $\beta_n$ converges in probability to $\beta / [1 + \sigma_u^2 / a]$ $\neq \beta$, where $a = \lim_{n \to \infty} (1/n) \sum_{i=1}^{n} Z_i^2$, in probability.

## 7.2.4 Censored data

Survival analysis is an area concerning lifetimes of subjects (humans, electronic components, etc.). Thus the unknown distribution function $F$ of the variable of interest $X$ (lifetime) or its hazard rate $\lambda(t) = f(t)/[1 - F(t)]$, where $f(.)$ is the derivative of $F$, is to be estimated from data. Now if the data are a *complete* random sample $X_1, X_2, ..., X_n$, then $F$ can be estimated consistently, pointwise, by the *empirical distribution function*. However, in medical statistics, we do not have, in general, complete data. Specifically some of the $X_i$ are observed, while the others are only *partially observed*. This is due to the nature of lifetimes!

For example, with $n$ subjects in a study, we need to conduct our analysis, say at a fixed time $T$. Thus the observed $X_i$ are those which are smaller than $T$, whereas the others are only known to be greater than $T$. Specifically, we observe the indicator function $I(X_i \leq T)$ and the $X_i$ for which the indicator function is 1. Thus, the available data are subject to a *censoring mechanism,* and the so obtained data are referred to as *censored data.* More generally, a random censoring mechanism is this.

Let $Y_1, Y_2, ..., Y_n$ be i.i.d. as the censoring times $(Y)$. What we are able to observe are the values of $Z_i = X_i \wedge Y_i$ and $\delta_i = 1_{(X_i \leq Y_i)}$. Without having the values of $X_1, X_2, ..., X_n$, we cannot even carry out the so-called "Fundamental Theorem of Statistics" (for statistical inference), namely estimating the distribution function $F$ of $X$, by the above usual procedure. Thus, the problem is how to estimate $F$ from the observed data $(Z_i, \delta_i, \ i = 1, 2, ..., n)$. Well, we rely on the concept of likelihood! For each $G \in \mathcal{F}$ (a class of distributions), the likelihood of getting the observed data under $G$ is

$$L(G|Z_i, \delta_i, i = 1, 2, ..., n) = \prod_{i=1}^{n} p^{\delta_{(i)}} \left( \sum_{j=i+1}^{n+1} p_j \right)^{1 - \delta_{(i)}}$$

where $p_i = dG(\{x_{(i)}\})$, $p_{n+1} = 1 - G(x_{(n)})$, with $x_{(i)}$ denoting order statistics. Maximizing this likelihood over $\mathcal{F}$, subject to $p_i \geq 0, \sum_{i=1}^{n+1} p_i = 1$, we obtain its MLE,

$$F_n^*(x) = 1 - \prod_{Z(i) \leq t} (1 - \delta_{(i)}/[n - i + 1]),$$

the famous *Kaplan-Meier product-limit estimator (1958).*

## Censored data in linear models

In many economic data, the censored data are in fact missing data of some special type. We illustrate this situation in regression models in which data on the response variables are censored in some specific way and cause selection bias. A popular linear regression model with censored data are the *Tobit model* [155]. The kind of censored data in the Tobit model is a special case of sample selection models of James Heckman which we will discuss later in the context of missing data. The treatment of the Tobit model will illustrate how we should modify existing statistical procedures to cope with new types of data.

Consider a standard linear model

$$Y_i = \alpha + \beta X_i + \varepsilon_i.$$

In contrast to measurement-error data in which the explanatory variable $X$ is coarsed in the sense that its values are obtained with errors, we consider now the case where $X_i$ are all observed with accuracy, for all individual $i$, in a sample of size $N$, but the response variable $Y$ is partially latent (unobserved) in that some of the values $Y_i$ are missing. The basic question is: Can we just use the available (observed) data (i.e., ignoring the missing ones) to estimate parameters by ordinary least squares (OLS) method? In fact, the precise question is: Are OLS estimators of the parameters, based only on the subsample of observed data, *consistent*? This clearly depends on "how data are missing."

We consider a common situation in economics where missing data are in fact censored of a special type. In other words, data are not missing at random, but rather in a "self-selected" manner. In such cases, the observed subsample is biased, and hence standard statistical procedures in linear models will lead to inconsistent estimators of model parameters. In order to use only the observed subsample to obtain consistent estimators of parameters, we need to find ways to "correct the sample bias," and that will lead to new statistical procedures. For concreteness, consider the problem investigated by Tobin himself.

We wish to regress the response variable $Y$, the amount a household would like to spend on new cars, on some explanatory variable $X$, say, household income, by using the linear model (to analyze car purchases in a population, namely, at each level of the covariate $X$, estimate how much households are willing to spend to buy a new car),

$$Y_i = \alpha + \beta X_i + \varepsilon_i, \tag{7.38}$$

using a random sample of $N$ households $(i = 1, 2, ..., N)$. As usual, we will collect data $(X_i, Y_i)$, $i = 1, 2, ..., N$, on these households and estimate the model parameters $\alpha, \beta$ (of course, assuming that our linear model is adequate). Once these estimates of parameters are obtained, prediction on $Y_j$ (at a given $X_i$) can be obtained. How accurate are our predictions clearly depend on how "good" are our parameters' estimates. Now, since we are in a linear model setting, with, say, standard assumptions about the random error term $\varepsilon$, we would like to use the ordinary least squares (OLS) method to estimate the parameters. It is well known that with full sample data, OLS estimators are consistent (since $E(\varepsilon_i) = 0$), a property of estimators that we consider "good."

Suppose the least expensive car costs $c$ dollars. A household $i$ whose desired level of car expenditure is less than $c$ will not be able to buy a new car, and hence $Y_i$ is not available. In such a situation, we know the source of missing data. The observed data in the sample are obtained by the *sample selection rule*: $Y_i$ is selected (observed) from the sample if $Y_i \geq c$. In terms of censoring, the data $Y_i$ are censored by itself, i.e., the value of $Y$ is observed only when some criterion in term of $Y$ itself is met. Note that, regardless of whether $Y$ is observed or not, the regressor $X$ is observed for the entire sample of size $N$.

The regression model is used to explain the desired expenditures in the population under study as a whole. Now, the missing data are "censored" so that the observed subsample would overrepresent the population, since it contains information only on households whose desired level of expenditures was sufficiently high (for a transaction to occur).

Specifically, in the above model, the $Y_i$ are observed when $Y_i \geq c$. Suppose we just consider the model with the data available in the observed subsample $(Y_i, X_i)$, say, $i = 1, 2, ..., n < N$. Doing that, we are in fact considering the variable $Y$ *truncated* by the sample selection rule $Y \geq c$. In other words, using only the observed subsample is using a sample from the truncated $Y$. We have

$$E(Y_i|X_i, Y_i \geq c) = \alpha + \beta X_i + E(\varepsilon_i|X_i, Y_i \geq c).$$

Now

$$E(\varepsilon_i|X_i, Y_i \geq c) = E(\varepsilon_i|X_i, \varepsilon_i \geq c - (\alpha + \beta X_i)).$$

The error term $\varepsilon_i|X_i$, $\varepsilon_i \geq c - (\alpha + \beta X_i)$, has a non-zero mean. Indeed, since $\varepsilon_i \geq c - (\alpha + \beta X_i)$, we see that, for any particular value of $X_i$, its mean can be negative, positive or zero. As a result, OLS estimators will be inconsistent. Note that consistency of OLS estimators of parameters in a linear model with random error term $u_i$ follows if $u_i$ has zero mean and is uncorrelated with the regressors.

In view of the above, we ask, "Is it possible to estimate consistently the model parameters using only the observed subsample?" because, after all, that's all the available data we have! Since OLS fails in this linear model with censored data, let's turn to another estimation method! Tobin suggested to estimate the parameters by MLE by assuming a normal distribution $N(0, \sigma^2)$ for the error $\varepsilon_i$, where the likelihood is computed as follows:

$$L_N(\alpha, \beta|(X_i, Y_i), i = 1, 2, ..., N) = L_n L_*$$

where $L_n$ denotes the part with observed $Y_i$, and $L_*$ is the part corresponding to unobserved $Y_i$. Specifically, denoting by $\varphi$ and $\Phi$ the density and distribution of the standard normal random variable, respectively, the density of the uncensored observations $Y_i$ is

$$f(y_i) = (1/\sigma)\varphi[(y_i - (\alpha + \beta x_i)/\sigma]$$

and, for censored $Y_i$, $L_*$ is a product of

$$
\begin{aligned}
P(Y_i < c | X_i) &= P(\varepsilon_i < [c - (\alpha + \beta X_i)]/\sigma) \\
&= \Phi([c - (\alpha + \beta X_i)]/\sigma).
\end{aligned}
$$

In 1979, James Heckman [70] proposed a technique to use OLS by correcting the sample bias. We discuss it here for the Tobit model. The more general *sample selection models* will be discussed in the next section on missing data.

First, observe that $E(\varepsilon_i | X_i, Y_i \geq c) \neq 0$ is the source of inconsistency. So let's remove it! Let $E(\varepsilon_i | X_i, Y_i \geq c) = \zeta_i$ (an unknown non-random term, unknown since it involves $\alpha, \beta$), and $u_i = \varepsilon_i - \zeta_i$. We have clearly

$$
E(u_i | X_i, Y_i \geq c) = 0.
$$

So consider

$$
Y_i = \alpha + \beta X_i + \zeta_i + u_i. \tag{7.39}
$$

Suppose we view $\zeta_i$ as a second regressor for $Y_i$, let's examine the correlation between the random error term $u_i$ and the regressors $X_i$ and $\zeta_i$. We have

$$
\begin{aligned}
E(X_i u_i | X_i, Y_i \geq c) &= EE(X_i u_i | X_i, Y_i \geq c) | X_i, Y_i \geq c) \\
&= E(X_i E(u_i | X_i, Y_i > c)) = 0
\end{aligned}
$$

and also

$$
E(\zeta_i u_i | X_i, Y_i \geq c) = 0.
$$

Thus, $u_i$ is uncorrelated with the regressors $X_i$ and $\zeta_i$. Therefore, after correcting the sample bias by normalizing the error term $\varepsilon_i$ to have zero mean under the sample selection rule, for using the observed subsample data, we are in standard setting to apply OLS to obtain consistent estimators in Equation (7.39) provided we can "observe" the second regressor variable $\zeta_i$. We will discuss this issue in the more general framework of sample selection models.

**Remark 7.4** *The censoring is governed by the response variable itself, while in Heckman's model, the missing data are governed by another variable.*

Just like in measurement-error models, ordinary least squares estimators of parameters are biased and inconsistent since $E(\varepsilon_i) \neq 0$. Indeed, since $Y_i \geq 0$, it follows that $\varepsilon_i \geq -(\alpha + \beta X_i)$ and hence for any particular value of $X_i$, $E(\varepsilon_i)$ can be negative, positive or zero. Let's turn to maximum likelihood estimation. Suppose $\varepsilon_i$ is distributed as $N(0,1)$ and let $\varphi, \Phi$ denote its density and distribution functions, respectively.

Since $\varepsilon_i \geq -(\alpha + \beta X_i)$, $\varepsilon_i^*$ is a truncated random variable of $\varepsilon_i$, i.e., the $\varepsilon_i^* = \varepsilon_i | \varepsilon_i \geq -(\alpha + \beta X_i)$ so that its density is

$$\begin{aligned} f(\varepsilon_i^*) &= \varphi(\varepsilon_i^*)/P(\varepsilon_i \geq -(\alpha + \beta X_i)) \\ &= \varphi(\varepsilon_i^*)/[1 - \Phi(-(\alpha + \beta X_i))]. \end{aligned}$$

Thus

$$E(\varepsilon_i | \varepsilon_i \geq -(\alpha + \beta X_i)) = \sigma \varphi(\alpha + \beta X_i)/\Phi(\alpha + \beta X_i)$$

where $\sigma$ is the standard deviation of $\varepsilon_i^*$. The quantity $\lambda_i = \varphi(\alpha + \beta X_i)/\Phi(\alpha + \beta X_i)$ is called the inverse Mills' ratio. If we have estimates of the $\lambda_i$, then we can use them to normalize $E(\varepsilon_i)$ to zero (considering the error $\varepsilon_i - \sigma \lambda_i$) and hence obtaining consistent estimators for the parameters. The famous two-stage estimation procedure of James Heckman is this:

1. We estimate the $\lambda_i$ by estimating $\alpha, \beta$ and plug in $\lambda_i = \varphi(\alpha + \beta X_i)/\Phi(\alpha + \beta X_i)$. Consider the probit model with $Z_i = 1$ if $Y_i^* > 0$ and $= 0$ if $Y_i^* \leq 0$,

$$p_i = \Phi(\alpha + \beta X_i).$$

   The parameters $\alpha, \beta$ are estimated by MLE, using the whole sample.

2. Since the estimates $\hat{\lambda}_i$ of $\lambda_i$ converge to $\lambda_i$ as $n \to \infty$, we consider

$$Y_i = \alpha + \beta X_i + \sigma \hat{\lambda}_i + \varepsilon_i - \sigma \hat{\lambda}_i = \alpha + \beta X_i + \sigma \hat{\lambda}_i + u_i.$$

## 7.2.5   Missing data

Missing data occur frequently in studies involving humans. Not accounting for missing data properly could lead to incorrect inferences. The most common occurrence of missing data is in survey data, in which some respondents fail to answer questions. As James Heckman has pointed out to us, when facing missing data, we need to ask the basic question: *Why are data missing?* The answer to this question is a must before we use the available data (i.e., the non-missing observations) for statistical inference. This is so because, as we will see, the source of missing data will dictate how we should design our statistical procedures appropriately to yield meaningful results. Failure to recognize this issue could lead to erroneous statistical conclusions from our analysis. Basically, there are two cases:

(i) Some are simply unavailable for reasons unknown to the analyst, so we can just "ignore it" [62]. The observed subsample is a usable data set. Perhaps we might wonder whether we can extract some useful information from the "incomplete data."

(ii) The missing data are systematically related to the phenomenon being modeled.

For example, in surveys data may be "self-selected" or "self-reported" (such as high-income individuals tending to withhold information about their income). In this case, the gaps in the data set would represent more than just missing information, and hence the observed subsample would be qualitatively different.

Recall the censored data in the Tobit model. We know that the response variable $Y_i$, for individual $i$, is missing when it is below some threshold $c$. Thus, the observed data are from a truncated distribution of $Y$. Two things come out of this: on one hand, we can describe the "sample selection rule," namely, the sample "selects" the data in the sample according to a rule to reveal to us the observed ones, and on the other hand, this sample selection rule provides us with the correct distributional information to analyze our observed data, leading to correct inferences.

This situation is general. Figuring out the source of the missing data, we will describe it as a sample selection rule. Once we have this rule, we proceed as in the Tobit special type of missing data. As we will see in Heckman's sample selection models, this will lead to two equations to represent the situation:

(i) An *outcome* equation, which is our main equation of interest, describing the relation between our response variable and its determinants (covariates),

(ii) A *selection* equation, which describes the sample selection rule, leading to the observed data.

We are interested in case (ii) as James Heckman has emphasized in his fundamental work [70]. Missing data occur frequently in statistical analysis of economic data for forecasting. We will elaborate on it in the context of regression. But since we will introduce Heckman sample selection models and his two-step estimation procedure in linear regression models, we need to review some special *generalized linear models*, namely, the *probit* and *logit models*.

## Regression for discrete response variables

Consider the so-called "linear probability model," or models for qualitative choices. This is the case where our response variable $Y$ is *discrete*, but the regressor $X$ could be continuous. For example, consider the labor force participation problem, in which $Y_i = 1$ or $0$ according to the respondent either works or does not work. We believe that the decision to work or not can be explained by a set of covariates such as age, martial status, education, etc., gathered in a vector $X$, i.e., we can write

$$P(Y = 1|X) = \Psi(X, \beta) = 1 - P(Y = 0|X),$$

where $\Psi$ is some function, and $\beta$ is a vector of parameters.

A simplest model (i.e., specifying the function $\Psi$) is the linear regression model where we take $\Psi$ as a linear function in $X$, i.e., $\Psi(X, \beta) = X'\beta$. In this model, the mean of $Y$ is predicted by the linear predictor $X'\beta$, since $E(Y|X) = P(Y = 1|X) = X'\beta$. From that, we can write

$$Y = E(Y|X) + Y - E(Y|X) = X'\beta + \varepsilon.$$

However this linear model for $Y$ has a number of shortcomings. First, the variance of $\varepsilon$ will depend on $\beta$ ($\varepsilon$ is heteroskedastic). Indeed, as $Var(\varepsilon) = Var(Y)$ and since $Y$ is Bernoulli, we have

$$Var(\varepsilon|X) = X'\beta(1 - X'\beta).$$

More importantly, recalling that $E(Y|X) = P(Y = 1|X) = X'\beta$, there is no guarantee that $X'\beta \in [0, 1]$ for all possible values of $X$, since we cannot constraint $X'\beta$ to the interval $[0, 1]$. This failure could lead to both nonsense probabilities and negative variance! Thus we should replace the linear probability model by another model which makes sense! One way is to "scale" the value $X'\beta$ so that it will always be in $[0, 1]$, in a suitable fashion. This can be achieved by using an appropriate function $F$, like a distribution function of a random variable, and write

$$P(Y = 1|X) = F(X'\beta).$$

Each choice of $F$ leads to a (nonlinear) model. For example, if we choose $F(x)$ to be the distribution function of the standard normal random variable, then we get the so-called probit model. The probit model was developed by Bliss in 1935.

How do we estimate $\beta$ from our observed data $(X_i, Y_i), i = 1, 2, ..., n$? Well, conditional on $X_i$, $Y_i$ is Bernoulli with parameter $p_i = F(X_i'\beta)$, and hence the likelihood of $\beta$ given the observations is

$$\prod_{i=1}^{n} p_i^{Y_i}(1-p_i)^{1-Y_i} = \prod_{i=1}^{n}(F(X_i'\beta))^{Y_i}(1-F(X_i'\beta))^{1-Y_i},$$

so that we are using maximum likelihood method for estimation.

Now the use of $F$ as the normal distribution function in the above probit model presents difficulties in computation since there is no closed form for $F^{-1}$. Another much simpler $F$ which is quite "close" to the normal distribution function is the *logistic* distribution function, namely

$$F(x) = 1/(1+e^{-x})$$

for which $F^{-1}(t) = \log(t/(1-t))$. Using the logistic $F$, the associated model is called the logit model, and we are using *logistic regression models.*

## Sample selection models

As stated before, we are concerned with statistical inference, especially in regression models, with missing data where we can describe the missing data mechanism. Facing missing observations in our survey data, we seek to discover why our data are missing. A full understanding of the situation under study will help in answering this question. With that qualitative information, we specify the way data are missing. Each such description of the sample selection rule gives rise to a *model*, and hence the term "sample selection models." More general than the Tobit censored regression model, in Heckman's sample selection models, the response variable is observed only when some criterion based on some other random variable is met. We could classify samples (in regression with response variable $Y$, and regressor $X$) as follows:

(i) *Censored samples*: $Y$ is observed only if some criterion defined in terms of $Y$ is met, $X$ is observable in the whole sample (regardless of whether $Y$ is observed or not),

(ii) *Truncated samples:* $Y$ is observed if some criterion defined in terms of $Y$ is met, $X$ is observed only if $Y$ is observed,

(iii) *Selected samples:* $Y$ is observed only if some criterion defined in terms of some other random variable ($Z$) is met, $X$ and another variable $W$ (which forms the determinants of whether $Z = 1$) are observable for the whole sample (regardless of whether $Y$ is observed or not).

Let's describe now the setting of *linear regression in a model of se-lected samples*. To understand the set-up we are going to put down, it is helpful to have a simple example. Consider the classic example of female labor supply. We wish to study the determinants of market wages of female workers. The market wage level for each female worker $(Y_i)$ could depend on observable characteristics (determinants) of the worker $(X_i)$ such as age, education, experience, marital status). Each woman $i$ sets a reservation wage level to accept to work, so let $Z_i$ denote the difference between the market wage offered to $i$ and her reservation wage. When $Z_i > 0$, the woman $i$ is employed and her wage is observed. Suppose linear models are appropriate for the study! The model of interest is:

$$Y_i = \alpha X_i + \varepsilon_i. \tag{7.40}$$

In the literature, this *outcome equation* is also written with $Y_i$ replaced by $Y_i^*$ to remind us that in fact $Y_i$ is a latent variable.

Suppose that $Z$ can also be regressed linearly by some covariate $W$, i.e.,

$$Z_i = \beta W_i + u_i. \tag{7.41}$$

This is the *selection equation* of the model. Again, in fact $Z$ is a latent variable (representing the "propensity" to work). The *sample selection rule* is this: $Y_i$ is observed only when $Z_i > 0$.

But $Z_i$ is not observable, what is observed is whether it is positive or negative, i.e., only the sign of $Z_i$ is observed. If we let the dummy variable $\delta_i$ taking values 1 or 0 according to $Z_i$ is positive or negative, then, given a sample of $n$ individuals, we actually "observe" $(X_i, W_i, \delta_i), i = 1, 2, ..., n$, whereas we observe $Y_i$ only for $i$ for which $\delta_i = 1$. Note that the Tobit model is a special case in which $Z = Y$.

Using only the subsample of available data (i.e., only for the individ-uals $i$ for which $Y_i$ is observed, equivalently, for which $\delta_i = 1$, by sample selection rule) to estimate the primary parameter $\alpha$ in equation (7.40) will result in biased estimates (i.e., the OLS fails like in the Tobit model), since $E(\varepsilon_i|\text{sample selection rule}) \neq 0$.

Indeed,

$$E(\varepsilon_i|X_i, W_i, \delta_i = 1) = \sigma\lambda(\beta W_i) \neq 0$$

where $\sigma = Cov(\varepsilon_i, u_i)$, assuming that $(\varepsilon_i, u_i)$ is bivariate normal with $Var(\varepsilon_i) = Var(u_i) = 1$ (for simplicity), and $\lambda(t) = \varphi(t)/[1 - \Phi(t)]$ which is the hazard rate of the standard normal random variable, which is also called the reciprocal of the Mills ratio. We denote as before by $\varphi$ and $\Phi$ the density function and distribution function of the standard normal random variable. Now, with the specification of error distribution, the parameters $\alpha$ and $\beta$ can be estimated at the same time by maximum likelihood method which is complicated. James Heckman proposed a *two-step estimation procedure* to avoid complications of maximum likelihood method. There are two steps since his method is sequential: Estimate $\beta$ first, then use it to estimate $\alpha$.

**Step 1**  Estimate $\beta$ by probit analysis, applying to the selection equation alone. We have

$$Z_i = \beta Wi + u_i$$

where $u_i$ is $N(0,1)$, and

$$\delta_i = 1 \text{ or } 0 \text{ according to } Z_i > 0 \text{ or } Z_i \le 0.$$

We then have

$$
\begin{aligned}
\pi_i &= P(\delta_i = 1) \\
&= P(Z_i > 0) \\
&= P(\beta W_i + u_i > 0) \\
&= P(u_i > -\beta W_i) \\
&= 1 - \Phi(-\beta W_i) \\
&= \Phi(\beta W_i).
\end{aligned}
$$

Thus, we are in the setting of probit analysis.

The likelihood of $\beta$ given the observations $(W_i, \delta_i)$, $i = 1, 2, ..., n$ (i.e., the whole sample) is

$$
\begin{aligned}
L_n(\beta | (W_i, \delta_i), i = 1, 2, ..., n) &= \prod_{i=1}^{n} \pi^{\delta_i}(1 - \pi)^{1-\delta_i} \\
&= \prod_{i=1}^{n} (\Phi(\beta W_i))^{\delta_i}(1 - \Phi(\beta W_i))^{1-\delta_i}.
\end{aligned}
$$

Maximizing this marginal likelihood, we obtain the estimator $\dot{\beta}$ of $\beta$.

**Step 2**  Estimate $\alpha$ as follows. Since $E(\varepsilon_i | X_i, W_i, \delta_i = 1) = \sigma \lambda(\beta W_i) = \sigma \lambda_i$ where we set $\lambda_i = \lambda(\beta W_i)$, we could normalize the error under the sample selection rule, i.e., $(\varepsilon_i | X_i, W_i, \delta_i = 1)$ to have zero mean by changing it to $u_i - \sigma \lambda_i$. For that, we rewrite (7.40) as

$$Y_i = \alpha X_i + \sigma \lambda_i + (u_i - \sigma \lambda_i).$$

Now if we estimate $\lambda_i$ by the plug-in estimator $\hat{\lambda}_i = \lambda(\hat{\beta}W_i)$, for all $i = 1, 2, ..., n$, and consider

$$Y_i = \alpha X_i + \sigma\hat{\lambda}_i + (u_i - \sigma\hat{\lambda}_i) \tag{7.42}$$

then, asymptotically, the error $(u_i - \sigma\hat{\lambda}_i)$ under the sample selection rule has zero mean and uncorrelated with $X_i$ and $\hat{\lambda}_i$. Thus, we could consider (7.42) as a linear model with two regressors, $X_i$ and $\hat{\lambda}_i$.

Applying only the observed data to (7.42) and using OLS estimation procedure, we get a consistent estimator for $\alpha$.

## 7.3   Modeling dependence structure

We will touch upon the issue of modeling data structure in the context of stochastic dominance. In view of the restricted context, we will not discuss dependence structures of economic time series data.

### 7.3.1   Copulas

Let $\mathbf{X} = (X_1, X_2, ..., X_n)$ be a random vector with joint distribution function

$$\begin{aligned}
F &: \quad \mathbb{R}^n \to [0, 1] \\
F(x_1, x_2, ..., x_n) &= \quad P(X_1 \le x_1, X_2 \le x_2, ..., X_n \le x_n) \\
&= \quad P_{\mathbf{X}}\left[\prod_{i=1}^{n}(-\infty, x_i]\right].
\end{aligned}$$

The joint distribution contains all the statistical information about $(X_1, X_2, ..., X_n)$. In particular, we can derive marginal distributions of the components as

$$F_j(x_j) = P(X_j \le x_j) = F(\infty, \infty, ..., x_j, \infty, ..., \infty).$$

The converse problem is this. Suppose we know all marginal distributions $F_j, j = 1, 2, ..., n$. Can we "guess" their joint distribution? Clearly, the answer is no. There are many different joint distributions which admit the same marginal distributions. With some additional information about the *dependence structure* of the components $X_j$'s, the answer could be yes. For example, if the components are mutually independent, then

$$F(x_1, x_2, ..., x_n) = \prod_{i=1}^{n} F_i(x_i).$$

Fréchet observed that if given the marginals $F_i$, then the joint $F$ cannot exceed $F(x_1, x_2, ..., x_n) = \wedge_{i=1}^{n} F_i(x_i)$. That is clear by the very definition of $F$ in terms of $P$, noting that

$$(X_1 \leq x_1, X_2 \leq x_2, ..., X_n \leq x_n) = (X_1 \leq x_1) \cap (X_2 \leq x_2) \cap ... \cap (X_n \leq x_n)$$

and

$$P[(X_1 \leq x_1) \cap (X_2 \leq x_2) \cap ... \cap (X_n \leq x_n)] \leq \wedge_{i=1}^{n} P(X_i \leq x_i).$$

In the case $n = 2$, Fréchet also obtained the lower bound, namely

$$F(x_1, x_2) \geq \max\{F_1(x_1) + F_2(x_2) - 1, 0\}$$

again by manipulating probabilities of events. In other words, given the marginals $F_i$, all possible forms of binary $F$ lie between the two bounds

$$\max\{F_1(x_1) + F_2(x_2) - 1, 0\} \leq F(x_1, x_2) \leq \min\{F_1(x_1), F_2(x_2)\}.$$

We need (economic) data to validate economic relationships between economic variables for applications. Data can be experimental or observational. In any case, the dependence structures of data $(X_1, X_2, ..., X_n)$ is essential as they dictate how inference procedures should be developed. Put it differently, inference procedures depend on the "type" of data available. It is not clear until the work of Sklar [148] that Fréchet's question is precisely about *modeling dependence structures of random variables*. As we will see, not only copulas describe dependence structures, they can be used to generate new multivariate distributions, a problem of great importance for applications. As such, it is not surprising that, after recognizing that concept of copula is fundamental for the above issues in around 1999, copulas appear, at an increasing speed, in economic research.

The key point is the observation that each $F_i(x_i)$ is a number in $[0, 1]$ and that a situation like

$$F(x_1, x_2, ..., x_n) = \wedge_{i=1}^{n} F_i(x_i)$$

can be rewritten as

$$F(x_1, x_2, ..., x_n) = C(F_1(x_1), F_2(x_2), ..., F_n(x_n)),$$

where

$$C : [0, 1]^n \to [0, 1]$$
$$C(a_1, a_2, ..., a_n) = \wedge_{i=1}^{n} a_i,$$

expressing explicitly $F$ as a function of its marginals. In other words, it suffices to find $C$ to arrive at a relation between $F$ and its marginals. So the first task is to find out basic properties of $C$ in

$$F(x_1, x_2, ..., x_n) = C(F_1(x_1), F_2(x_2), ..., F_n(x_n))$$

in order to *axiomatize it*, just as we did for the concept of distribution functions.

To figure out the appropriate conditions for $C(\cdot, \cdot)$, let's consider the case where we have two random variables $U, V$, each uniformly distributed on $(0, 1)$.

Now let $C(\cdot, \cdot)$ be the joint distribution of $(U, V)$, i.e., $C(a, b) = P(U \leq a, V \leq b)$, then, as a joint distribution, $C(\cdot, \cdot)$ satisfies the following:

(i)  $C(a, b) = 0$ if at least $a$ or $b = 0$,

(ii)  $C(a, 1) = a$, $C(1, b) = b$, and

(iii)  For $a, b, c, d \in [0, 1]$, such that $a \leq b \leq c \leq d$,

$$C(b, d) - C(b, c) - C(a, d) + C(a, c) \geq 0.$$

Note that the condition (iii) is a condition of a joint distribution function (as defined axiomatically to generate a probability measure on $\mathcal{B}(\mathbb{R}^2)$) which follows from properties of the probability measure $P$. It means simply that the probability for $(X, Y)$ to fall into any rectangle is nonnegative!

If $X$ and $Y$ have continuous distribution functions $F_X, F_Y$, respectively, then $U = F_X(X)$ and $V = F_Y(Y)$ are each uniformly distributed on $(0, 1)$. Let $C(\cdot, \cdot)$ be the joint distribution function of $(U, V)$. Then necessarily

$$H(x, y) = P(X \leq x, Y \leq y) = C(F_X(x), F_Y(y)).$$

It can be checked that the following are 2-copulas:

(i)  $C_\alpha(x, y) = x + y - 1 + [(1 - x)^{-1/\alpha} + (1 - y)^{-1/\alpha}]^{-\alpha}$, $\alpha \neq 0$,

(ii)  $C_\theta(x, y) = xy + \theta xy(1 - x)(1 - y)$, $\theta \in [-1, 1]$, and

(iii)  $C_\theta(x, y) = [\min(x, y)]^\theta [xy]^{1-\theta}$, $\theta \in [0, 1]$.

Note that Fréchet's bounds now mean that if $C$ is a 2-copula, then

$$(x + y - 1) \vee 0 \leq C(x, y) \leq x \wedge y.$$

An interesting copula in survival analysis is the Marshall-Olkin copula which is a *parametric* copula (i.e., depending on a finite dimensional parameter):

$$C_\theta(x, y) = (xy)(x^{-\theta} \wedge y^{-\theta}), \theta \in [0, 1].$$

For $\theta = 0$, we get $C_0(xy) = xy$ and for $\theta = 1$, we get $C_1(x, y) = x \wedge y$.

These copulas are joint distributions on $\mathbb{R}^2$ (with appropriate extension of its domain) having uniform marginals on $[0, 1]$. With these copulas, we obtain new bivariate distributions, for example, using exponential margins or Weibull margins.

Now, an important corollary to Sklar's theorem is this. Let $H(\cdot, \cdot)$ be any bivariate distribution function with *continuous* marginal distributions $F, G$. Then there exists a unique copula $C$ such that

$$C(u, v) = H(F^{-1}(u), G^{-1}(v))$$

where $F^{-1}(u) = \inf\{x \in \mathbb{R} : F(x) \geq u\}$.

This corollary says that from $H$, we can extract its $C$, after, of course, deriving the marginals. This result is essential for simulations (generating copulas) as well as for copula estimation. Also, this corollary allows us to propose lots of multivariate models: after extracting a copula $C$ from $H, F$ and $G$, we can use $C$ to create new joint distributions, independently of the previous marginals. This is a method for generating copulas called the *inversion method*.

For example, if $X, Y$ are independent with $F, G$ continuous, then

$$
\begin{aligned}
H(F^{-1}(u), G^{-1}(v)) &= P(X \leq F^{-1}(u), Y \leq G^{-1}(v)) \\
&= P(X \leq F^{-1}(u))P(Y < G^{-1}(v)) \\
&= P(F(X) \leq F \circ F^{-1}(u))P(G(Y) \leq G \circ G^{-1}(v)) \\
&= P(F(X) \leq u)P(G(Y) \leq v) \\
&= uv = C(u, v)
\end{aligned}
$$

by noting that when $F$ is continuous, $F(X)$ is uniformly distributed on $[0, 1)$ and $F \circ F^{-1}$ is the identity function. A Gaussian copula is derived from multivariate normal distribution using the above inversion method.

## 7.3.2   Copulas for sampling designs

We illustrate here an example where we can choose copulas for specific purposes. This is an application of copulas to a basic problem in collecting data, namely, *sampling designs for finite populations*. It also illustrates the case of discrete random vectors. The problem in sampling from a finite population $U$ of size $|U| = n$ is this.

Let $\theta : U \to \mathbb{R}$ be a quantity if interest, e.g., $\theta(u)$ is the income of the individual $u \in U$. Suppose we cannot conduct a census and are interest in the population total $T(\theta) = \sum_{u \in U} \theta(u)$. To obtain an estimate of $T(\theta)$, we could rely on the values of $\theta(.)$ on some part $A$ of $U$ and use $\sum_{u \in A} \theta(u)$ as an estimator. This is the essence of *inductive logic*, namely "making statements about the whole population from the knowledge of a part of it." To make this inductive logic reliable (valid), we create a chance model for selecting subsets of $U$. This probabilistic approach not only has public acceptance (!) but is also essential in avoiding biases as well as providing necessary ingredients for assessing errors and the performance of the estimation procedure.

A man-made chance model is then specified by a set-function

$$f : 2^U \to [0, 1]$$

such that $\sum_{A \subseteq U} f(A) = 1$, where $f(A)$ is the probability of selecting the subset $A$ of $U$.

In practice, it is desirable to specify the *inclusion function* $\pi : U \to [0, 1]$, where $\pi(u)$ is the probability that the individual $u$ will be included in the selected sample. The set-function $f$ is called the sampling design. With $\pi$ specified, we are looking for designs $f$ such that

$$\pi(u) = \sum_{u \in A} f(A).$$

Let's number elements of $U$ as $\{1, 2, ..., n\}$. Then, for each $j \in U$, consider the Bernoulli random variable

$$I_j \quad : \quad 2^U \to \{0, 1\}$$

$$I_j(A) \quad = \quad \begin{cases} 1 & \text{if } j \in A \\ 0 & \text{if not} \end{cases}$$

with parameter

$$\pi(j) = P(I_j = 1) = \sum_{I_j(A)=1} f(A).$$

The sampling design $f$ is equivalent to the joint distribution of the Bernoulli random vector $(I_1, I_2, ..., I_n)$. This can be seen as follows. Making the bijection between $2^U$ and $\{0, 1\}^n$:

$$A \longleftrightarrow (\varepsilon_1, \varepsilon_2, ..., \varepsilon_n) = \varepsilon \in \{0, 1\}^n$$

with

$$A(\varepsilon) = \{j \in U : \varepsilon_j = 1\}.$$

We have

$$f(A(\varepsilon)) = P(I_1 = \varepsilon_1, I_2 = \varepsilon_2, ..., I_n = \varepsilon_n).$$

Thus, if we specify $\pi$, then $(I_1, I_2, ..., I_n)$ has *fixed marginals*. As such, this joint distribution is determined by some $n$-copula, according to Sklar's theorem. Specifically, let $F_j$ be the distribution function of $I_j$, i.e.,

$$F_j(x) = \begin{cases} 0 & \text{if } x < 0 \\ 1 - \pi(j) & \text{if } 0 \le x < 1 \\ 1 & \text{if } x \ge 1. \end{cases}$$

Let $C$ be an $n$-copula. Then the joint distribution function of $(I_1, I_2, ..., I_n)$ is of the form

$$F(x_1, x_2, ..., x_n) = C(F_1(x_1), F_2(x_2), ..., F_n(x_n)).$$

Hence each choice of $C$ leads to a possible sampling design compatible with the specified inclusion function $\pi$. For example, with the choice of

$$C(u_1, u_2, ..., u_n) = \prod_{i=1}^n u_i$$

we obtain the well-known Poisson sampling design

$$f(A) = \prod_{i \in A} \pi(i) \prod_{i \notin A} (1 - \pi(i)).$$

If we choose

$$C(u_1, u_2, ..., u_n) = \wedge_{i=1}^n u_i$$

we get

$$f(A) = \sum_{B \subseteq A} (-1)^{|A \setminus B|} [1 - \vee_{j \notin B} \pi(j)].$$

## 7.3.3  Estimation of copulas

A copula captures the dependence structure of a random vector of interest. Rather than proposing a copula, we could use data to make an "educated guess" about its true structure.

The problem of copula estimation is still in a research phase as it is more complicated than joint density functions. Suppose we wish to estimate the copula $C(u, v)$ of a bivariate variable $(X, Y)$, with joint distribution function $H$ and marginals $F, G$.

**Nonparametric estimation**  Since

$$H(x, y) = C(F(x), G(y))$$

and

$$C(u, v) = H(F^{-1}(u), G^{-1}(v))$$

we can use a plug-in estimator. Let $(x_i, y_i), i = 1, 2, ..., n$ be the sample data, we can first estimate $F, G$ by their empirical distribution functions $F_n$, $G_n$, then use $F_n^{-1}, G_n^{-1}$. For an estimator of $H$, kernel estimator could be used.

**Parametric estimation**  Here, we consider a family of copulas $C_\theta, \theta \in \Theta$ for a random vector $(X, Y)$. Suppose the marginals $F, G$ is also parametric with parameters $\alpha, \beta$, respectively. Let $\lambda = (\theta, \alpha, \beta)$. We have

$$H(x, y) = C_\theta(F(x; \alpha), G(y; \beta)).$$

Differentiating this equation, we get

$$h(x, y) = c_\theta(F(x), G(y))f(x)g(y),$$

where the copula density $c_\theta(\cdot, \cdot)$ is

$$c_\theta(u, v) = \partial^2 C_\theta(u, v) / \partial u \partial v.$$

The log-likelihood function is

$$L(\theta) = \sum_{i=1}^{n} \log c_\theta(F(x_i; \alpha), G(y_i; \beta)) + \sum_{i=1}^{n} [\log f(x_i) + \log g(y_i)]$$

and the MLE of $\lambda$ can be obtained.

## 7.4 Some additional statistical tools

As a complement, we mention below Bayesian statistics followed by an engineering tool (the Kalman filter) which has become a useful tool in econometrics.

### 7.4.1 Bayesian statistics

If we allow all possible estimators (unbiased or not) into the competition, then the UMVUE (uniformly minimum variance unbiased estimator) might not be the best, as some biased estimator could have smaller mean squared error. Note that error is what we are mostly concerned about! One way to deal with this issue is this.

The risk at $\theta$ of the estimator $T$ for the parameter $\varphi(\theta)$, is

$$R(\theta, T) = E_\theta(T - \varphi(\theta))^2.$$

An estimator $T^*$ is called a *minimax estimator* if

$$\sup_{\infty \in \Theta} R(\theta, T^*) = \inf_T \sup_{\theta \in \Theta} R(\theta, T).$$

One way to get minimax estimators is using *Bayes' estimators*. Viewing the unknown parameter $\theta$ as a random variable, we assign to it a distribution $\pi$, called the *prior distribution* (before collecting data) expressing our probabilistic (subjective) information about possible location of $\theta$. Then we go out collecting data $X_1, X_2, ..., X_n$. Then we revise our belief about $\theta$ by a *posterior distribution*, i.e., after getting data, by using Bayes' formula,

$$f(\theta|X_1, X_2, ..., X_n) = f(X_1, X_2, ..., X_n|\theta)\pi(\theta)/\int f(X_1, X_2, ..., X_n|\theta)\pi(\theta)d\theta,$$

and the Bayes' estimator is the posterior conditional mean, i.e.,

$$T^* = \int_{\Theta k} \theta f(\theta|X_1, X_2, ..., X_n)d\pi(\theta).$$

Under suitable conditions, Bayes' estimators are *admissible*, i.e., there are no better estimators.

### 7.4.2 From linear regression to filtering models

The Kalman filter is a statistical tool (filter) useful for estimating (recursively) unobserved and time-varying parameters or variables in econometric models. We will elaborate on its formulation with indications on applications in econometrics.

We are familiar with the following situation in econometrics. Let $Y$ be a variable of interest, and $X$ be its covariate. We wish to regress $Y$ on $X$. The simplest model for the link between $Y$ and $X$ is a linear model of the form

$$Y = \theta X + \varepsilon$$

where $\theta$ is an unknown parameter (which could be a scalar or vector) and $\varepsilon$ is the random error term, usually, in a classical setting, normally distributed with zero mean and constant variance $\sigma^2$ (unknown). The objective is to arrive at the conditional mean $E(Y|X = x) = \theta x$. For that purpose, we need to estimate $\theta$ from the data $(X_i, Y_i), i = 1, 2, ...n$, a random sample from $(X, Y)$.

In this standard linear regression model, the unknown parameter $\theta$ is not viewed as a random variable (i.e., no Bayesian philosophy involved) and independent of the "time" $i$ of the observations $(X_i, Y_i)$, i.e., we have

$$Y_i = \theta X_i + \varepsilon_i, i = 1, 2, ..., n.$$

Note right away that this model is appropriate when the relationship between $Y$ and $X$ seems *constant*, at least during the estimation period.

Otherwise, a *time-varying parameter linear model* should be considered. For example, in macroeconomics, relationships for countries that have undergone structural reforms during the sample period, the $\theta$'s should be indexed by "time" $i$. As we will see, in such models, the Kalman filter can be used to estimate the time-dependent parameters of the model, provided, as in this example, we can postulate structural changes in demand for international reserves, which forms an additional equation (called "system equation") needed to carry out the Kalman filter. Here $Y_t$ is the reserve demand of a country, and $X_t$ is its imports.

Soon, even in the context of econometrics (where we make observations sequentially in *time*, e.g., data as a time series), we will replace the index $i$ to $t$ to emphasize the fact that we do make observations at various points in time. Note that although time is continuous in nature, we will use discrete time, as in practical applications (quality control or radar tracking of moving targets) we sample at discrete times, i.e., we discretize the time interval. Thus, we will write $1, 2, ..., t-1, t$ as observation time periods. In this setting, our regression coefficient $\theta$ in the standard linear model is *time-invariant*. Thus, the linear regression is described by one *equation* (which is our linear model above) with time-invariant parameters, and the available data are $(X_i, Y_i)$, $i = 1, 2, ..., t$.

We are interested in estimating $\theta$ based on the data up to time $t$ by some statistic $\hat{\theta}_t$, but $\theta$ itself does not depend on time $t$. Thus, the problem is just to estimate the constant (but unknown) $\theta$ from our available data. Of course, if we happen to have one more observation $(X_{t+1}, Y_{t+1})$, we will estimate $\theta$ based on $(X_i, Y_i)$, $i = 1, 2, ..., t, t + 1$. That could be done *recursively*, i.e., using $\hat{\theta}_t$ and $(X_{t+1}, Y_{t+1})$ to arrive at $\hat{\theta}_{t+1}$, rather than recompute it from the statistic leading to $\hat{\theta}_t$. Again, it could be an improved estimate for *the same* $\theta$ since we have more information about it.

Consider now a linear regression model in which the regression coefficient $\theta$ depends on time, i.e., the model is

$$Y_i = \theta_i X_i + \varepsilon_i, i = 1, 2, ..., t \qquad (7.43)$$

termed a *time-varying parameter* linear model, where the variance of $\varepsilon_t$ is now depending on $t$, say $\sigma_t^2$. Then, what is our goal, i.e., what do we want to estimate and what for, from the same type of data $(X_i, Y_i), i = 1, 2, ..., t$? Well, among other things, perhaps we would like to estimate the value of $\theta_t$. Now, in this time-varying parameter model, the variables $X, Y$, without time index, do not make sense any more, as in the standard time-invariant model. For example, if $Y$ is the position of an object, it only makes sense to talk about "position at some point in time." Thus, $\theta_t$ is a link between $Y_t$ and $X_t$. But if we are going to use the whole data set $(X_i, Y_i)$, $i = 1, 2, ..., t$, to estimate $\theta_t$, these past data should bear some information on $\theta_t$, in other words, the $\theta_t$'s should be related somehow.

For example, suppose

$$\theta_t = G_t \theta_{t-1} + w_t \qquad (7.44)$$

where $G_t$ is known, and the error term $w_t$ is normally distributed with zero mean and variance $\sigma^2(w_t)$, time dependent. Thus, we have two equations forming our new model. The first equation 7.43 is called the *observation (or measurement) model* since observed data come from it. We could also call it the "observed model," whereas the second equation 7.44 is called the *structural or system equation*. We could also call it the "unobserved model," since no observed data come out from this model. The above should remind you of many similar situations in statistics or econometrics. This is the case of *coarse data* (i.e., data of low quality, such as data with measurement error, censored data, grouped data, incomplete data, imprecise data, approximate data and indirect measurements). For example, see the auction data in the previous lecture. The simplest case is *measurement-error data in linear models*. Here we have two equations describing the situation:

(i) A standard linear model $Y = \theta X + \varepsilon$, and

(ii) A measurement model $Z = X + u$.

The covariate $X$ is unobserved. It is observed by another variable $Z$. The goal is to estimate $\theta$ from the data $(Y_i, Z_i), i = 1, 2, ..., n$. As we will see, Kalman filter was developed for control engineering problems. Then the above structure becomes apparent (specific examples in subsequent sections). Equation 7.43 is the observation equation where $X_t$ plays the role of a (linear) operator representing our capability of $\theta_t$ resulting in $Y_t$. Another way of saying this is "the observed value $Y_t$ is the 'output' of the 'hidden' $\theta_t$," i.e., $Y_t$ depends on the unobservable quantity $\theta_t$ (see the hidden Markov model below).

Equation 7.44 is the dynamics of a system whose "state" at time $t$ is $\theta_t$. It is a linear dynamical system since $G_t$ is a matrix (time-dependent) in discrete-time. The fact that $\theta_t$ only depends on $\theta_{t-1}$ and not on previous states is an assumption known as *Markov* assumption "the future depends only on the present, and not the past," or put it differently "given the present, the future and the past are independent." Thus, if we want to predict $\theta_t$, we only need an estimate of $\theta_{t-1}$ and measurement obtained at time $t$, i.e., $Y_t$. That is why *recursive estimation* comes into the picture. Putting it more nicely, "The Markov property dictates recursive estimation." It is a dynamical system since it is the discrete version of a continuous linear system of the form $\dot{\theta} = A\theta$, where derivative is with respect to time.

In engineering applications, such as tracking a moving target (details below), the "quantity" of interest is the *state of the system*, here that is the position, speed and acceleration of the target at a time $t$. Since these components cannot be measured (observed) with accuracy, it is necessary to estimate the states using both measurement data (via an observed equation) and the dynamics of the target which supplies Equation 7.44. By the very nature of problems of this type, successive measurements $Y_i, i = 1, 2, ..., t$, up to time $t$, should be used to estimate $\theta_t$. While $E(Y|X = x)$ makes no sense, an expression like $E(Y_t|X_t = x)$ does make sense!

Now if we look at Equation 7.44, we see that while the state $\theta_t$ is in fact nonrandom (only unknown), but the error term $w_t$ makes it a stochastic (random) equation. In other words, $\theta_t$ becomes a random variable. This is precisely the Bayesian point of view! And hence, the problem of estimation of the state $\theta_t$ will be carried out within the framework of Bayesian statistics (using Bayes' formula). So let us formulate the *filtering* problem with the Kalman filter (details will be given in subsequent sections). We want to filter out the noise to arrive at a good estimate of $\theta_t$ using information up to time $t$ (the present). The Kalman filter is a recursive algorithm to achieve this in an optimal fashion, i.e., minimizing the mean squared error.

It works as follows: We start out by making our "best guess" of $\theta_0$ (say, its prior mean and variance). At time $t = 1$, we get a measurement $(Y_1, X_1)$ and we produce $\hat{\theta}_1$ by a "formula" for which we will give details later. At time $t = 2$, we get a new observation $(Y_2, X_2)$. We follow a recursive procedure, using $\hat{\theta}_1$ and $(Y_2, X_2)$ to produce $\hat{\theta}_2$, and so on, until getting $\hat{\theta}_t$.

### 7.4.3 Recursive estimation

Since the computational efficiency of the estimation procedure that we are going to discuss in this lecture is recursiveness, let's start out by illustrating it briefly.

Consider the problem of estimating a quantity by successive measurements. Let the measurements at times $1, 2, ..., n$ be $x_1, x_2, ..., x_n$. We use their average as our best guess for the true value of the quantity, i.e.,

$$\mu_n = (x_1 + x_2 + ... + x_n)/n.$$

If we get a new measurement $x_{n+1}$, we should use it to update our guess. Of course, we can recompute the new sample mean as

$$\mu_{n+1} = (x_1 + x_2 + ... + x_n + x_{n+1})/(n+1).$$

But it is more efficient to use our old $\mu_n$ (already computed) and make a small correction using the new measurement $x_{n+1}$. Such a strategy, if possible, is called *recursive computation*.

Here, we have

$$
\begin{aligned}
\mu_{n+1} &= (n/(n+1))[(1/n)\sum_{i=1}^{n} x_i + (1/n)x_{n+1}] \\
&= (n/(n+1))\mu_n + (1/(n+1))x_{n+1} \\
&= \mu_n + K(x_{n+1} - \mu_n),
\end{aligned}
$$

where the "gain factor" $K = 1/(n+1)$.

Thus, the update $\mu_{n+1}$ is computed using quickly $\mu_n$ and $x_{n+1}$ after figuring out the gain factor. As an exercise, you are asked to verify the recursive computation of the variance, using the above recursive computation of the mean.

Let

$$\sigma_n^2 = (1/n) \sum_{i=1}^{n} (x_i - \mu_n)^2$$

and verify that

$$
\begin{aligned}
\sigma_{n+1}^2 &= (1/(n+1)) \sum_{i=1}^{n+1} (x_i - \mu_{n+1})^2 \\
&= (1/(n+1)) \sum_{i=1}^{n+1} [x_i - \mu_n - K(x_{n+1} - \mu_n)^2] \\
&= (1-K)[\sigma_n^2 + K(x_{n+1} - \mu_n)^2]
\end{aligned}
$$

with $K = 1/(n+1)$.

## 7.4.4   What is a filter?

Next, since we are going to talk about *filters*, let's spell it out and ask when do we need filters?

Consider a time series $X_1, X_2, ..., X_n$ as our current data (observations).

(i) *Forecaster*: find a best guess for the *future* $X_{n+1}$ based, of course, on the data.

(ii) *Smoother (or interpolation)*: look back at the *past* data and compute the best possible $X_i$ based on the data.

(iii) *Filter*: provide a correction for a *present* $X_n$ based on its past $X_1, X_2, ..., X_{n-1}$ and an inexact measurement of $X_n$.

In general we need a filter to filter out the noise from our actual observation, such as to filter out the noise in an audio or video system. The filtering process is needed especially when our models contain *unobservable* variables. In other words, there are economic models in which the variables of interest are *hidden*, and we can only observe their outputs. Thus, there is a need to estimate the *states* of the system. Clearly, we need filtering also when we can only make *indirect measurements* on states of nature, as it happens often in physical sciences. We have run into such a situation in auction theory in a previous lecture. Another well-known situation is in biology, e.g., the identification of DNA, called the hidden Markov model. For your information, here is a tutorial.

In many physical domains, the observed data are not necessarily independent. A type of dependence which is not "too far" from independence is "conditional independence." A sequence of random variables $(X_n, n \geq 0)$, where each $X_n$ take values in a common discrete (state) space $S$, is called a (discrete-time) *Markov chain* when it satisfies the following *Markov property:*

For any $i_0, \ldots, i_{n-1}, i, j \in S$, and $n \geq 0$,

$$
\begin{aligned}
P(X_{n+1} &= j | X_0 = i_0, ..., X_{n-1} = i_{n-1}, X_n = i) \\
&= P(X_{n+1} = j | X_n = i),
\end{aligned}
$$

i.e., the future is independent of the past, given the present. Of course, i.i.d. sequences of (discrete) random variables are (special) Markov chains. Markov chains are *stochastic processes*. Its random evolution is characterized by the transition probabilities $P(X_{n+1} = j | X_n = i) = P_{ij}$. What we are going to present is the situation where the data come from a Markov chain observed in noise, i.e., they are *hidden Markov data.*

For an illustration, consider a simple situation. Let $(X_n, n \geq 0)$ be a discrete-time stochastic process with state space $S$ (finite). The objective of studies is to answer questions about this process. If the $X_n$ are observable, then the problem is simply statistical inference with Markov chains where research literature exists. Suppose we cannot directly observe the $X_n$, and instead, we can observe another process $Y_n$ which is "linked" to $X_n$ in some probabilistic way. How do we carry out the studies of the process $X_n$ in such a situation? Motivated by speech recognition and biology, let's consider the following.

(i) The process $(X_n, n \geq 0)$ is a stationary Markov chain, so that it is characterized by an initial distribution $\pi$ on $S$, and a stationary one- step transition matrix $\mathbb{P} = [P_{ij}]$.

(ii) Each $X_n$, when in a state, emits symbols according to some probability law. Let $Y_n$ be the emitted symbol of $X_n$. Then $(Y_n, n \geq 0)$ is the observed process, with state space $A$, called an alphabet (assuming, for simplicity, finite).

(iii) Conditional upon the realizations of $X_n$ (unobservable), the distributions of $Y_n$ are completely specified (i.e., the distribution of $Y_n$ depends only on $X_n$). Specifically, each distribution $P(Y_n = a | X_n = i)$, $a \in A$, is known, for each $i \in S$.

**Example 7.5** Let $S = \{1, 2\}$ and $A = \{a, b\}$. And let $\pi(1) = \pi(2) = 1/2$.

$$P = \begin{bmatrix} 0.7 & 0.3 \\ 0.6 & 0.4 \end{bmatrix}$$

$$P(Y_n = a|X_n = 1) = 1/4 = 1 - P(Y_n = b|X_n = 1)$$
$$P(Y_n = a|X_n = 2) = 1/6 = 1 - P(Y_n = b|X_n = 2).$$

*One question about $(X_n, n \geq 0)$ could be: What are the true underlying states at $X_n, X_{n+1}, X_{n+2}$, given the observation $Y_n = a$, $Y_{n+1} = a$, $Y_{n+2} = b$? This question reminds us of the maximum likelihood principle in classical statistics! Thus, we look at*

$$P(X_n = i, X_{n+1} = j, X_{n+2=k}|Y_n = a, Y_{n+1} = a, Y_{n+2} = b)$$

*for all eight triples $(i, j, k) \in \{1, 2\}^3$. The most plausible triple of states which produces (emits) the observed $(a, a, b)$ is the one that maximizes the above conditional probability function of $(i, j, k)$.*

## 7.4.5   The Kalman filter

Basically, the Kalman filter is a recursive estimator. In fact, it is a minimum mean-squared error estimator. A filter is an estimator for the present state of nature (or of a system) based upon past observations and current measurements. As its name indicates, a filter aims at improving the accuracy of the present estimation. The filter is named after its inventor, Rudolf Emil Kalman (Hungary, 1960). It is incorporated in the navigation computer of the Apollo trip to the moon (for trajectory estimation).

The filter was designed for control engineering. It is a set of mathematical equations that provides an efficient (recursive) computational means to estimate the state of a process in a way that minimizes the mean squared error. To put it differently, the Kalman filter is an efficient recursive filter that estimates the state of a dynamical system from a series of noisy measurements. The original set-up of the Kalman filter (for control engineering) involves control variables in the dynamics (system equation) of the moving objects, and its derivation is a straightforward mean-squared optimization.

In the following tutorial exposition, we ignore the control variable in the "system equation" (as it does not appear in statistical problems) and we will exploit the sequential *recursive Bayes' estimation* flavor of the

Kalman filter to explain it from a purely statistical viewpoint. In other words, the optimal Kalman filter can be viewed as a Bayes' estimator. As a final note, originally, the Kalman filter is derived for linear discrete-time systems with Gaussian noise. We will take this specific context to introduce it. Extensions of the Kalman filter to more general situations are possible, but will not be discussed here.

So, why do economists want to learn about the Kalman filter? Well, from previous examples, it becomes clear that, besides the meanings of the equations (models), there are similarities between physical systems and econometric models. Thus, facing the same type of problems, economists could use the same tool as engineers. The Kalman filter was invented for control engineering and became known to statisticians, then to economists! In fact, several basic tools used in contemporary economics are from engineering, such as Ito Calculus. That should not be a surprise, since after all, we are all studying dynamical systems, and hence we should use the same tools!

We realize, from above examples, that there are economic problems whose models comprise two equations: the system equation corresponds to the standard model between variables, and the measurement or observation equation corresponds to the measurement error model. For example, consider a linear model

$$Y_t = \beta_t X_t + \varepsilon_t$$

where the $\beta_t$ are unknown, time-varying parameters, as opposed to a classical linear model in which $\beta$ does not depend on $t$. In financial econometrics, stochastic volatility models, such as GARCH, are such examples.

Given a sequence of noisy observation $Y_0, Y_1, ..., Y_{t-1}$ for a linear dynamical system, estimate the internal state $X_t$. Specifically, these are observed values of $Y$ which depends on an unobserved quantity $X$, called the *state of nature*. Our goal is to make inference about $X_t$ in this hidden model. Suppose the link between $Y_t$ and $X_t$ is linear, so resulting in an observation equation

$$Y_t = F_t X_t + v_t$$

where $F_t$ is known and $v_t$ is the observation error. This is different than a conventional linear model, as $X_t$ (playing the role of regression coefficients) now changes with time. This dynamical feature is expressed in a system equation of the form

$$X_t = G_t X_{t-1} + w_t$$

where $G_t$ is known. With all the previous motivations and heuristics, we are ready to describe the Kalman filter and see how it works. The idea of the Kalman filter is to arrive at a conditional distribution of the unobservables using Bayes' theorem, link with observables, an equation of motion, and of course, assumptions about error terms.

The filter works as follows:

1. Use the current observation to predict the next period value of the unobservable,

2. Use the new observation to update.

**Outline of the derivation of the Kalman filter**   The *observation equation* is

$$Y_t = F_t \theta_t + \varepsilon_t$$

where $F_t$ is known (i.e., observed at time $t$), and $\varepsilon_t$ is $N(0, V_t)$ with $V_t$ known.

The *system equation* is

$$\theta_t = G_t \theta_{t-1} + w_t$$

where $G_t$ is known, and $w_t$ is $N(0, W_t)$ with $W_t$ known. Also, $\varepsilon_t$ and $w_t$ are assumed to be independent. The goal is to estimate $\theta_t$ given the data $(Y_1, Y_2, ..., Y_t) = \mathbf{Y}_t$. This can be achieved by a direct application of Bayes' theorem:

$$P(\text{state|data}) \varpropto P(\text{data|state})$$

where the symbol $\varpropto$ means "proportional to," i.e.,

$$P(\theta_t | \mathbf{Y}_t) \varpropto P(Y_t | \theta_t, \mathbf{Y}_{t-1}) P(\theta_t | \mathbf{Y}_{t-1}).$$

In this expression, the left-hand side is the *posterior distribution* of $\theta_t$, whereas the first term and the second term on the right-hand side are the *likelihood* and the *prior distribution* for $\theta_t$.

At time $t - 1$, our knowledge about $\theta_{t-1}$ is the posterior distribution of $\theta_{t-1}$ given $\mathbf{Y}_{t-1}$ with $P(\theta_{t-1} | \mathbf{Y}_{t-1})$ distributed as $N(\hat{\theta}_{t-1}, \Sigma_{t-1})$. The recursive procedure is started at time 0 by choosing $\hat{\theta}_0$ and $\Sigma_0$ to be our best guesses about the mean and the variance of $\theta_0$, respectively. Then we look forward to time $t$ in two stages.

**Stage 1 (Prior to observing $Y_t$)** Prior to observing $Y_t$, our best choice of $\theta_t$ is governed by the system equation, and is given as $G_t\theta_{t-1} + w_t$. Since $P(\theta_{t-1}|\mathbf{Y}_{t-1})$ is distributed as $N(\hat{\theta}_{t-1}, \Sigma_{t-1})$, our knowledge about $\theta_t$ is in the *prior distribution* of $\theta_t$ given $\mathbf{Y}_{t-1}$ which is $P(\theta_t|\mathbf{Y}_{t-1})$ distributed as $N(G_t\hat{\theta}_{t-1}, R_t = G_t\Sigma_{t-1}G_t' + W_t)$, which is derived by observing that, for any constant $c$, if $X$ is $N(\mu, \Sigma)$ then $cX$ is $N(c\mu, c\Sigma c')$ where $c'$ denotes the transpose of $c$.

**Stage 2 (After observing $Y_t$)** On observing $Y_t$, our goal is to compute the posterior distribution of $\theta_t$ using Bayes' theorem. This is carried out as follows. Let $e_t$ be the error in predicting $Y_t$ from the time point $t-1$, i.e.,

$$e_t = Y_t - \hat{Y}_t = Y_t - F_tG_t\hat{\theta}_{t-1},$$

noting that $F_t$, $G_t$ and $\hat{\theta}_{t-1}$ are known up to time $t$, observing $Y_t$ is equivalent to observing $e_t$. Thus, $P(\theta_t|\mathbf{Y}_t) \propto P(Y_t|\theta_t, \mathbf{Y}_{t-1})P(\theta_t|\mathbf{Y}_{t-1})$ can be written as

$$P(\theta_t|\mathbf{Y}_{t-1}, Y_t) = P(\theta_t|\mathbf{Y}_{t-1}, e_t) \propto P(e_t|\theta_t, \mathbf{Y}_{t-1})P(\theta_t|\mathbf{Y}_{t-1}), \quad (7.45)$$

noting that $P(e_t|\theta_t, \mathbf{Y}_{t-1})$ is (equivalent to) the likelihood of $\theta_t$ given $\mathbf{Y}_t$.
Now using the observation equation, we get

$$e_t = F_t(\theta_t - G_t\hat{\theta}_{t-1}) + \varepsilon_t,$$

so that

$$E(e_t|\theta_t, \mathbf{Y}_{t-1}) = F_t(\theta_t - G_t\hat{\theta}_{t-1}).$$

Since $\varepsilon_t$ is $N(0, V_t)$, it follows that the likelihood is described by the conditional distribution of $(e_t|\theta_t, \mathbf{Y}_{t-1})$ which is $N(F_t(\theta_t - G_t\hat{\theta}_{t-1}), V_t)$. From (7.45), we have that

$$P(\theta_t|\mathbf{Y}_{t-1}, Y_t) = [P(e_t|\theta_t, \mathbf{Y}_{t-1})P(\theta_t|\mathbf{Y}_{t-1})] / \left[ \int P(e_t, \theta_t|\mathbf{Y}_{t-1})d\theta_t \right]$$

which is the posterior distribution of $\theta_t$ and best describes our knowledge about $\theta_t$.

## 7.5 Exercises

1. Recall that the Dirac measure $\delta_x$ at a point $x \in \mathbb{R}$ is a probability measure on $\mathcal{B}(\mathbb{R})$ defined by $\delta_x(A) = 1$ or $0$ according to $x \in A$ or $x \notin A$. Let $X_1, X_2, ..., X_n$ be a random sample from $X$

with unknown distribution function $F$. The empirical distribution function, based on $X_1, X_2, ..., X_n$, is

$$F_n(x; X_1, X_2, ..., X_n) = (1/n) \sum_{i=1}^{n} 1_{(X_i \leq x)}$$

and the empirical measure is $dF_n(.) = (1/n) \sum_{i=1}^{n} \delta_{X_i}(.)$.

(a) Let $\hat{F}(t) = \int_{-\infty}^{t} F(x)dx$, $t \in \mathbb{R}$. Show that

$$\int_{-\infty}^{t} F_n(x)dx = \int_{-\infty}^{t} (t-x)dF_n(x) = (1/n) \sum_{i=1}^{n} (t-X_i)1_{(X_i \leq t)}.$$

(b) What is the (a.s.) limit of $\int_{-\infty}^{t} F_n(x)dx$ as $n \to \infty$?

2. Consider the Wang's family of distortion functions

$$g_\lambda(x) = \Phi(\Phi^{-1}(x) + \lambda) \text{ for } \lambda \in \mathbb{R}.$$

Verify that this family forms a commutative group under composition of functions, i.e.,

(a) $g_0(x) = x$, for all $x \in \mathbb{R}$,

(b) $g_\alpha \circ g_\beta = g_\beta \circ g_\alpha = g_{\alpha+\beta}$, and

(c) $g_\alpha \circ g_{-\alpha} = g_{-\alpha} \circ g_\alpha = $ identity.

3. Let $F$ be the distribution function of a random variable $X$. As in Exercise 1, let

$$\hat{F}(t) = \int_{-\infty}^{t} F(x)dx, t \in \mathbb{R}.$$

Is $\hat{F}(.)$ a distribution function?

4. Let $X, Y$ be random variables with distribution functions $F, G$, respectively. The empirical investigation of second stochastic dominance is based on the function

$$t \in \mathbb{R} \to \hat{F}(t) - \hat{G}(t) = \int_{-\infty}^{t} [F(x) - G(x)]dx.$$

Assuming that $E|X| < \infty$ and $E|Y| < \infty$, find $\lim_{t \to -\infty}[\hat{F}(t) - \hat{G}(t)]$ and $\lim_{t \to \infty}[\hat{F}(t) - \hat{G}(t)]$.

5. Let $X_1, X_2, ..., X_n$ be a random sample from $X$ (with distribution function $F$), and $Y_1, Y_2, ..., Y_m$ be a random sample from $Y$ (with distribution function $G$), both with finite means. Let $F_n$, $G_m$ be the empirical distribution functions based on the above samples. A natural point estimator of $\int_{-\infty}^{t} [F(x) - G(x)]dx$ is

$$\int_{-\infty}^{t} [F_n(x) - G_m(x)]dx = D_{n,m}(t).$$

(a) Verify that

$$D_{n,m}(t) = (1/n) \sum_{i=1}^{n} (t - X_i) 1_{(X_i \leq t)} - (1/m) \sum_{j=1}^{m} (t - Y_j) 1_{(Y_j \leq t)}.$$

(b) What are the (a.s.) limits of $D_{n,m}(t)$ as $t \to -\infty$, and $t \to \infty$?

(c) Show that, as $n, m \to \infty$, $D_{n,m}(t)$ converges almost surely to $\int_{-\infty}^{t} [F(x) - G(x)]dx$, i.e., $D_{n,m}(t)$ is a strongly consistent estimator of $\int_{-\infty}^{t} [F(x) - G(x)]dx$, for each $t \in \mathbb{R}$.

6. Let $X$ be a random variable with distribution function $F$. Its survival function is $S(x) = 1 - F(x)$. Let $g : [0, 1] \to [0, 1]$ be a distortion function, i.e., nondecreasing with $g(0) = 0, g(1) = 1$.

(a) Find the conditions on $g$ so that $S^*(.) = (g \circ S)(.)$ is a distribution function.

(b) The dual distortion of $g$ is defined to be

$$g^*(x) = 1 - g(1 - x)$$

and let $F^*(x) = 1 - S^*(x-)$. Verify that $S^*(.) = (g \circ S)(.)$ if and only if $F^*(.) = g^*(F(.))$, except at a countable number of points.

7. Let $X_1, X_2, ..., X_n$ be a random sample from $X$ with distribution function $F$. Estimate $F^{-1}(\alpha)$, for $\alpha \in (0, 1)$ by $F_n^{-1}(\alpha)$, where $F_n(.)$ is the empirical distribution function based on $X_1, X_2, ..., X_n$. Determine $F_n^{-1}(\alpha)$.

8. Let the distribution function $F$ of $X$ be absolutely continuous with density function $f$. Given a random sample $X_1, X_2, ..., X_n$ from $X$, $f(x)$ is estimated by the kernel estimator

$$f_n(x) = (1/nh_n) \sum_{i=1}^{n} K((x - X_i)/h_n),$$

where $K : \mathbb{R} \to \mathbb{R}^+$, symmetric, with $\lim_{x \to \infty} xK(x) = 0$, and $h_n \to 0$, $nh_n \to \infty$, as $n \to \infty$.

Consider the smooth estimator of $F(x)$

$$F_n^*(x) = \int_{-\infty}^{x} f_n(y)dy.$$

(a) Verify that $F_n^*(x) = (1/n)\sum_{i=1}^{n} K^*((x-X_i)/h_n)$, where $K^*(x) = \int_{-\infty}^{x} K(y)dy$.

(b) Show that $F_n^*(x)$ is a strongly consistent estimator of $F(x)$.

9. Let $g$ be a concave distortion function. The distorted probability risk measure

$$C_v(X) = \int_0^{\infty} v(X > t)dt = \int_0^{\infty} g(1 - F_X(t))dt$$

can be written as

$$\int_0^{\infty} g(1 - F_X(t))dt = \int_0^1 \varphi_g(s)F_X^{-1}(s)ds,$$

where $\varphi_g(s) = g'(1 - s)$.

Find $\varphi_g$ when

(a) $g(s) = s^{1-\theta}$, for $0 < \theta < 1$,

(b) $g(.) = g_\alpha(.)$, the distortion function of $TVaR_\alpha(.)$, and

(c) $g(.) = g_\lambda$, the Wang distortion function with parameter $\lambda > 0$.

# Chapter 8

# Applications to Finance

*In this chapter and the next two we illustrate some applications to finance, risk management and economics which are drawn mainly from our research interests.*

## 8.1 Diversification

### 8.1.1 Convex stochastic dominance

The following notation is used throughout this chapter and the next two. Various classes of utility functions and stochastic orders are defined as follows. For $n = 1, 2, 3,$

$$
\begin{aligned}
\mathcal{U}_n^A &= \{u : \mathbb{R} \to \mathbb{R} : (-1)^i u^{(i)} \leq 0, \quad i = 1, ..., n\} \\
\mathcal{U}_n^{SA} &= \{u : \mathbb{R} \to \mathbb{R} : (-1)^i u^{(i)} < 0, \quad i = 1, ..., n\} \\
\mathcal{U}_n^D &= \{u : \mathbb{R} \to \mathbb{R} : u^{(i)} \geq 0, \quad i = 1, ..., n\} \\
\mathcal{U}_n^{SD} &= \{u : \mathbb{R} \to \mathbb{R} : u^{(i)} > 0, \quad i = 1, ..., n\}
\end{aligned}
$$

where $u^{(i)}$ denotes the $i$th derivative of $u$. Also,

$$
\begin{aligned}
\mathcal{U}_1^{EA} \ (\mathcal{U}_1^{ESA}) &= \{u : \mathbb{R} \to \mathbb{R} : u \text{ is (strictly) increasing}\} \\
\mathcal{U}_2^{EA} \ (\mathcal{U}_2^{ESA}) &= \{u : \mathbb{R} \to \mathbb{R} : u \text{ is increasing, and (strictly) concave}\} \\
\mathcal{U}_2^{ED} \ (\mathcal{U}_2^{ESD}) &= \{u : \mathbb{R} \to \mathbb{R} : u \text{ is increasing, and (strictly) convex}\} \\
\mathcal{U}_3^{EA} \ (\mathcal{U}_3^{ESA}) &= \{u : \mathbb{R} \to \mathbb{R} : u \in \mathcal{U}_2^{EA}, u' \text{ is (strictly) convex}\} \\
\mathcal{U}_3^{ED} \ (\mathcal{U}_3^{ESD}) &= \{u : \mathbb{R} \to \mathbb{R} : u \in \mathcal{U}_2^{ED}, u' \text{ is (strictly) convex}\}
\end{aligned}
$$

Let $\overline{\mathbb{R}}$ be the set of extended real numbers and $\Omega = [a, b]$ be a subset of $\overline{\mathbb{R}}$ in which $a < 0$ and $b > 0$. Let $B$ be the Borel $\sigma$-field of $\Omega$ and $\mu$

be a measure on $(\Omega, B)$. For random variables $Y, Z$ with corresponding distribution functions $F, G$, respectively,

$$F_1^A(x) = P_X[a, x] \quad \text{and} \quad F_1^D(x) = P_X[x, b] \quad \text{for all} \quad x \in \Omega, \qquad (8.1)$$

and

$$\mu_F = \mu_Y = E(Y) = \int_a^b x\, d\,F(x)$$

$$\mu_G = \mu_Z = E(Z) = \int_a^b x\, d\,G(x)$$

$$H_j^A(x) = \int_a^x H_{j-1}^A(y)\, dy$$

$$H_j^D(x) = \int_x^b H_{j-1}^D(y)\, dy, \quad j = 2, 3, \qquad (8.2)$$

where $H = F$ or $G$. We assume

$$F_1^A(a) = 0 \quad \text{and} \quad F_1^D(b) = 0. \qquad (8.3)$$

For $H = F$ or $G$, we define the following functions for *Markowitz Stochastic Dominance* (MSD) and *Prospect Stochastic Dominance* (PSD):

$$H_1^a(x) = H(x) = H_1^A(x)$$
$$H_1^d(x) = 1 - H(x) = H_1^D(x)$$
$$H_j^d(y) = \int_y^0 H_{j-1}^d(t)dt, \quad y \le 0$$
$$H_j^a(x) = \int_0^x H_{j-1}^a(t)dt, \quad x \ge 0, \quad j = 2, 3. \qquad (8.4)$$

See Definition 8.5 (and Section 8.3). In order to make the computation easier, we further define

$$H_j^M(x) = \begin{cases} H_j^A(x) & \text{for} \quad x \le 0 \\ H_j^D(x) & \text{for} \quad x > 0 \end{cases}$$

$$H_j^P(x) = \begin{cases} H_j^d(x) & \text{for} \quad x \le 0 \\ H_j^a(x) & \text{for} \quad x > 0 \end{cases} \qquad (8.5)$$

where $H = F$ and $G$ and $j = 1, 2$ and $3$.

**Definition 8.1** $Y$ *dominates* $Z$ *and* $F$ *dominates* $G$ *in the sense of FASD (SASD, TASD), denoted by*

$$Y \succeq_1^A Z \text{ or } F \succeq_1^A G \quad (Y \succeq_2^A Z \text{ or } F \succeq_2^A G, \quad Y \succeq_3^A Z \text{ or } F \succeq_3^A G),$$

*if and only if*

$$F_1^A(x) \leq G_1^A(x) \quad (F_2^A(x) \leq G_2^A(x), \ F_3^A(x) \leq G_3^A(x))$$

*for each* $x$ *in* $[a, b]$, *where FASD, SASD and TASD stand for first, second and third-order* ascending stochastic dominance, *respectively.*

**Definition 8.2** $Y$ *dominates* $Z$ *and* $F$ *dominates* $G$ *in the sense of FDSD (SDSD, TDSD), denoted by*

$$Y \succeq_1^D Z \text{ or } F \succeq_1^D G \quad (Y \succeq_2^D Z \text{ or } F \succeq_2^D G, \quad Y \succeq_3^D Z \text{ or } F \succeq_3^D G),$$

*if and only if*

$$F_1^D(x) \geq G_1^D(x) \quad (F_2^D(x) \geq G_2^D(x), \ F_3^D(x) \geq G_3^D(x))$$

*for each* $x$ *in* $[a, b]$, *where FDSD, SDSD and TDSD stand for first, second and third-order* descending stochastic dominance, *respectively.*

**Definition 8.3** $Y$ *dominates* $Z$ *and* $F$ *dominates* $G$ *in the sense of FMSD (SMSD, TMSD), denoted by*

$$Y \succeq_1^M Z \text{ or } F \succeq_1^M G \quad (Y \succeq_2^M Z \text{ or } F \succeq_2^M G, \quad Y \succeq_3^M Z \text{ or } F \succeq_3^M G),$$

*if and only if*

$$F_1^M(-x) \leq G_1^M(-x) \quad (F_2^M(-x) \leq G_2^M(-x), \ F_3^M(-x) \leq G_3^M(-x))$$

*and*

$$F_1^M(x) \geq G_1^M(x) \quad (F_2^M(x) \geq G_2^M(x), \ F_3^M(x) \geq G_3^M(x))$$

*for each* $x \geq 0$; *where FMSD, SMSD and TMSD stand for first, second and third-order MSD, respectively.*

**Definition 8.4** $Y$ *dominates* $Z$ *and* $F$ *dominates* $G$ *in the sense of FPSD (SPSD, TPSD), denoted by*

$$Y \succeq_1^P Z \text{ or } F \succeq_1^P G \quad (Y \succeq_2^P Z \text{ or } F \succeq_2^P G, \ Y \succeq_3^P Z \text{ or } F \succ_3^P G),$$

*if and only if*

$$F_1^P(-x) \geq G_1^P(-x) \quad (F_2^P(-x) \geq G_2^P(-x), \quad F_3^P(-x) \geq G_3^P(-x))$$

*and*

$$F_1^P(x) \leq G_1^P(x) \quad (F_2^P(x) \leq G_2^P(x), \quad F_3^P(x) \leq G_3^P(x))$$

*for each $x \geq 0$; where FPSD, SPSD and TPSD stand for first, second and third-order PSD, respectively.*

We note that in Definitions 8.1 to 8.4, if in addition there exists a strict inequality for any $x$ in $[a, b]$, we say that $Y$ dominates $Z$ and $F$ dominates $G$ in the sense of SFXSD, SSXSD and STXSD, denoted by

$$Y \succ_1^X Z \text{ or } F \succ_1^X G, \quad Y \succ_2^X Z \text{ or } F \succ_2^X G, \quad \text{and } Y \succ_3^X Z \text{ or } F \succ_3^X G,$$

respectively, where SFXSD, SSXSD and STXSD stand for strictly first, second and third-order XSD, respectively, with X = A, D, M or P.

We also note that in Definitions 8.1 and 8.4, we use

$$Y \succeq_j (\succ_j) Z \quad [F \succeq_j (\succ_j) G]$$

to represent

$$Y \succeq_j^A (\succ_j^A) Z \quad [F \succeq_2^A (\succ_j^A) G]$$

and use

$$Y \succeq^j (\succ^j) Z \quad [F \succeq^j (\succ^j) G]$$

to represent

$$Y \succeq_j^D (\succ_j^D) Z \quad [F \succeq_j^D (\succ_j^D) G)$$

for $j = 1, 2$ and $3$.

Levy and Levy [95] define the MSD and PSD functions as

$$H^M(x) = \begin{cases} \int_a^x H(t)\,dt & \text{for} \quad x < 0 \\ \int_x^b H(t)\,dt & \text{for} \quad x > 0 \end{cases} \quad \text{and}$$

$$H^P(x) = \begin{cases} \int_x^0 H(t)\,dt & \text{for} \quad x < 0 \\ \int_0^x H(t)\,dt & \text{for} \quad x > 0, \end{cases} \quad (8.6)$$

where $H = F$ and $G$. Then MSD and PSD are expressed in the following definition:

**Definition 8.5** $F \succeq_{MSD} G$ if $F^M(x) \leq G^M(x)$ for all $x$, and $F \succeq_{PSD} G$ if $F^P(x) \leq G^P(x)$ for all $x$.

One can easily show that $F \succeq_{MSD} G$ if and only if $F \succeq_2^M G$ and $F \succeq_{PSD} G$ if and only if $F \succeq_2^P G$. Hence, the MSD and PSD defined in [93] are the same as the second-order MSD and PSD we have defined here. We note that Levy and Wiener [92] and Levy and Levy [93] define PSD as $F \succeq_{PSD} G$ if and only if

$$0 \leq \int_{x_1}^{x_2} [G(z) - F(z)]\, dz \quad \text{for all} \quad x_1 \leq 0 \leq x_2$$

with at least one strict inequality.

**Definition 8.6** *For $j = 1,\ 2,\ 3,\ \mathcal{U}_j^A,\ \mathcal{U}_j^{SA},\ \mathcal{U}_j^D$ and $\mathcal{U}_j^{SD}$ are the sets of the utility functions[1] $u$ such that*

$$
\begin{aligned}
\mathcal{U}_j^A\ (\mathcal{U}_j^{SA}) &= \{u : (-1)^i u^{(i)} \leq (<)\, 0, \quad i = 1, \ldots, j\} \\
\mathcal{U}_j^D\ (\mathcal{U}_j^{SD}) &= \{u : u^{(i)} \geq (>)\, 0, \quad i = 1, \ldots, j\} \\
\mathcal{U}_j^S\ (\mathcal{U}_j^{SS}) &= \{u : u_+ \in \mathcal{U}_j^A\ (U_j^{SA}) \quad and \quad u_- \in \mathcal{U}_j^D\ (U_j^{SD})\} \\
\mathcal{U}_j^R\ (\mathcal{U}_j^{SR}) &= \{u : u_+ \in \mathcal{U}_j^D\ (U_j^{SD}) \quad and \quad u_- \in \mathcal{U}_j^A\ (U_j^{SA})\},
\end{aligned}
$$

*where $u^{(i)}$ is the $i$th derivative of the utility function $u$, $u_+ = max\{u, 0\}$, and $u_- = min\{u, 0\}$.*

First, we look at the concept of convex stochastic dominance for risk averters and risk seekers, which is used to compare a convex combination of several continuous distributions and a single continuous distribution. We state theoretical results as theorems whose proofs will be left as solutions to exercises. For more details, see relevant references cited in the Bibliography.

**Theorem 8.7** *Let $F_1, \ldots, F_n, G_1, \ldots, G_n$ be distribution functions. The following hold for $m = 1,\ 2$ or $3$, and*

$$\Lambda_n = \{(\lambda_1, \ldots, \lambda_n) : \lambda_i \geq 0, \quad i = 1, \ldots, n, \quad and \quad \sum_{i=1}^n \lambda_i = 1\}.$$

*1. There exists $(\lambda_1, \ldots, \lambda_n) \in \Lambda_n$ such that*

$$\sum_{i=1}^n \lambda_i F_i \succeq_m (\succ_m) \sum_{i=1}^n \lambda_i G_i \tag{8.7}$$

---

[1]We note that the theory can be easily extended to satisfy utilities defined to be non-differentiable and/or non-expected utility functions. In this paper, we will skip the discussion of non-differentiable utilities or non-expected utility functions. Readers may refer to Wong and Ma [174] for the discussion.

*if and only if for every u in $\mathcal{U}$ such that*

$$\mathcal{U}_m^A \subseteq \mathcal{U} \subseteq \mathcal{U}_m^{EA} \quad (\mathcal{U}_m^{SA} \subseteq \mathcal{U} \subseteq \mathcal{U}_m^{ESA})$$

*there exists $i \in \{1, \ldots, n\}$ such that $F_i$ is preferred to $G_i$ for u.*

2. *There exists $(\lambda_1, \ldots, \lambda_n) \in \Lambda_n$ such that*

$$\sum_{i=1}^{n} \lambda_i F_i \succeq^m (\succ^m) \sum_{i=1}^{n} \lambda_i G_i \tag{8.8}$$

*if and only if for every u in $\mathcal{U}$ such that*

$$\mathcal{U}_m^D \subseteq \mathcal{U} \subseteq \mathcal{U}_m^{ED} \quad (\mathcal{U}_m^{SD} \subseteq \mathcal{U} \subseteq \mathcal{U}_m^{ESD}),$$

*there exists $i \in \{1, \ldots, n\}$ such that $F_i$ is preferred to $G_i$ for u.*

Using Theorem 8.7, we can compare a convex combination of several continuous distributions with a single continuous distribution for risk averters and risk seekers as follows.

**Corollary 8.8** *Let $F_1, \ldots, F_n, F_{n+1}$ be distribution functions. For $m = 1, 2$ and $3$,*

1. *There exists $(\lambda_1, \ldots, \lambda_n) \in \Lambda_n$ such that*

$$\sum_{i=1}^{n} \lambda_i F_i \succeq_m (\succ_m) F_{n+1}$$

*if and only if for every utility function u in $\mathcal{U}_m^A$ ($\mathcal{U}_m^{SA}$), there exists $i \in \{1, \ldots, n\}$ such that $F_i$ is preferred to $F_{n+1}$ for u, and*

2. *There exists $(\lambda_1, \ldots, \lambda_n) \in \Lambda_n$ such that*

$$\sum_{i=1}^{n} \lambda_i F_i \geq^m (\succ^m) F_{n+1}$$

*if and only if for every utility function u in $\mathcal{U}_m^D$ ($\mathcal{U}_m^{SD}$) there exists $i \in \{1, \ldots, n\}$ such that $F_i$ is preferred to $F_{n+1}$ for u.*

The sufficient part of Theorem 8.7 allows us to draw conclusions about preferences of risk seekers that supplement those specified by Theorem 8.7. For example, if we assume that $u$ is in $\mathcal{U}_2^{SD}$, and if it is not true that $F_1 \succ_2 G_1$ or $G_1 \succ_2 F_1$, and not true that $F_2 \succ_2 G_2$ or $G_2 \succ_2 F_2$ and if $\lambda_1 F_1 + (1 - \lambda_1)F_2 \succ_2 \lambda_1 G_1 + (1 - \lambda_1)G_2$ for some $\lambda_1$ strictly between 0 and 1, then we know that either $F_1$ is preferred to $G_1$ or $F_2$ is preferred to $G_2$ for $u$.

Similarly, we can also draw some conclusions about the preferences of distribution functions by using the necessary part of Theorem 8.7. For example, assuming $u \in \mathcal{U}_2^{SD}$, if $\sum \lambda_i F_i \succ_2 \sum \lambda_i G_i$ is false for every $\lambda \in \Lambda_n$, then there exists $u \in \mathcal{U}_2^{SD}$ such that $u(G_i) \geq u(F_i)$ for all $i$, so that it is not possible to conclude that $F_i$ is preferred to $G_i$ for some $i$. Note that Corollary 8.8 is a special case of Theorem 8.7 in which all $G_i$ are identical. It can be used to compare a convex combination of distributions with a distribution for risk averters and risk seekers.

## 8.1.2 Diversification for risk averters and risk seekers

Note that for any pair of random variables $X$ and $Y$, the statements $X \succeq_m Y$, and $F \succeq_m G$ are equivalent. But for $n > 1$, the statements

$$\sum_{i=1}^{n} \alpha_i X_i \succeq_m \sum_{i=1}^{n} \alpha_i Y_i \quad \text{and} \quad \sum_{i=1}^{n} \alpha_i F_i \succeq_m \sum_{i=1}^{n} \alpha_i G_i$$

are different because the distribution functions of $\sum_{i=1}^{n} \alpha_i X_i$ and $\sum_{i=1}^{n} \alpha_i Y_i$ are different from $\sum_{i=1}^{n} \alpha_i F_i$ and $\sum_{i=1}^{n} \alpha_i G_i$. Therefore, we cannot apply the convex stochastic dominance theorem of the previous subsection to the convex combinations of random variables.

Consider first the case of $X$ and $Y$ which are in the same location and scale family such that $Y = p + qX$. The location parameter $p$ can be viewed as the random variable with degenerate distribution at $p$.

**Theorem 8.9** *Let $X$ be a random variable with range $[a, b]$ and finite mean $\mu_X$. Define the random variable $Y = p + qX$ with mean $\mu_Y$.*

1. *If $p + qy \geq y$ for all $y \in [a, b]$, then $Y \succeq_1 X$, equivalently $Y \succeq^1 X$.*

2. *If $0 \leq q < 1$ such that $p/(1-q) \geq \mu_X$, i.e., $\mu_Y \geq \mu_X$, then $Y \succeq_2 X$.*

3. *If $0 \leq q < 1$ such that $p/(1-q) \leq \mu_X$, i.e., $\mu_X \geq \mu_Y$, then $X \succeq^2 Y$.*

**Theorem 8.10** *Let $X$ and $Y$ denote two random variables with distribution functions $F$ and $G$, respectively, and assume that random variable $W$ is independent of both $X$ and $Y$. Let the distribution functions of the random variables $aX + bW$ and $aY + bW$ be denoted by $\hat{F}$ and $\hat{G}$, respectively, where $a > 0$ and $b \geq 0$. Then*

1. *If $G$ is larger than $F$ in the sense of FASD, then $\hat{G}$ is larger than $\hat{F}$ in the sense of FASD.*

2. *If $G$ is larger than $F$ in the sense of SASD, then $\hat{G}$ is larger than $\hat{F}$ in the sense of SASD.*

The above results could be further extended as follows.

**Theorem 8.11** *Let $\{X_1, \ldots, X_m\}$ and $\{Y_1, \ldots, Y_m\}$ be two sets of independent variables. For $n = 1$, $2$ and $3$, we have:*

1. *$X_i \succeq_n (\succ_n) Y_i$ for $i = 1, \ldots, m$ if and only if $\sum_{i=1}^{m} \alpha_i X_i \succeq_n (\succ_n) \sum_{i=1}^{m} \alpha_i Y_i$ for any $\alpha_i \geq 0$, $i = 1, \ldots, m$; and*

2. *$X_i \succeq^n (\succ^n) Y_i$ for $i = 1, \ldots, m$ if and only if $\sum_{i=1}^{m} \alpha_i X_i \succeq^n (\succ^n) \sum_{i=1}^{m} \alpha_i Y_i$ for any $\alpha_i \geq 0$, $i = 1, \ldots, m$.*

**Corollary 8.12** *Let $X, Y$ be random variables and $k \in \mathbb{R}$. For $n = 1$, $2$ and $3$.*

1. *If $X \succeq_n (\succ_n) Y$ then $X + k \succeq_n (\succ_n) Y + k$; and*

2. *If $X \succeq^n (\succ^n) Y$ then $X + k \succeq^n (\succ^n) Y + k$.*

*If $X_1$ and $X_2$ are independent and identically distributed, then*

$$\frac{1}{2}(X_1 + X_2) \succeq_2 \lambda_1 X_1 + \lambda_2 X_2 \succeq_2 X_1 \text{ for any } (\lambda_1, \lambda_2) \in \Lambda_2.$$

More generally,

**Theorem 8.13** *Let $n \geq 2$. If $X_1, \ldots, X_n$ are independent and identically distributed, then*

1. *$\frac{1}{n} \sum_{i=1}^{n} X_i \succeq_2 \sum_{i=1}^{n} \lambda_i X_i \succeq_2 X_i$   for any   $(\lambda_1, \ldots, \lambda_n) \in \Lambda_n$, and*

2. *$X_i \succeq^2 \sum_{i=1}^{n} \lambda_i X_i \succeq^2 \frac{1}{n} \sum_{i=1}^{n} X_i$   for any   $(\lambda_1, \ldots, \lambda_n) \in \Lambda_n$,*

*where $\Lambda_n = \{(\lambda_1, \ldots, \lambda_n) \lambda_i \geq 0 \text{ for } i = 1, \ldots, n, \text{ and } \sum_{i=1}^{n} \lambda_i = 1\}$.*

We call a person a *second-order ascending stochastic dominance* (SASD) risk averter if his/her utility function belongs to $U_2^{EA}$, and a *second-order descending stochastic dominance* (SDSD) risk seeker if his/her utility function belongs to $\mathcal{U}_2^{ED}$.

Investors could be confronted with the choice of transforming their current portfolio containing a random prospect into a diversified portfolio containing a sure prospect and a specified amount of the original random prospect. One could apply the findings stated in this chapter to compare the second-degree stochastic dominance of one portfolio over the other. One could further include the following:

**Property 8.14** *Information for risk averters or risk seekers in a single investment or gamble:*

1. *Let $X$ and $Y$ be the returns of two investments or gambles. If $X$ has the same distribution form as $Y$ but has a higher mean, then all risk averters and risk seekers will prefer $X$.*

2. *For an investment or gamble with the mean of return less than or equal to zero, the highest preference of SASD risk averters is not to invest or gamble.*

3. *For an investment or gamble with the mean of return which is greater than or equal to zero, SDSD risk seekers will prefer to invest or gamble as much as possible.*

4. *Let $X$ be the return of an investment or gamble with zero return, and $Y = qX$ with $0 \leq q < 1$, then SASD risk averters will prefer $Y$ while SDSD risk seekers will prefer $X$.*

A diversified portfolio can be larger in the sense of SASD than a specialized portfolio only if its constituent prospects have equal means. This result could be further extended for the portfolio of $n$ i.i.d. prospects as follows.

**Property 8.15** *For the portfolio of $n$ independent and identically distributed prospects with $n \geq 2$, SASD risk averters will prefer the equal weight portfolio whereas SDSD risk seekers will prefer a single prospect.*

## 8.2 Diversification on convex combinations

It is well known that the limitations of using the mean-variance (MV) portfolio selection are that it is derived by assuming the quadratic utility function and the distributions of the returns or assets being examined are required to be normally distributed or elliptically distributed. The stochastic dominance (SD) approach was proposed to circumvent the limitations of the MV criterion.

By combining majorization theory with stochastic dominance theory, the theory of portfolio diversification could be extended. The theory could also be used in determining the second-order stochastic dominance efficient set. Traditionally, there are two decision stages in determining the efficient set: investors could screen prospects or investments and check their stochastic dominance relation and classify the feasible set into the efficient and inefficient sets. Thereafter, investors can use the convex stochastic dominance approach to eliminate elements that are not optimal. In addition, investors can use the approach in this section to rank convex combinations of assets by majorization order to determine whether they are optimal.

Considering risk-averse investors with initial capital make a decision in the single-period portfolio selection to allocate their wealth to the $n$ ($n > 1$) risks without short selling in order to maximize their expected utilities from the resulting final wealth. Let random variable $X$ be an (excess) return of an asset or prospect. If there are $n$ assets $\vec{X}_n = (X_1, \ldots, X_n)'$, a portfolio of $\vec{X}_n$ without short selling is defined by a convex combination, $\vec{\lambda_n}'\vec{X}_n$, of the $n$ assets $\vec{X}_n$ for any $\vec{\lambda_n} \in S_n^0$ where

$$S_n^0 = \left\{ (s_1, s_2, \ldots, s_n)' \in \mathbb{R}^n,\, 0 \le s_i \le 1 \text{ for any } i,\, \sum_{i=1}^{n} s_i = 1 \right\}. \quad (8.9)$$

The $i$th element of $\vec{\lambda_n}$ is the weight of the portfolio allocation on the $i$th asset of return $X_i$. A portfolio will be equivalent to return on asset $i$ if $s_i = 1$ and $s_j = 0$ for all $j \ne i$. It is diversified if there exists $i$ such that $0 < s_i < 1$, and is completely diversified if $0 < s_i < 1$ for all $i = 1, 2, \ldots, n$. Without loss of generality, we further assume $S_n$ satisfy:

$$S_n = \left\{ (s_1, s_2, \ldots, s_n)' \in \mathbb{R}^n,\, 1 \ge s_1 \ge \cdots \ge s_n \ge 0,\, \sum_{i=1}^{n} s_i = 1 \right\}. \quad (8.10)$$

Note that the condition of $\sum_{i=1}^{n} s_i = 1$ is not necessary. It could be any positive number. For convenience, we set $\sum_{i=1}^{n} s_i = 1$ so that the sum of all relative weights is equal to one. We will mainly study the properties of majorization by considering $\vec{\lambda_n} \in S_n$ instead of $S_n^0$. Suppose that an investor has utility function $u$, and his/her expected utility for the portfolio $\vec{\lambda}_n' \vec{X}_n$ is $E[u(\vec{\lambda}_n' \vec{X}_n)]$. In this context, we study only the behavior of non-satiable and risk-averse investors whose utility functions are defined in Definition 8.1. There are many ways to order the elements in $S_n$. A popular one is to order them by majorization.

**Definition 8.16** *Let* $\vec{\alpha}_n, \vec{\beta}_n \in S_n$ *in which* $S_n$ *is defined in (8.10).* $\vec{\beta}_n$ *is said to majorize* $\vec{\alpha}_n$, *denoted by* $\vec{\beta}_n \succeq_M \vec{\alpha}_n$, *if* $\sum_{i=1}^{k} \beta_i \geq \sum_{i=1}^{k} \alpha_i$, *for all* $k = 1, 2, \ldots, n$.

Vectors that can be ordered by majorization have some interesting properties. One of them is a Pigou-Dalton transfer, as follows.

**Definition 8.17** *For any* $\vec{\alpha}_n, \vec{\beta}_n \in S_n$, $\vec{\alpha}_n$ *is said to be obtained from* $\vec{\beta}_n$ *by applying a single Pigou-Dalton transfer, denoted by* $\vec{\beta}_n \xrightarrow{d} \vec{\alpha}_n$, *if there exist* $h$ *and* $k$ $(1 \leq h < k \leq n)$ *such that* $\alpha_i = \beta_i$ *for any* $i \neq h, k$; $\alpha_h = \beta_h - \epsilon$; *and* $\alpha_k = \beta_k + \epsilon$ *with* $\epsilon > 0$.

The relationship between majorization and a Pigou-Dalton transfer is this.

**Theorem 8.18** *Let* $\vec{\alpha}_n, \vec{\beta}_n \in S_n$, $\vec{\beta}_n \succeq_M \vec{\alpha}_n$ *if and only if* $\vec{\alpha}_n$ *can be obtained from* $\vec{\beta}_n$ *by applying a finite number of Pigou-Dalton transfers, denoted by* $\vec{\beta}_n \xrightarrow{D} \vec{\alpha}_n$.

This theorem states that if $\vec{\beta}_n$ majorizes $\vec{\alpha}_n$, then $\vec{\alpha}_n$ can be obtained from $\vec{\beta}_n$ by applying a finite number of single Pigou-Dalton transfers, and vice versa. We link a Pigou-Dalton transfer and majorization order to stochastic dominance.

We now develop the theory of diversification for risk-averse investors to make comparisons among different portfolios by incorporating both majorization theory and stochastic dominance theory. We first state the diversification problem for the i.i.d. case in a bivariate setting as follows:

$$E[u(\vec{\alpha_2}' \vec{X}_2)] \geq E[u(\vec{\beta_2}' \vec{X}_2)] \tag{8.11}$$

whenever $|\alpha_1 - \alpha_2| \leq |\beta_1 - \beta_2|$ where $\vec{\alpha}_2 = (\alpha_1, \alpha_2)'$ and $\vec{\beta}_2 = (\beta_1, \beta_2)' \in S_2$, $u \in U_2$, and $\vec{X}_2 = (X_1, X_2)'$ in which $X_1$ and $X_2$ are nonnegative i.i.d. random variables.

Note that for any pair of random variables $X$ and $Y$, the statements $X \succeq_2 Y$ and $F \succeq_2 G$ are equivalent. But for $n > 1$, the statements

$\vec{\alpha}'_n \vec{X}_n \succeq_2 \vec{\beta}'_n \vec{Y}_n$ and $\vec{\alpha}'_n \vec{F}_n \succeq_2 \vec{\beta}'_n \vec{G}_n$ are different because the distribution functions of $\vec{\alpha}'_n \vec{X}_n$ and $\vec{\beta}'_n \vec{Y}_n$ are different from $\vec{\alpha}'_n \vec{F}_n$ and $\vec{\beta}'_n \vec{G}_n$, respectively. Thus, we cannot apply the convex stochastic dominance theorems to the convex combinations of random variables. To compare two sets of independent random variables in a multivariate setting, we need

**Theorem 8.19** *For* $n > 1$, *let* $\vec{X}_n = (X_1, \ldots, X_n)'$ *and* $\vec{Y}_n = (Y_1, \ldots, Y_n)'$ *where* $\{X_1, \ldots, X_n\}$ *and* $\{Y_1, \ldots, Y_n\}$ *be two sets of i.i.d. random variables. Then,* $X_i \succeq_2 Y_i$ *for any* $i = 1, 2, \ldots, n$ *if and only if* $\vec{\alpha}'_n \vec{X}_n \succeq_2 \vec{\alpha}'_n \vec{Y}_n$ *for any* $\vec{\alpha}_n \in S_n^0$.

Let $\left(\frac{\vec{1}}{n}\right) = (\frac{1}{n}, \ldots, \frac{1}{n})'$. In the bivariate case, for any $\left(\frac{\vec{1}}{2}\right)$ and $\vec{\alpha}_2 \in S_2^0$, if $X_1$ and $X_2$ are i.i.d., one can get

$$\left(\frac{\vec{1}}{2}\right)' \vec{X}_2 \succeq_2 \vec{\alpha}'_2 \vec{X}_2 \succeq_2 X_i \text{ for } i = 1, 2.$$

Furthermore,

**Theorem 8.20** *For* $n > 1$, *let* $\vec{X}_n = (X_1, \ldots, X_n)'$ *where* $X_1, \ldots, X_n$ *are i.i.d., then for any* $i = 1, 2, \ldots, n$ *and for any* $\left(\frac{\vec{1}}{n}\right)$, $\vec{\alpha}_n \in S_n^0$,

$$\left(\frac{\vec{1}}{n}\right)' \vec{X}_n \succeq_2 \vec{\alpha}'_n \vec{X}_n \succeq_2 X_i.$$

It is well known that $\left(\frac{1}{n}\right)' \vec{X}_n$ attains its maximum among $\vec{\alpha}'_n \vec{X}_n$ for any i.i.d. nonnegative $X_i$. This theorem verifies the optimality of diversification that the maximal expected utility will be achieved in an equally weighted portfolio of independent and identically distributed assets. The manifestation of the generality of the theorem is that it places very weak restrictions on the weights. This theorem implies that for i.i.d. assets, risk-averse investors will prefer the equally weighted portfolio to any convex combination portfolio, which, in turn, is preferred to any individual asset.

Nonetheless, Theorem 8.20 does not permit investors to compare the preferences of other different convex combinations of random variables.

**Theorem 8.21** *Let* $\vec{\alpha}_2, \vec{\beta}_2 \in S_2$ *and* $\vec{X}_2 = (X_1, X_2)'$ *where* $X_1$ *and* $X_2$ *are independent and identically distributed. Then,*

$$\vec{\beta}_2 \succeq_M \vec{\alpha}_2 \text{ if and only if } \vec{\alpha}'_2 \vec{X}_2 \succeq_2 \vec{\beta}'_2 \vec{X}_2.$$

This theorem provides a methodology for investors to make comparisons among a wide range of portfolios so that they make better choices in their investment decisions, especially the implications concerning the weights of allocations.

It is interesting to note from Theorem 8.18 that, if $\vec{\alpha}_2$ is majorized by $\vec{\beta}_2$, $\vec{\alpha}_2$ can be obtained from vector $\vec{\beta}_2$ by applying Pigou-Dalton transfer(s) and vice versa. Thus, incorporating Theorem 8.18 into Theorem 8.21, we get

**Corollary 8.22** *Let $\vec{\alpha}_2, \vec{\beta}_2 \in S_2$ and $\vec{X}_2 = (X_1, X_2)'$ where $X_1$ and $X_2$ are i.i.d. Then,*

$$\vec{\beta}_2 \xrightarrow{D} \vec{\alpha}_2 \quad \text{if and only if} \quad \vec{\alpha}_2' \vec{X}_2 \succeq_2 \vec{\beta}_2' \vec{X}_2 \,.$$

The multivariate version is

**Theorem 8.23** *For $n > 1$, let $\vec{\alpha}_n, \vec{\beta}_n \in S_n{}^2$ and $\vec{X}_n = (X_1, \ldots, X_n)'$ where $X_1, \ldots, X_n$ are i.i.d. If $\vec{\beta}_n \succeq_M \vec{\alpha}_n$, then $\vec{\alpha}_n' \vec{X}_n \succeq_2 \vec{\beta}_n' \vec{X}_n$.*

Incorporating Theorem 8.18 into Theorem 8.23, we get

**Corollary 8.24** *For $n > 1$, let $\vec{\alpha}_n, \vec{\beta}_n \in S_n$ and $\vec{X}_n = (X_1, \ldots, X_n)$ where $X_1, \ldots, X_n$ are i.i.d. If $\vec{\beta}_n \xrightarrow{D} \vec{\alpha}_n$, then $\vec{\alpha}_n' \vec{X}_n \succeq_2 \vec{\beta}_n' \vec{X}_n$.*

Can the i.i.d. assumption be dropped in the diversification problem and the completely diversified portfolio still be optimal? The answer is no, in general. In this section, we extend the results stated in the above theorems and corollaries by relaxing the i.i.d. condition as stated in the following corollaries:

**Corollary 8.25** *For $n > 1$, let $\vec{X}_n = (X_1, \ldots, X_n)'$ be a series of random variables that could be dependent. For any $\vec{\alpha}_n$ and $\vec{\beta}_n$,*

$$\vec{\alpha}_n' \vec{X}_n \succeq_2 \vec{\beta}_n' \vec{X}_n$$

*if there exist $\vec{Y}_n$ and $A_{nn}$ such that $\vec{Y}_n = (Y_1, \ldots, Y_n)'$ in which $\{Y_1, \ldots, Y_n\}$ are i.i.d., $\vec{X}_n = A_{nn} \vec{Y}_n$, and*

$$\vec{\beta}_n' A_{nn} \succeq_M \vec{\alpha}_n' A_{nn},$$

*where $\vec{\alpha}_n' A_{nn}, \vec{\beta}_n' A_{nn} \in S_n$.*

---

[2] We keep the condition $\sum_{i=1}^{n} s_i = 1$ in $S_n$ for convenience. One could exclude this condition and relax it to be $\vec{1}_n' \vec{\alpha}_n = \vec{1}_n' \vec{\beta}_n$.

**Corollary 8.26** *For $n > 1$, let $\vec{X}_n = (X_1, \ldots, X_n)'$ and $\vec{Y}_n = (Y_1, \ldots, Y_n)'$ be two series of random variables that could be dependent. For any $\vec{\alpha}_n$ and $\vec{\beta}_n$,*

$$\vec{\alpha}_n' \vec{X}_n \succeq_2 \vec{\beta}_n' \vec{Y}_n$$

*if there exist $\vec{U}_n = (U_1, \ldots, U_n)'$, $\vec{V}_n = (V_1, \ldots, V_n)'$, $A_{nn}$, and $B_{nn}$ in which $\{U_1, \ldots, U_n\}$ and $\{V_1, \ldots, V_n\}$ are two series of i.i.d. random variables such that $\vec{X}_n = A_{nn}\vec{U}_n$, $\vec{Y}_n = B_{nn}\vec{V}_n$, $U_i \succeq_2 V_i$ for all $i = 1, 2, \ldots, n$; and*

$$\vec{\beta}_n' B_{nn} \succeq_M \vec{\alpha}_n' A_{nn},$$

*where $\vec{\alpha}_n' A_{nn}$, $\vec{\beta}_n' B_{nn} \in S_n$.*

One could simply apply Theorem 8.23 to obtain the results of Corollary 8.25 and apply Theorem 8.19 to obtain the results of Corollary 8.26. One could then apply Theorem 8.18 to the above corollaries to obtain the following results:

**Corollary 8.27** *For $n > 1$, let $\vec{X}_n = (X_1, \ldots, X_n)'$ be a series of random variables that could be dependent. For any $\vec{\alpha}_n$ and $\vec{\beta}_n$,*

$$\vec{\alpha}_n' \vec{X}_n \succeq_2 \vec{\beta}_n' \vec{X}_n$$

*if there exist $\vec{Y}_n$ and $A_{nn}$ such that $\vec{Y}_n = (Y_1, \ldots, Y_n)'$ in which $\{Y_1, \ldots, Y_n\}$ are i.i.d. $\vec{X}_n = A_{nn}\vec{Y}_n$, and*

$$\vec{\beta}_n' A_{nn} \xrightarrow{D} \vec{\alpha}_n' A_{nn},$$

*where $\vec{\alpha}_n' A_{nn}$, $\vec{\beta}_n' A_{nn} \in S_n$.*

**Corollary 8.28** *For $n > 1$, let $\vec{X}_n = (X_1, \ldots, X_n)'$ and $\vec{Y}_n = (Y_1, \ldots, Y_n)'$ be two series of random variables that could be dependent. For any $\vec{\alpha}_n$ and $\vec{\beta}_n$,*

$$\vec{\alpha}_n' \vec{X}_n \succeq_2 \vec{\beta}_n' \vec{Y}_n$$

*if there exist $\vec{U}_n = (U_1, \ldots, U_n)'$, $\vec{V}_n = (V_1, \ldots, V_n)'$, $A_{nn}$, and $B_{nn}$ in which $\{U_1, \ldots, U_n\}$ and $\{V_1, \ldots, V_n\}$ are two series of i.i.d. random variables such that $\vec{X}_n = A_{nn}\vec{U}_n$, $\vec{Y}_n = B_{nn}\vec{V}_n$, $U_i \succeq_2 V_i$ for all $i = 1, 2, \ldots, n$; and*

$$\vec{\beta}_n' B_{nn} \xrightarrow{D} \vec{\alpha}_n' A_{nn},$$

*where $\vec{\alpha}_n' A_{nn}$, $\vec{\beta}_n' B_{nn} \in S_n$.*

In summary, by incorporating the majorization theory, we have presented several new results of interest on stochastic dominance. Specifically, we establish some basic relationships in the portfolio choice problem by using both majorization theory and stochastic dominance. We also provide the foundation for applying majorization theory and stochastic dominance to investors' choices under uncertainty. The results are general, but presumably they are applicable to investment decision theory and comparisons of diversification of assets in a multivariate setting. Thus, risk-averse and non-satiable investors will increase their expected utilities as the diversification of the portfolio increases.

These results permit comparisons among assets and portfolios. The results could also be used to demonstrate the optimality of diversification and to obtain the preference orderings of portfolios for different asset allocations. Also, the results also impose further restrictions on admissible portfolios on the efficient frontier, and thus, our findings could also be used in determining the second-order stochastic dominance efficient set.

Nonetheless, it is well known that for any two distributions with the same mean, the mean-preserving spread and second-order stochastic dominance (SSD) are equivalent, whereas under some conditions, marginal conditional stochastic dominance is equivalent to SSD. Thus, incorporating the theory stated in this section, one could conclude that under some regularity conditions, the preferences obtained from a Pigou-Dalton transfer, majorization and stochastic dominance could be equivalent to those obtained from the mean-preserving spread and marginal conditional stochastic dominance. Unlike the stochastic dominance approach, which is consistent with utility maximization, the dominance findings using the mean-variance criterion may not be consistent with utility maximization if the asset returns are not normally distributed or investors' utilities are not quadratic. However, under some specific conditions, the mean-variance optimality could be consistent with the stochastic dominance approach with utility maximization.

As is well known, if the returns of assets follow the same location-scale family, then a mean-variance domination could infer preferences by risk averters for the dominant fund over the dominated one. In addition, the Markowitz mean-variance optimization is equivalent to minimizing variance subject to obtaining a predetermined level of expected gain. Thus, by incorporating the results stated in this section and under some regularity conditions, the efficient set derived from a Pigou-Dalton transfer, majorization and stochastic dominance could belong to the same efficient set obtained from the mean-variance criterion and risk minimization.

## 8.3   Prospect and Markowitz SD

According to the von Neumann and Morgenstern expected utility theory, the utility functions of risk averters and risk seekers are nondecreasing concave and convex, respectively. Examining the relative attractiveness of various forms of investments, it is found that the strictly concave utility functions may not be able to explain the behavior when investors buy insurance or lottery tickets. To address this issue, a utility function that has convex and concave regions in both the positive and the negative domains is proposed. In addition, prospect theory studies the (value) utility function which is concave for gains and convex for losses, yielding an S-shaped function and studies the theory in which investors can maximize the expectation of the S-shaped utility function.

In addition, a theory for the reverse S-shaped utility functions for investors is developed. The criterion MSD is used to determine the dominance of one investment alternative over another for all reverse S-shaped functions, and another criterion called PSD is used to determine the dominance of one investment alternative over another for all prospect theory S-shaped utility functions. These theories, introduced in Section 8.1, could be linked to the corresponding S-shaped and reverse S-shaped utility functions for the first three orders.

**Theorem 8.29** *Let $X$ and $Y$ be random variables with probability distribution functions $F$ and $G$, respectively. Suppose $u$ is a utility function. For $i = 1, 2$ and $3$, we have $F \succeq_i^M (\succ_i^M) G$ if and only if $u(F) \geq (>) u(G)$ for any $u$ in $\mathcal{U}_i^R$ $(\mathcal{U}_i^{SR})$.*

**Theorem 8.30** *Let $X$ and $Y$ be random variables with probability distribution functions $F$ and $G$, respectively. Suppose $u$ is a utility function. For $i = 1, 2$ and $3$, we have $F \succeq_i^P (\succ_i^P) G$ if and only if $u(F) \geq (>) u(G)$ for any $u$ in $\mathcal{U}_i^S$ $(\mathcal{U}_i^{SS})$.*

The above theorems enable investors to examine the compatibility of the MSD and PSD of any order with the expected utility theory and demonstrates the MSD and PSD of any order to be consistent with the expected utility paradigm. It is well known that hierarchy exists in SD relationships for risk averters and risk seekers: the first-order SD implies the second-order SD which, in turn, implies the third-order SD in the SD rules for risk averters as well as risk seekers. For MSD and PSD, their hierarchical relationships are shown in the following:

**Corollary 8.31** *For any random variables* $X$ *and* $Y$, *for* $i = 1$ *and* 2, *we have the following:*

1. *If* $X \succeq_i^M (\succ_i^M) Y$, *then* $X \succeq_{i+1}^M (\succ_{i+1}^M) Y$; *and*

2. *If* $X \succeq_i^P (\succ_i^P) Y$, *then* $X \succeq_{i+1}^P (\succ_{i+1}^P) Y$.

This corollary suggests that practitioners report the MSD and PSD results to the lowest order in empirical analyses. It is possible for MSD to be "the opposite" of PSD in their second orders and that $F$ dominates $G$ in SPSD, but $G$ dominates $F$ in SMSD. This result could be extended to include MSD and PSD to the second and third orders:

**Corollary 8.32** *For any random variables* $X$ *and* $Y$, *if* $F$ *and* $G$ *have the same mean which is finite, then,*

1. *We have*

$$F \succeq_2^M (\succ_2^M) G \quad \textit{if and only if} \quad G \succeq_2^P (\succ_2^P) F; \textit{ and} \qquad (8.12)$$

2. *If, in addition, either* $F \succeq_2^M (\succ_2^M) G$ *or* $G \succeq_2^P (\succ_2^P) F$ *holds, we have*

$$F \succeq_3^M (\succ_3^M) G \quad \textit{and} \quad G \succeq_3^P (\succ_3^P) F. \qquad (8.13)$$

However, there are cases when distributions $F$ and $G$ have the same mean and do not satisfy (8.12) yet satisfying (8.13). One could find an example in which there is no SMSD and no SPSD dominance but $F \succeq_3^M G$ and $G \succeq_3^P F$. The above corollary provides the conditions in which $F$ is "the opposite" of $G$. On the other hand, under some regularities, $F$ becomes "the same" as $G$ in the sense of TMSD and TPSD as shown in the corollary below:

**Corollary 8.33** *If* $F$ *and* $G$ *satisfy the conditions*

$$\begin{aligned} F_2^A(0) &= G_2^A(0), \quad F_3^A(0) = G_3^A(0), \\ F_2^a(b) &= G_2^a(b), \quad \textit{and } F_3^a(b) = G_3^a(b), \end{aligned} \qquad (8.14)$$

*then*

$$F \succeq_3^M (\succ_3^M) G \quad \textit{if and only if} \quad F \succeq_3^P (\succ_3^P) G.$$

Note that the assumptions in (8.14) are very restrictive. In fact, if some of the assumptions are not satisfied, there exists $F$ and $G$ such that $G \succeq_3^P F$ but neither $F \succeq_3^M G$ nor $G \succeq_3^M F$ holds. Hence, under some regularities, $F$ is "the same" as $G$ in the sense of TMSD and TPSD. One may wonder whether this "same direction property" could appear in FMSD versus FPSD and SMSD versus SPSD. The following corollary shows that this is possible:

**Corollary 8.34** *If the random variable $X = p+qY$ and if $p+qx \geq (>)\,x$ for all $x \in [a,b]$, then we have $X \succeq_i^M (\succ_i^M) Y$ and $X \succeq_i^P (\succ_i^P) Y$ for $i = 1, 2$ and $3$.*

MSD is generally not "the opposite" of PSD. In other words, if $F$ dominates $G$ in PSD, it does not necessarily mean that $G$ dominates $F$ in MSD. This is easy to see because having a higher mean is a necessary condition for dominance by both rules. Therefore, if $F$ dominates $G$ in the sense of PSD, and $F$ has a higher mean than $G$, $G$ cannot possibly dominate $F$ in the sense of MSD. The above corollary goes one step further and shows that they could be "the same" in the sense of MSD and PSD. In addition, we derive the following corollary to show the relationship between the first-order MSD and PSD.

**Corollary 8.35** *For any random variables $X$ and $Y$, we have*

$$X \succeq_1^M (\succ_1^M)Y \quad \text{if and only if} \quad X \succeq_1^P (\succ_1^P)Y.$$

One can show that $X$ stochastically dominates $Y$ in the sense of FMSD or FPSD if and only if $X$ stochastically dominates $Y$ in the sense of the first-order SD (FSD). This result can be extended to yield the following corollary:

**Corollary 8.36** *If the market is complete, then for any random variables $X$ and $Y$, $X \succ_1^M Y$ or if $X \succ_1^P Y$ if and only if there is an arbitrage opportunity between $X$ and $Y$ such that one will increase one's wealth as well as one's utility if one shifts the investments from $Y$ to $X$.*

A "complete" market could be defined as "an economy where all contingent claims on the primary assets trade." It is known that if the market is complete, then $X$ stochastically dominates $Y$ in the sense of FSD if and only if there is an arbitrage opportunity between $X$ and $Y$. As $X \succeq_1^M Y$ is equivalent to $X \succeq_1^P Y$ (see Corollary 8.35), both are equivalent to $X$ stochastically dominating $Y$ in the sense of FSD.

The safety-first rule for decision-making under uncertainty stipulates choosing an alternative that provides a target mean return while minimizing the probability of the return falling below some threshold of disaster. The idea could be extended to include both MSD and PSD.

Using the results in Theorems 8.29 and 8.30, we can call a person a *first-order-MSD* (FMSD) investor if his/her utility function $u$ belongs to $U_1^R$, and a *first-order-PSD* (FPSD) investor if his/her utility function $U$ belongs to $U_1^S$. A second-order-MSD (SMSD) risk investor, a second-order-PSD (SPSD) risk investor, a third-order-MSD (TMSD) risk investor and a third-order-PSD (TPSD) risk investor can be defined in the same way. It is easy to find that the risk aversion of an SPSD investor is positive in the positive domain and negative in the negative domain and an SMSD investor's risk aversion is negative in the positive domain and positive in the negative domain. If one's risk aversion is positive and decreasing in the positive domain and negative and decreasing in the negative domain, then one is a TPSD investor; but the reverse is not true. Similarly, if one's risk aversion is negative and decreasing in the positive domain and positive and decreasing in the negative domain, then one is a TMSD investor. These results could be summarized as follows.

**Corollary 8.37** *For an investor with an increasing utility function $u$ and risk aversion $r$,*

1. *She is an SPSD investor if and only if her risk aversion $r$ is positive in the positive domain and negative in the negative domain;*

2. *She is an SMSD investor if and only if her risk aversion $r$ is negative in the positive domain and positive in the negative domain;*

3. *If her risk aversion $r$ is always decreasing and is positive in the positive domain and negative in the negative domain, then she is a TPSD investor; and*

4. *If her risk aversion $r$ is always decreasing and is negative in the positive domain and positive in the negative domain, then she is a TMSD investor.*

Corollary 8.37 states the relationships between different types of investors and their risk aversions. Note that the converses of (c) and (d) are not true.

## 8.3.1　Illustration

In the experiment, $10,000 is invested in either Stock $F$ or Stock $G$ with the following dollar gain one month later with probabilities $f$ and $g$, respectively, as shown in Table 8.1.

Table 8.1: The Distributions for Investments $F$ and $G$

| Investment $F$ | | Investment $G$ | |
|---|---|---|---|
| Gain | Probability ($f$) | Gain | Probability ($g$) |
| −1,500 | 1/2 | −3,000 | 1/4 |
| 4,500 | 1/2 | 3,000 | 3/4 |

See [95]. We use the MSD and PSD integrals, $H_i^M$ and $H_i^P$, for $H = F$ and $G$ and $i = 1$, 2 and 3 as defined in (8.5).

Table 8.2: The MSD Integrals and Their Differentials for $F$ and $G$

| Gain | First Order | | | Second Order | | | Third Order | | |
|---|---|---|---|---|---|---|---|---|---|
| $X$ | $F_1^M$ | $G_1^M$ | $GF_1^M$ | $F_2^M$ | $G_2^M$ | $GF_2^M$ | $F_3^M$ | $G_3^M$ | $GF_3^M$ |
| −3 | 0 | 0.25 | 0.25 | 0 | 0 | 0 | 0 | 0 | 0 |
| −1.5 | 0.5 | 0.25 | −0.25 | 0 | 0.375 | 0.375 | 0 | 0.28125 | 0.28125 |
| $0^-$ | 0.5 | 0.25 | −0.25 | 0.75 | 0.75 | 0 | 5.625 | 1.125 | 0.5625 |
| $0^+$ | 0.5 | 0.75 | 0.25 | 2.25 | 2.25 | 0 | 5.0625 | 3.375 | −1.6875 |
| 3 | 0.5 | 0.75 | 0.25 | 0.75 | 0 | −0.75 | 0.5625 | 0 | −0.5625 |
| 4.5 | 0.5 | 0 | −0.5 | 0 | 0 | 0 | 0 | 0 | 0 |

See [171]. The functions $F_i^M$ and $G_i^M$ are defined in (8.6) and $GF_i^M = G_i^M - F_i^M$ for $i = 1$, 2 and 3.

Table 8.3: The PSD Integrals and Their Differentials for $F$ and $G$

| Gain | First Order | | | Second Order | | | Third Order | | |
|---|---|---|---|---|---|---|---|---|---|
| $X$ | $F_1^P$ | $G_1^P$ | $GF_1^P$ | $F_2^P$ | $G_2^P$ | $GF_2^P$ | $F_3^P$ | $G_3^P$ | $GF_3^P$ |
| −3 | 1 | 1 | 0 | 2.25 | 2.25 | 0 | 2.8125 | 3.375 | 0.5625 |
| −1.5 | 1 | 0.75 | −0.25 | 0.75 | 1.125 | 0.375 | 0.5625 | 0.84375 | 0.28125 |
| $0^-$ | 0.5 | 0.75 | 0.25 | 0 | 0 | 0 | 0 | 0 | 0 |
| $0^+$ | 0.5 | 0.25 | −0.25 | 0 | 0 | 0 | 0 | 0 | 0 |
| 3 | 0.5 | 1 | 0.5 | 1.5 | 0.75 | −0.75 | 2.25 | 1.125 | −1.125 |
| 4.5 | 1 | 1 | 0 | 2.25 | 2.25 | 0 | 5.0625 | 3.375 | −1.6875 |

See [171]. The functions $F_i^P$ and $G_i^P$ are defined in (8.6) and $GF_i^P = G_i^P - F_i^P$ for $i = 1$, 2 and 3.

To make the comparison easier, we define their differentials

$$GF_i^M = G_i^M - F_i^M \quad \text{and} \quad GF_i^P = G_i^P - F_i^P \tag{8.15}$$

for $i = 1, 2$ and 3 and present the results of the MSD and PSD integrals with their differentials for the first three orders in Tables 8.2 and 8.3.

In this example, we find that $F \succeq_i^M G$ and $G \succeq_i^P F$ for $i = 2$ and 3. From Corollary 8.31, we know that hierarchy exists in both MSD and PSD such that $F \succeq_2^M G$ implies $F \succeq_3^M G$ while $G \succeq_2^P F$ implies $G \succeq_3^P F$. Hence, one only has to report the lowest SD order.

The advantage of the stochastic dominance approach is that we have a decision rule which holds for all utility functions of certain class. Specifically, PSD (MSD) of any order is a criterion which is valid for all S-shaped (reverse S-shaped) utility functions of the corresponding order. Moreover, the SD rules for S-shaped and reverse S-shaped utility functions can be employed with mixed prospects. Note that we do not restrict the S-shaped utility functions to be steeper than their shapes for gains as the restricted set in value functions.

## 8.4 Market rationality and efficiency

Without the need to identify a risk index or a specific model, using SD rules could test market rationality and market efficiency. In applying these rules, we look at the whole distribution of returns and not only at certain parameters, such as mean and variance. In examining market data the criterion that SD employs are: (a) Can some investors switch their portfolio choice, say from $X$ to $Y$ and increase their (expected) wealth? (b) Can some investors switch their portfolio choice, say from $X$ to $Y$ and increase their expected utility given that others have no utility loss or even increase their expected utility by switching their portfolio choice from $Y$ to $X$? We first discuss (a): suppose that there are two portfolios or assets $X$ and $Y$. If $X$ dominates $Y$ by FSD, there exists an arbitrage opportunity and all types of investors will increase their wealth and utility by switching from $Y$ to $X$. If this is the case, is there any market irrationality? Jarrow [72] and Falk and Levy [42] claim that if FSD exists, under certain conditions arbitrage opportunity exists and investors will increase their wealth and expected utilities if they shift from holding the dominated hedge fund to the dominant one. One may believe that in this situation, there is a market irrationality and one may conclude that the market is inefficient.

Firstly, if FSD exists and the cumulative distribution function (CDF) of $X$ lies below that of $Y$, then $X$ dominates $Y$ mathematically. However, the dominance may not be significant when the SD tests are applied. In this case, the FSD does not exist. In another situation, if FSD exists statistically, arbitrage opportunity may not exist but investors could increase their expected wealth as well as their expected utilities if they shift from holding the dominated asset to the dominant one. See [42], [72] and [171] for more discussion. For example, Jarrow [72] discovered the existence of the arbitrage opportunities by SD rules. He defined a "complete" market as "an economy where all contingent claims on the primary assets trade." His arbitrage versus SD theorem says that when the market is complete, $X$ stochastically dominates $Y$ in the sense of FSD if and only if there is an arbitrage opportunity between $X$ and $Y$. In addition, if the test detects the first-order dominance of a particular fund over another but the dominance does not last for a long period, these results cannot be used to reject market efficiency or market rationality, see [12], [42], [85] and [175] for more discussion.

In general, the first-order dominance should not last for a long period of time because market forces cause adjustment to a condition of no FSD if the market is rational and efficient. For example, if asset $X$ dominates asset $Y$ at FSD, then all investors would buy asset $X$ and sell asset $Y$. This will continue, driving up the price of asset $X$ relative to asset $Y$, until the market price of asset $X$ relative to asset $Y$ is high enough to make the marginal investor indifferent between $X$ and $Y$. In this case, we infer that the market is still efficient and rational. In the conventional theories of market efficiency, if one is able to earn an abnormal return for a considerable length of time, the market is considered inefficient. If new information is either quickly made public or anticipated, the opportunity to use the new information to earn an abnormal return is of limited value. On the other hand, in the case where the first-order dominance holds for a long time and all investors increase their expected wealth by switching their asset choice, we claim that the market is neither efficient nor rational.

Another possibility for the existence of FSD to be held for a long period is that investors do no realize such dominance exists. It would be interesting to find out if FSD relationships among some funds would disappear over time. If they do not, then this would be considered a financial anomaly. Note that there are some financial anomalies which do not disappear after discovered. For example, Jegadeesh and Titman [73] first documented the momentum profit in stock markets that for a longer holding period former winners are still winners and former losers are still losers. After many years, many studies, for example, Chan et al. [21] still found momentum profits empirically.

On the other hand, the SSD criterion assumes that all individuals have increasing utility with decreasing marginal utility functions. Given a market only with these investors, if asset $X$ dominates asset $Y$, then all investors would buy $X$ and (short) sell $Y$. This will continue, driving up the price of asset $X$ relative to $Y$, until the market price of $X$ relative to $Y$ is high enough to make the marginal investor indifferent between $X$ and $Y$. But the higher price for $X$ (and lower price for $Y$) implies lower expected return for $X$ (and higher expected return for $Y$). This has the effect of altering the equilibrium distributions of returns until there is no SSD for the two assets. This means that in a rational market in which all individuals have increasing utility but decreasing marginal utility, we should not observe SSD between any two distinct assets, as well as any two different portfolios, continue to exist for a considerable length of time.

We note that it should be possible in a rational market to have two assets or portfolios where one SD dominates another for a short period of time. However, we claim that this phenomenon should not last for an extended period of time because market forces cause adjustment to a condition of no SSD if the market is rational. One could still claim that the market is inefficient if the market contains only risk averse investors. However, if the markets also contain other types of investors, for example, risk seekers (see [5], [66], [113], [131], [151], [173] and [170]), could one claim that the markets are inefficient?

Risk averters could prefer to invest $X$ than $Y$ while risk seekers could prefer to invest in $Y$ than $X$. Both parties could get what they want and there is no pressure to push up the price of $X$ or pull down the price of $Y$. Risk averters could prefer to invest $X$ than $Y$ while risk seekers could prefer to invest in $Y$ than $X$. Both parties could get what they and we claim that this could not infer markets are inefficient.

In addition, the markets could contain other types of investors, for example, investors with S-shaped utility functions or reverse S-shaped utility functions, see for example, [75], [92, 95, 93], [104], [158] and [171]. This will contribute to the equilibrium of stock prices by different groups of investors. This could also help to explain why some investors prefer to buy whereas others prefer to sell as well as the existence of many financial phenomena and the complexity of the financial markets.

A similar argument can be made for the TSD criterion which assumes that all investors' utility functions exhibit non-satiation, risk aversion, with decreasing absolute risk aversion. If a market only contains these types of investors, we should not observe TSD between any two distinct assets continuing to exist for a considerable length of time. However, if the market contains both investors with decreasing absolute risk aversion and investors with non-decreasing absolute risk aversion, then it is possible to have two different assets or portfolios, where one dominates another in the sense of TSD for a considerable length of time. It should be noted that the SSD and TSD between any two distinct assets or different portfolios do not imply return arbitrage opportunities.

Here we emphasize utility improvement or welfare gain for the higher order stochastic dominance, rather than risk-free profit opportunities. Essentially, we claim that there should be no opportunities for further welfare improvement or gain in a rational market so that we should not observe the SSD and TSD for any two different assets or portfolios for a considerable length of time.

Conceptually, market rationality within the SD framework is not different from the conventional understanding with some rational asset pricing models, such as the capital asset pricing model (CAPM). The only difference is that the latter approach defines an abnormal return as excess return adjusted to some risk measure, while SD market rationality tests employ the whole distribution of returns. In particular, both SD and the asset pricing model based residual analyses are consistent with the concept of expected utility maximization. Given that no model is true, the SD approach with less restrictions on both investors' preferences and return distributions seems to help us better understand any financial anomalies.

### 8.4.1   Applications of SD to calendar anomalies

The day-of-the-week effect was first observed in 1931, when the U.S. stock market consistently experienced significant negative returns on Mondays and significant positive returns on Fridays. Explanations for the existence of a day-of-the-week effect include statistical errors, micro-market effects, information-flow effects and order-flow effects. There is a large empirical literature documenting a day-of-the-week effect with low returns on Monday for U.S. stock markets.

In addition, many studies examine the day-of-the-week effect for other developed markets. They observe strong negative returns on Monday and Tuesday in some developing countries while some observe Monday had the lowest negative return and Friday had the highest positive return. Later on, it was observed that the day-of-the-week effect has largely disappeared in many countries.

On the other hand, the January effect is the anomaly that stock returns are larger in January than in other months. The January effect in the Dow Jones Industrial average was first observed in 1942. It was found that stock returns for January were significantly higher than the other 11 months for several indices for NYSE stocks. Research also confirms the existence of a January effect in some emerging markets but not other markets. After calendar anomaly is found, recent studies reveal that calendar effects have been weakening and/or have disappeared. Most of the existing literature has employed the mean-variance criterion or CAPM statistics. As there are many limitations on using these approaches, to address these shortcomings, we recommend to apply a non-parametric stochastic dominance approach to examine calendar effects.

## 8.4.2 Data

We examine the calendar anomalies by using daily stock indices for the period January 1, 1988 to December 31, 2002, including the Hang Seng Index for Hong Kong, Jakarta Composite Index for Indonesia, Kuala Lumpur Composite Index for Malaysia, Nikkei Index for Japan, Straits Times Index for Singapore, Taiwan Stock Exchange Index for Taiwan and the SET Index for Thailand. The daily log-return

$$R_{it} = log_e P_{i,t} - log_e P_{i,t-1} \tag{8.16}$$

is calculated based on the closing values $P_{i,t}$ of the stock index $i$ on days $t$ and $t-1$, respectively, and exclude any week with fewer than five trading days in our study for the day-of-the-week effect. Similarly, we only examine returns for the first 20 calendar days in our test for the January effect so as to fulfill the requirement of equal sample size. We form the portfolio of each weekday (month) by grouping the returns of the same weekday (month) over the entire sample period. Following grouping, we adopt the SD approach to perform a pair-wise comparison for all portfolios in our study.

## 8.4.3 Results

**Day-of-the-week effect** We exhibit in Table 8.4 the mean return, standard deviation, skewness and kurtosis of the returns for each day of the week and for each country. Consistent with most previous studies for the day-of-the-week effect, our results display a tendency for the lowest mean return to be on Monday and the highest mean return on Friday.

Table 8.4: *Weekday Returns for Asian Countries (1988–2002)*

|  | Mon | Tue | Wed | Thu | Fri |
|---|---|---|---|---|---|
| | Hong Kong | | | | |
| Mean (%) | −0.0801 | 0.0944c | 0.1148c | −0.0567 | 0.1054c |
| Std Dev (%) | 2.15 | 1.48 | 1.78 | 1.61 | 1.51 |
| Skewness | −2.41 | −0.92 | 1.01 | −0.89 | 0.71 |
| Kurtosis | 29.31 | 17.59 | 14.68 | 5.75 | 4.29 |
| | Taiwan | | | | |
| Mean (%) | 0.0816 | −0.1152 | 0.0013 | 0.0223 | 0.0983 |
| Std Dev (%) | 2.62 | 1.87 | 1.93 | 1.96 | 1.9 |
| Skewness | 0.13 | 0.19 | −0.26 | −0.31 | −0.17 |
| Kurtosis | 2.45 | 2.71 | 1.45 | 2.21 | 1.9 |
| | Japan | | | | |
| Mean (%) | −0.1736a | 0.0382 | 0.0208 | 0.0189 | −0.0654 |
| Std Dev (%) | 1.6 | 1.35 | 1.46 | 1.37 | 1.41 |
| Skewness | −0.07 | 1.04 | 0.21 | −0.04 | 0.35 |
| Kurtosis | 2.97 | 10.93 | 2.81 | 2.35 | 3.4 |
| | Malaysia | | | | |
| Mean (%) | −0.1720a | 0.025 | 0.1008c | 0.0099 | 0.1020c |
| Std Dev (%) | 1.88 | 1.83 | 1.42 | 1.43 | 1.36 |
| Skewness | 1.85 | −1.3 | 0.52 | −1.11 | 1.13 |
| Kurtosis | 26.28 | 69.33 | 8.94 | 10.42 | 13.97 |
| | Singapore | | | | |
| Mean (%) | −0.1477b | 0.0055 | 0.0597 | 0.0548 | 0.0916c |
| Std Dev (%) | 1.72 | 1.23 | 1.31 | 1.25 | 1.22 |
| Skewness | 0.34 | 0.21 | −0.13 | −0.12 | 0.32 |
| Kurtosis | 13.72 | 11.35 | 5.34 | 4.77 | 6.95 |
| | Indonesia | | | | |
| Mean (%) | −0.0047 | −0.034 | 0.0015 | 0.1027 | 0.1538c |
| Std Dev (%) | 1.78 | 1.42 | 1.73 | 1.84 | 2.14 |
| Skewness | 1.93 | 0.31 | 0.75 | 1.24 | 9.92 |
| Kurtosis | 24.14 | 9.81 | 18.34 | 28.51 | 184.96 |
| | Thailand | | | | |
| Mean (%) | −0.2333a | −0.086 | 0.0936 | −0.0016 | 0.2430a |
| Std Dev (%) | 1.95 | 1.75 | 1.82 | 1.77 | 1.64 |
| Skewness | 0.43 | 0.12 | −0.52 | 0.31 | 0.86 |
| Kurtosis | 5.03 | 6.21 | 3.42 | 4.76 | 6.52 |

See [86].

In contrast to the mean returns, the standard deviation of returns generally decreases as the week progresses as the volatility tends to be highest on Monday and lowest on Friday in all countries except Indonesia. While not reported, pair-wise t-tests show that some weekdays have statistically different mean returns than others whereas F-statistics indicate that some standard deviations are statistically higher at the 5% level. In Table 8.4, a, b and c denote significance differences from zero at the 1%, 5% and 10% level, respectively.

For comparative purposes, we first applied the MV approach. Taking Japan as an example, from Table 8.4 it can be seen that in Japan, Monday's mean return is −0.1736%, which is lower than all other weekdays and its standard deviation is 1.6%, which is significantly higher than all other weekdays. Hence, applying the MV criterion, Monday is dominated by Tuesday, Wednesday, Thursday and Friday by the MV rule.

The results using the MV approach are summarized in Table 8.5. All results in Table 8.5 use a 5% significance level. We find that Monday is dominated by four other weekdays for Hong Kong, Japan, Malaysia, Singapore and Thailand. However, Monday is dominated by Friday for Taiwan only and is dominated by Tuesday and Wednesday for Indonesia. Friday also dominates other weekdays for each of the countries in the sample except Japan and Indonesia.

*Table 8.5: MV Test Results of Day-of-the-Week Effect for Asian Countries*

| Hong Kong | Tue ≻ Mon | Wed ≻ Mon | Thu ≻ Mon | Fri ≻ Mon |
|---|---|---|---|---|
|  | Tue ≻ Thu | Fri ≻ Thu |  |  |
| Taiwan | Fri ≻ Mon | Fri ≻ Wed | Fri ≻ Thu |  |
| Japan | Tue ≻ Mon | Wed ≻ Mon | Thu ≻ Mon | Fri ≻ Mon |
|  | Tue ≻ Wed | Tue ≻ Thu | Tue ≻ Fri | Thu ≻ Fri |
| Malaysia | Tue ≻ Mon | Wed ≻ Mon | Thu ≻ Mon | Fri ≻ Mon |
|  | Wed ≻ Tue | Wed ≻ Thu | Fri ≻ Tue | Fri ≻ Wed |
|  | Fri ≻ Thu |  |  |  |
| Singapore | Tue ≻ Mon | Wed ≻ Mon | Thu ≻ Mon | Fri ≻ Mon |
|  | Fri ≻ Tue | Fri ≻ Wed | Fri ≻ Thu |  |
| Indonesia | Tue ≻ Mon | Wed ≻ Mon |  |  |
| Thailand | Tue ≻ Mon | Wed ≻ Mon | Thu ≻ Mon | Fri ≻ Mon |
|  | Fri ≻ Tue | Fri ≻ Wed | Fri ≻ Thu |  |

Results of the MV test are reported for day-of-the-week effect for seven Asian countries. The sample period is January 1988 to December 2002. The inequality $X \succ Y$ means $X$ dominates $Y$ by MV rule if $\mu_x \geq \mu_y$ and $\sigma_x \leq \sigma_y$ and at least one of them is significant at the 5% significance level. See [86].

In Japan, Friday is dominated by Tuesday and Thursday. Overall, the findings from the MV criterion are inconsistent with a diminishing weekday effect that has been suggested by recent studies. However, if normality does not hold then the MV rule may lead to paradoxical results. Table 8.4 shows that weekday returns in the countries under consideration are non-normal, as evidenced by higher kurtosis and the highly significant unreported Kolmogorov-Smirnov statistics.

While not reported, the CDF plots show that there is no FSD between the Monday and Friday stock returns for Indonesia as their CDFs cross. For Malaysia, the CDF plot for Friday is below that of Monday, thus Friday FSD Monday. To verify this inference formally we apply the Davidson-Duclos (DD) test to the series. To minimize a Type II error of finding dominance when there is none, and to take care of the almost SD effect, a conservative 5% cutoff point is used in this study. The values of the first three orders of DD statistics over the entire distribution of returns for Indonesia and Malaysia show that in general the first-order DD statistics, $T_1$, moves from negative to positive along the distribution of returns. This implies that Friday FSD Monday in the lower range of returns (negative returns) while Monday FSD Friday in the upper range (positive returns). However, the difference could be significant or insignificant.

We found that no $T_1$ is significantly negative and positive for Indonesia while 10% of $T_1$ is significantly negative and no $T_1$ is significantly positive for Malaysia. All second- and third-order DD statistics ($T_2$ and $T_3$) are negative along the distribution of returns.

*Table 8.6: DD Test Results of Day-of-the-Week Effect for Asian Countries*

| Hong Kong | Tue $\succ_1$ Mon | | | |
|-----------|-------------------|---|---|---|
| Taiwan    | Tue $\succ_2$ Mon | Wed $\succ_2$ Mon | Thu $\succ_2$ Mon | Fri $\succ_2$ Mon |
|           | Tue $\succ_1$ Mon | Thu $\succ_2$ Mon | | |
| Japan     | Tue $\succ_1$ Mon | Thu $\succ_2$ Mon | | |
| Malaysia  | Wed $\succ_2$ Mon | Fri $\succ_1$ Mon | | |
| Singapore | Tue $\succ_1$ Mon | Wed $\succ_2$ Mon | Thu $\succ_1$ Mon | Fri $\succ_1$ Mon |
| Indonesia | ND | | | |
| Thailand  | Wed $\succ_1$ Mon | Fri $\succ_1$ Mon | Fri $\succ_1$ Tue | Fri $\succ_1$ Thu |

See [86]. Results of the Davidson-Duclos (DD) test of stochastic dominance are reported for day-of-the-week effect for seven Asian countries [29]. The sample period is January 1988 to December 2002. The DD test statistics are computed over a grid of 100 daily different weekday portfolio returns. ND denotes FSD, SSD & TSD do not exist; $X \succ_j Y$ means $X$ dominates $Y$ at the 5% significance level and at the $j$ order stochastic dominance where $j = 1, 2,$ and 3.

Most are found to be significant at the 5% level for Malaysia (50%-SSD, 63%-TSD) but not for Indonesia (0%-SSD, 0%-TSD). Thus, we conclude that Friday dominates Monday for the first three orders for Malaysia but not for Indonesia at the 5% SMM significance level. This infers that any risk-averse investor will prefer Friday to Monday in the Malaysian stock market as they will increase their wealth as well as their expected utility, subject to trading costs, by switching their investments from Monday to Friday.

We display in Table 8.6 the dominance among different weekday returns for each country in our study using the DD test. We find that Monday stock returns are stochastically dominated by at least one of the other weekday returns at the first order in all countries except Indonesia and Taiwan. For example, Tuesday returns FSD Monday returns in Hong Kong, Japan and Singapore; Friday returns FSD Monday returns in Malaysia, Singapore and Thailand; Thursday returns FSD Monday returns in Singapore and Wednesday returns FSD Monday returns in Thailand. Moreover, our results also show that Friday returns FSD both Tuesday and Thursday returns in Thailand. Monday returns are stochastically dominated by Thursday returns in Japan and dominated by Wednesday returns in Malaysia and Singapore at second- and third-order. Moreover, Monday returns are dominated by all other weekdays by SSD and TSD in Taiwan. Thus, we conclude that there is a day-of-the-week effect in some Asian markets.

Comparing Tables 8.5 and 8.6, we find that the number of cases where pairwise dominance is found using the MV approach (42 pairs) is much larger than the number of cases where pairwise dominance is found using the DD approach (17 pairs). Except for Taiwan, each time that pairwise dominance was found with the DD test, it was also found with the MV approach. Thus, while sometimes the conclusions drawn from the MV and SD approaches are the same, there are several instances where they differ. The SD and MV approaches could be the same when the returns being studied belong to the same location-scale family.

However, it is not surprising that there are several cases where the two approaches give different results. First, the MV criterion is only applicable when the decision-maker's utility function is quadratic or the probability distribution of return is normal while these assumptions are not required by the SD criterion. Second, even when all the distributions of the underlying returns being studied are normal, it is still possible that the conclusion drawn from the SD approach will be different from that drawn from the MV approach.

## January effect

We present in Table 8.7 summary statistics of monthly returns for each country.

Table 8.7: Summary Statistics of Weekday Returns for
Asian Countries (1988-2002)

| | Hong Kong | Taiwan | Japan | Malay-sia | Sing-apore | Indo-nesia | Thai-land |
|---|---|---|---|---|---|---|---|
| | January | | | | | | |
| Mean (%) | -0.0097 | 0.2258c | 0.1122 | 0.0502 | 0.07 | 0.1877b | 0.2685b |
| Std Dev (%) | 1.91 | 2.16 | 1.6 | 1.75 | 1.88 | 1.73 | 2.19 |
| Skewness | -0.65 | 0.02 | 0.87 | 0.51 | -0.12 | -0.37 | 0.72 |
| Kurtosis | 3.69 | 2.2 | 5.02 | 6.1 | 5.58 | 12.34 | 4.25 |
| | February | | | | | | |
| Mean (%) | 0.2586a | 0.3400a | 0.0325 | 0.2406b | 0.0924 | 0.0017 | -0.0351 |
| Std Dev (%) | 1.66 | 1.88 | 1.16 | 1.75 | 1.34 | 1.52 | 2.06 |
| Skewness | 1.27 | 0.35 | -0.17 | 4.57 | 3.89 | 1.02 | 0.23 |
| Kurtosis | 11.6 | 2.67 | 1.42 | 55.99 | 42.83 | 21.51 | 6.54 |
| | March | | | | | | |
| Mean (%) | -0.0169 | 0.0613 | -0.0314 | -0.0674 | -0.0592 | 0.0711 | -0.0565 |
| Std Dev (%) | 1.51 | 1.68 | 1.56 | 1.06 | 1.15 | 1.31 | 1.39 |
| Skewness | -0.57 | -0.02 | 0.49 | -0.14 | 0.15 | 1.8 | 0.06 |
| Kurtosis | 2.8 | 1.99 | 2.37 | 2.71 | 2.24 | 15.49 | 1.35 |
| | April | | | | | | |
| Mean (%) | 0.0902 | 0.0206 | 0.0683 | 0.078 | 0.1396b | 0.0191 | 0.1225 |
| Std Dev (%) | 1.43 | 1.85 | 1.49 | 1.35 | 1.21 | 1.47 | 1.4 |
| Skewness | -0.41 | -0.15 | -0.35 | -0.36 | -1.1 | -0.77 | 0.9 |
| Kurtosis | 5.35 | 0.69 | 4.67 | 3.6 | 10.23 | 28.26 | 5.09 |
| | May | | | | | | |
| Mean (%) | 0.063 | -0.1205 | 0.0105 | 0.0224 | -0.0184 | 0.1838b | -0.0841 |
| Std Dev (%) | 1.66 | 1.93 | 1.13 | 1.2 | 0.98 | 1.36 | 1.77 |
| Skewness | -1.03 | -0.47 | -0.02 | 0.12 | -0.18 | 0.05 | 0.26 |
| Kurtosis | 10.75 | 1.99 | 1.75 | 3.59 | 1.87 | 6.32 | 6.04 |
| | June | | | | | | |
| Mean (%) | -0.0082 | -0.0546 | -0.0535 | -0.0168 | -0.0013 | 0.0728 | 0.0409 |
| Std Dev (%) | 1.94 | 1.88 | 1.26 | 1.16 | 1.08 | 1.31 | 1.57 |
| Skewness | -5.62 | -0.57 | 0.15 | -0.18 | 0.29 | 1.96 | 0.4 |
| Kurtosis | 74.36 | 2.54 | 0.83 | 1.94 | 3.25 | 13.51 | 4.14 |

Table 8.7: Summary Statistics of Weekday Returns for
Asian Countries (1988-2002) (continued)

| | Hong Kong | Taiwan | Japan | Malay-sia | Sing-apore | Indo-nesia | Thai-land |
|---|---|---|---|---|---|---|---|
| | July | | | | | | |
| Mean (%) | 0.0781 | 0.0271 | −0.0005 | −0.0003 | 0.0143 | −0.0607 | −0.071 |
| Std Dev (%) | 1.23 | 2.12 | 1.31 | 1.14 | 0.96 | 1.07 | 1.63 |
| Skewness | 0.06 | 0.09 | 0.18 | −0.44 | 0.28 | −0.53 | 0.94 |
| Kurtosis | 0.55 | 1.14 | 1.53 | 3.05 | 1.98 | 3.8 | 4.31 |
| | August | | | | | | |
| Mean (%) | −0.2061b | −0.1424 | −0.1216 | −0.2115b | −0.1824b | −0.1331 | −0.1725 |
| Std Dev (%) | 1.81 | 2.18 | 1.58 | 1.73 | 1.47 | 2.42 | 1.96 |
| Skewness | −0.65 | −0.15 | 0.1 | −0.29 | −0.97 | −0.39 | −0.43 |
| Kurtosis | 5.14 | 1.68 | 2.3 | 5.81 | 4.1 | 35.73 | 5.66 |
| | September | | | | | | |
| Mean (%) | −0.0156 | −0.2452b | −0.1644b | −0.1091 | −0.1290c | −0.1995b | −0.1235 |
| Std Dev (%) | 1.62 | 1.87 | 1.54 | 2.7 | 1.29 | 1.75 | 1.82 |
| Skewness | −0.09 | −0.47 | −0.34 | −0.02 | −0.32 | 0.27 | −0.07 |
| Kurtosis | 5.41 | 3.32 | 2.15 | 33.81 | 7.45 | 8.45 | 3.7 |
| | October | | | | | | |
| Mean (%) | 0.2253c | −0.0936 | 0.0277 | 0.0793 | 0.1306 | −0.0132 | 0.0814 |
| Std Dev (%) | 2.22 | 2.3 | 1.56 | 1.48 | 1.64 | 1.81 | 1.83 |
| Skewness | 0.17 | −0.08 | 1.47 | −1.35 | −0.65 | 0.18 | 0.41 |
| Kurtosis | 17.79 | 2.07 | 12.23 | 16.41 | 10.43 | 9.97 | 3.48 |
| | November | | | | | | |
| Mean (%) | 0.0135 | 0.111 | 0.0225 | −0.005 | 0.1231c | 0.0316 | −0.0681 |
| Std Dev (%) | 1.47 | 2.25 | 1.46 | 1.44 | 1.24 | 1.56 | 1.8 |
| Skewness | 0.07 | 0.58 | 0.3 | −1.4 | 0.66 | 0.67 | −0.29 |
| Kurtosis | 1.52 | 5.39 | 3.46 | 16.59 | 4.95 | 7.53 | 2.45 |
| | December | | | | | | |
| Mean (%) | 0.1053 | 0.1326 | −0.0369 | 0.2799a | 0.1869a | 0.3573b | 0.1733b |
| Std Dev (%) | 1.54 | 2.19 | 1.36 | 1.4 | 1.07 | 3.06 | 1.43 |
| Skewness | −0.56 | −0.03 | −0.11 | 0.54 | 0.05 | 7.42 | 0.4 |
| Kurtosis | 4 | 2.19 | 1.9 | 14.67 | 1.18 | 92.89 | 3.21 |

Refer to [86]. a, b, c denote significance difference from zero at the 1%, 5% and 10% level, respectively.

January returns are positive, although not necessarily significant, in all countries for our sample except Hong Kong. However, in general,

January does not have the highest returns of the month (except in Japan and Thailand) and Hong Kong even has very low returns in January. Table 8.7 suggests that January returns exhibit a high standard deviation. Japan and Thailand are the two countries with the highest mean and standard deviation for January returns. The higher kurtosis than normal and the highly significant unreported Kolmogorov-Smirnov statistics show that January returns in each of the seven Asian countries in this study are non-normal.

*Table 8.8: MV Test Results of January Effect for Asian Countries*

| Hong Kong | Feb ≻ Jan | Apr ≻ Jan | May ≻ Jan | Jul ≻ Jan |
|---|---|---|---|---|
|  | Nov ≻ Jan | Dec ≻ Jan |  |  |
| Taiwan | Feb ≻ Jan | Jan ≻ Aug | Jan ≻ Oct | Jan ≻ Nov |
|  | Jan ≻ Dec |  |  |  |
| Japan | ND |  |  |  |
| Malaysia | Apr ≻ Jan | Oct ≻ Jan | Dec ≻ Jan | Jan ≻ Aug |
| Singapore | Feb ≻ Jan | Apr ≻ Jan |  |  |
| Indonesia | Jan ≻ Aug | Jan ≻ Sep | Jan ≻ Oct |  |
| Thailand | ND |  |  |  |

Refer to [86]. Results of the MV test are reported for January effect for seven Asian countries. The sample period is January 1988 to December 2002. $X \succ Y$ means $X$ dominates $Y$ by MV rule if $\mu_x \succ \mu_y$ and $\sigma_x \succ \sigma_y$ at the 5% significance level.

We display in Table 8.8 dominance between January and non-January returns using the MV approach at the 5% level. The table shows that there is no MV dominance for Japan and Thailand because these two countries have the highest mean and standard deviation for their January returns. Hong Kong has the most MV dominance pairs (six pairs) with February, April, May, July, November and December returns preferred to January returns. Singapore has the least MV dominance pairs (two pairs) with returns in February and April preferred to January returns. For Taiwan, January dominates four months (August, October, November and December) but is only dominated by February. In Indonesia, January dominates August, September and October and is not dominated by any months.

We report in Table 8.9 the DD test results for January returns with each of the non-January months for the whole sample period. We find that July returns FSD January returns in Hong Kong. January returns are dominated by July, November and December at second- and third-order in Singapore. For other markets, the January returns do not dominate any of the non-January months and vice versa by FSD, SSD or

TSD. Thus, we conclude that there is no monthly seasonality effect in the Asian markets except in Hong Kong.

Table 8.9: DD Test Results of January Effect for Asian Countries

| Hong Kong | Jul $\succ_1$ Jan | | |
|-----------|-----------|-----------|-----------|
| Taiwan | ND | | |
| Japan | ND | | |
| Malaysia | ND | | |
| Singapore | Jul $\succ$ Jan | Nov $\succ$ Jan | Dec $\succ$ Jan |
| Indonesia | ND | | |
| Thailand | ND | | |

Refer to [86]. Results of the Davidson-Duclos (DD) test of stochastic dominance are reported for January effect for seven Asian countries [29]. The sample period is January 1988 to December 2002. The DD test statistics are computed over a grid of 100 daily different month portfolio returns. ND denotes FSD, SSD & TSD do not exist; $X \succ_j Y$ means $X$ dominates $Y$ at the 5% significance level and at the $j$ order stochastic dominance where $j = 1, 2$ and 3.

Comparing Tables 8.8 and 8.9, with the MV approach there are 20 pair-wise cases of dominance while there are only four using the DD approach. While there is no evidence of dominance for two countries (Japan and Thailand) using the MV approach, there is no evidence of dominance for five countries (Taiwan, Japan, Malaysia, Indonesia and Thailand) using the DD approach. July dominates January for Hong Kong using the MV and DD approaches, but for Singapore, the SSD of July, November and December to January is not picked up with the MV approach. The possible reasons for the different results are the same as the day-of-the-week effect that are discussed above.

Recent studies have suggested a weakening and/or disappearance of the day-of-the-week effect in non-U.S. markets over the course of the 1990s. However these findings are still tentative due to the differing statistical tools used in the studies, some of which may have been mis-specified or suffering serious measurement problems. As stock returns in Asian markets are not normally distributed by nature, the parametric MV approach is of limited value.

Another limitation is that findings using the MV approach cannot be used to conclude whether investors' portfolio preferences will increase expected wealth or, in the case of risk-averse investors, lead to an increase in expected utility without an increase in wealth. Thus, this study has used the SD approach, which is not distribution-dependent and can shed light on the utility and wealth implications of portfolio preferences through exploiting information in higher order moments to test for day-of-the-week and January effects in Asian markets.

The findings that Monday returns are dominated by other weekdays and Friday dominates other weekdays, applying the MV criterion, suggests that the diminishing of a weekday effect claimed by recent studies is questionable. Our DD test results for the day-of-the-week effect also indicate there is FSD of other weekdays over Monday returns in the Asian countries studied. Moreover, the existence of SSD and TSD in some of the markets suggests that risk-averse individuals would prefer (or not prefer) certain weekdays in some of the Asian markets to maximize their expected utility.

The existence of a weekday effect raises the possibility that investors could exploit this to earn abnormal returns. On the other hand, the DD test results for the January effect suggest that it has largely disappeared from Asian markets and that only in Singapore is January dominated by some other months at SSD and TSD. The reason for the re-appearance of the day-of-the-week and disappearance of the January effects from Asian markets is an interesting topic for future research.

## 8.5    SD and rationality of momentum effect

We apply SD tests to momentum strategies implemented on international stock market indices. The momentum effect is first observed in 1993 that the tendency for portfolios of stocks that performed well (poorly) in the past 3 to 12 months to continue earning positive (negative) returns over the next 12 months.

This pattern appears to be an anomaly from the perspective of efficient markets as empirical studies show that the effect cannot be explained away using standard asset pricing models such as the CAPM and Fama-French three-factor models. Nonetheless, the search for more general asset pricing models continues on the premise that existing models may be inadequate because of omitted risk factors.

Stochastic dominance theory is used to distinguish between the hypothesis that there exists some (more general) asset pricing models that can explain the momentum effect versus the alternative hypothesis that there are no asset pricing models consistent with risk-averse investors that can rationalize that effect. If winner portfolios stochastically dominate loser portfolios at second or higher orders, then it is unlikely that the problem is due to omitted risk factors but more likely that momentum reflects market inefficiency.

We apply SD tests to analyze momentum strategies implemented on 24 international stock indices for the period 1989 to 2000. Previous studies of international momentum strategies discover significant risk-adjusted profits for different sample periods. Our stochastic dominance approach yields consistent but more general results: We find that winners dominate losers by second- and third-order stochastic dominance, implying that all investors with strictly increasing concave utility functions prefer to buy winners and sell losers over the sample period.

### 8.5.1  Evidence on profitability of momentum strategies

It was observed that a simple strategy of buying stocks with high returns over 3 to 12 months and selling stocks with low returns over the same period produces annualized returns of about 12% for the following year.

The momentum effect also appears in many non-U.S. markets. Much research has focused on whether momentum profits can be explained by risk. Some suggest that momentum cannot be explained by exposure to market risk alone. Standard risk factors do not explain moments. The first is a high minus low (HML) book-to-value risk factor, which is usually interpreted as a firm distress proxy. The second is a small minus large (SML) firm size risk factor to proxy for the higher risk and lower liquidity of small firms. It is found that their unconditional three-factor model cannot account for momentum profits.

However, some research findings also show that neither can a conditional three-factor model explain momentum, they conjecture that the three-factor model may be mis-specified and that momentum returns may reflect an omitted component of returns that differs across stocks. In other words, momentum profits may simply be due to cross-sectional dispersions in unconditional expected returns.

However, studies also show that differences in unconditional expected returns cannot explain momentum profits. Some argue that momentum profits may also be due to industry risk exposures that are not captured by standard factor models, but studies find that it cannot fully explain the momentum effect. In particular, a random industry strategy still earns statistically significant returns in months other than January.

Business cycle risk is another potential explanation of the momentum effect. If the expected return to the momentum strategy is related to business cycle risk, then one would expect momentum profits to be lower in poor economic states than in good economic states.

Some studies found evidence against the hypothesis: momentum profits appear to be large and statistically significant across good and bad economic states. They also show that the multifactor macroeconomic model cannot explain momentum across markets. In short, current risk-based explanations fail to account fully for the momentum effect. A full understanding of the source of the risk-adjusted profitability of the momentum strategy is still an open question.

The failure of risk-based models has led to research which attempts to explain momentum in terms of behavioral inconsistencies of investors. Studies attribute momentum to the fact that investors are overconfident and tend to overreact to confirming news while attributing wrong forecasts to external noise. An increase in overconfidence following the arrival of confirming news leads to further overreaction, generating momentum in stock prices.

Other studies attribute momentum to under-reaction, and add that under-reaction is a delayed response to the slow diffusion of private information. They conjecture that the less risk-averse investors are, the greater is the degree of overreaction. This implies that momentum should be stronger in good economic states than in bad economic states.

There is increasing evidence supporting some aspects of behavioral explanations of momentum. For example, behavioral models imply that momentum profits should be better explained in terms of stock-specific risks than systematic risks. Consistent with this hypothesis, some studies show that a momentum strategy that defines winners and losers based on stock-specific returns (residuals from the three-factor model) is significantly more profitable than a strategy that ranks stocks on total returns.

Other studies focus on more specific aspects of idiosyncratic risk. For example, the literature shows that momentum profits coexist with earnings momentum, and document stronger momentum effects in stocks with high turnover.

By conditioning momentum profits on the state of the market, studies find that momentum profits arise only following up-market periods. They interpret this finding as consistent with the hypothesis that there is more delayed overreaction and hence stronger momentum following periods when the market is up and investors are less risk averse.

Although the evidence points to investor irrationality as the primary cause of momentum, many aspects of behavioral theories are intrinsically hard to test. If momentum is due to overreaction, momentum returns must eventually reverse as investors correct the mispricing. Unfortunately, this hypothesis is difficult to refute since behavioral theories do not specify a precise time frame for such reversals to occur. Thus, to behavioral skeptics, the issue of whether momentum profits are compensation for risk or reflect investor irrationality remains an open question.

## 8.5.2   Data and methodology

We implement momentum strategies on international stock market indices, including 24 indices from January 1989 through December 2001 compiled by Morgan Stanley Capital International (MSCI). Every MSCI index comprises large capitalization and actively traded stocks, thus mitigating biases due to non-synchronous trading or bid-ask spreads.

All data are obtained from Datastream. The 24 countries in our sample are Australia, Austria, Belgium, Canada, Denmark, France, Germany, Hong Kong, Indonesia, Italy, Japan, Korea, Malaysia, Netherlands, Norway, Singapore, South Africa, Spain, Sweden, Switzerland, Taiwan, Thailand, U.K. and U.S. This sample includes all the well-established markets as well as a few emerging markets.

The momentum strategy of international indices is implemented by buying stock indices with high returns over the previous 1 to 12 months and selling stock indices with low returns over the same period. On each day, $t$, all 24 markets are ranked based on their compounded returns from $t - J$ to $t - 1$, where $J = 22, 66, 132, 198$ and $264$ days (approximately 1, 3, 6, 9 and 12 months, respectively). This is known as the ranking period. A winner ($W$) portfolio is then formed by comprising the four markets with the highest ranking period returns and a loser ($L$) portfolio comprising the four markets with the lowest ranking period returns.

To mitigate microstructure biases due to bid-ask bounce effects, we skip day $t$ and invest in a momentum portfolio on day $t+1$ to buy $W$ and short sell $L$. The momentum portfolio is held for $K$ days (the holding period) and its geometric mean daily return is computed assuming equal weights. To reduce the number of portfolio combinations, we let $K$ be the same as $J$.

## 8.5.3   Profitability of momentum strategy

We report in Table 8.10 the ranking period returns of winner $W$, loser $L$ and momentum ($W - L$) portfolios for the overall sample period. As all returns are in local currency, we take the perspective of multinational investors. Refer to [50] for more details.

For all ranking periods, winner returns are positive while loser returns are negative. The mean daily return across all horizons is 0.216% for winners and $-0.174\%$ for losers, which translates to annualized returns of 54% and $-43.5\%$, respectively. The distribution of momentum portfolio returns is right-skewed. Together, these results hint of potentially large profits from a strategy of buying past winners and shorting past losers.

Table 8.11 reports holding period returns of such a strategy.

Results for the full sample period are shown in Panel A of Table 8.11. In contrast to the ranking period, winners show continuation in returns but losers experience reversals. Thus, a strategy of buying winners without short selling losers earned larger profits than a more costly strategy that requires short selling losers.

Nonetheless, based on the $t$-test, the momentum strategy is profitable across all holding periods. The 1-month and 6-month momentum portfolios produced the highest average daily return of 0.036% and 0.035%, respectively, while the 12-month portfolio has the lowest mean daily return (0.0053%).

*Table 8.10: Ranking Period Average Returns of Winner, Loser and Momentum Portfolios*

| Ranking period returns | Winner $(W)$ | Loser $(L)$ | $W - L$ |
|---|---|---|---|
| 1-month | 0.378 | −0.340 | 0.718 |
| $t$-stat | (29.49) | (−26.99) | (65.37) |
| Skewness | 0.83 | −1.14 | 1.12 |
| 3-month | 0.236 | −0.199 | 0.435 |
| $t$-stat | (28.96) | (−23.53) | (62.99) |
| Skewness | 0.74 | −1.06 | 1.00 |
| 6-month | 0.176 | −0.136 | 0.312 |
| $t$-stat | (30.64) | (−23.10) | (66.36) |
| Skewness | 0.23 | −0.98 | 0.82 |
| 9-month | 0.152 | −0.104 | 0.256 |
| $t$-stat | (32.12) | (−22.50) | (63.54) |
| Skewness | 0.64 | −0.46 | 1.07 |
| 12-month | 0.140 | −0.089 | 0.229 |
| $t$-stat | (33.32) | (−21.25) | (62.15) |
| Skewness | 0.57 | −0.47 | 1.18 |

The $t$-test assumes that returns are i.i.d. normal. To check whether our results are affected by violations of this assumption, we performed a bootstrap test whereby we scramble the stock market indices simultaneously to remove any dependence structure for each market while preserving cross-market correlations. We then execute our momentum strategy on 5,000 bootstrap samples and compute the empirical $p$-values, i.e., the percentage of the simulations generating higher mean returns than the actual sample.

Table 8.11: *Holding Period Average Returns of Winner, Loser and Momentum Portfolios*

| Holding Period | W | L | W − L | | |
|---|---|---|---|---|---|
| | | | All Months | January | Non-January |
| Panel A | Full sample | | | | |
| 1-month | 0.0505 | 0.0148 | 0.0357 | 0.0391 | 0.0426 |
| | (4.12) | (1.11) | (3.3) | (−1.22) | (3.83) |
| 3–month | 0.0332 | 0.0292 | 0.0040 | −0.0051 | 0.0049 |
| | (4.01) | (3.04) | (0.49 | (−0.30) | (0.56) |
| 6–month | 0.0521 | 0.0173 | 0.0348 | 0.0344 | 0.0349 |
| | (9.09) | (2.58) | (4.91) | (2.2) | (4.66) |
| 9–month | 0.0365 | 0.0124 | 0.0241 | 0.0225 | 0.0243 |
| | (8.04) | (2.18) | (3.57) | (1.62) | (3.38) |
| 12–month | 0.0239 | 0.0187 | 0.0053 | −0.0104 | 0.0067 |
| | (5.99) | (3.65) | (0.84) | (−0.67) | (1.02) |
| Panel B | Subperiod 1 | | | | |
| 1-month | 0.0511 | 0.0272 | 0.0239 | −0.0235 | 0.0283 |
| | (4.09) | (2.07) | (2.08) | (−0.64) | (2.41) |
| 3 month | 0.0274 | 0.0387 | −0.0113 | −0.0230 | 0.0102 |
| | (3.17) | (4.17) | (−1.36) | (−1.22) | (1.16) |
| 6 month | 0.0453 | 0.0326 | 0.0127 | −0.0138 | 0.0151 |
| | (6.63) | (4.89) | (1.69) | (0.82) | (1.93) |
| 9–month | 0.0406 | 0.0214 | 0.0193 | 0.0112 | 0.0200 |
| | (7.65) | (4.08) | (3.01) | (0.94) | (2.93) |
| 12–month | 0.0341 | 0.0261 | 0.008 | −0.0228 | 0.0109 |
| | (7.27) | (5.81) | (1.37) | (−1.57) | (1.79) |
| Panel C | Subperiod 2 | | | | |
| 1-month | 0.0496 | −0.0050 | 0.0545 | −0.0643 | 0.0655 |
| | (2.17) | (−0.19) | (2.79) | (−1.25) | (3.40) |
| 3–month | 0.0425 | 0.0139 | 0.0286 | 0.0240 | 0.0290 |
| | (2.85) | (0.76) | (1.91) | (0.87) | (1.92) |
| 6–month | 0.0631 | −0.0072 | 0.0703 | 0.1122 | 0.0664 |
| | (7.16) | (−0.59) | (5.78) | (6.7) | (5.43) |
| 9–month | 0.0299 | −0.0019 | 0.0319 | 0.0409 | 0.0310 |
| | (4.17) | (0.17) | (2.45) | (1.50) | (2.40) |
| 12–month | 0.0076 | 0.0067 | 0.0009 | 0.0095 | 0.0001 |
| | (1.25) | (0.66) | (0.08 ) | (0.33) | (0.01) |

Refer to [50] for more details. The full sample period is January 1, 1989 to December 31, 2001. Subperiod 1 is from January 1, 1989 through December 31, 1996, and subperiod 2 is from January 1, 1997 through December 31, 2001.

If the $p$-value is above 5%, we infer that momentum profits detected by the $t$-test are due to chance or are biased by non-normality. The bootstrap results (not reported) show that the $p$-values are uniformly below 5% across all time horizons. We conclude that the documented momentum profits are not spurious.

To test whether momentum profits for our sample persisted through time, we implement our momentum strategy over two subperiods with different return and volatility characteristics. The first subperiod is from January 1989 through December 1996. This is a bullish period for most stock markets. Of the 24 stock markets in the sample, 22 registered positive holding period returns. The mean geometric average return across all the markets is 0.024% per day. The second subperiod, from January 1997 through December 2001, is relatively more bearish and volatile than the first subperiod. Major events which contributed to this volatility include the Thai baht devaluation (July 1997) which adversely affected many other South East Asian stock markets, the Russian debt crisis (August 1998) and the bursting of the Internet stock bubble (March 2000). Only 16 markets registered positive holding period returns in this period. The mean geometric average return for all the markets was $-0.031$%.

Panel B reports momentum results for the first subperiod. Both winners and losers have positive returns, indicating momentum for winners but reversals for losers. Despite this, the momentum strategy of longing winners and shorting losers is generally profitable (the 3- and 12-month holding periods being the exceptions). The 1- and 9-month momentum portfolios generated the highest mean daily return of 0.0239% and 0.0193%, respectively.

Results for the second subperiod are shown in Panel C. The mean return for winners across all holding periods is 0.039%, which is only slightly below their mean return in the first subperiod (0.04%). Therefore, winner momentum has clearly persisted over the two periods. On average, returns for losers are significantly smaller than in the first subperiod, and are in fact negative for the 1-, 6- and 9-month holding periods. Consequently, the momentum strategy is more profitable in the second subperiod than in the first. Across all horizons, the mean daily return of the momentum portfolio is 0.0372%, with the 1- and 6-month strategy performing best.

In summary, the momentum strategy has been highly profitable over the entire sample period. This profit is mainly due to winners, which showed consistently strong returns continuation. Could this persistence in returns be explained by macroeconomic risks? If investing in past winners is a risky strategy, we would expect returns to winners to be positive when the market is bullish but negative when the market is bearish. The data does not support this hypothesis. If anything, winner profits were higher in the more bearish second subperiod, suggesting that the profits cannot be explained in terms of economy-wide systematic risk. Our results are consistent with those of Griffin et al. [61] who find that momentum profits are positive in all economic states, regardless of whether economic states are defined by world GDP growth rate or aggregate stock market returns.

### 8.5.4   Results of stochastic dominance tests

Results of the KS test are shown in Table 8.12.

*Table 8.12: Results of KS Test for Stochastic Dominance*

| Holding Period | $H_{02}$ | | $H_{03}$ | |
|---|---|---|---|---|
| | $W \succ_2 L$ | $L \succ_2 W$ | $W \succ_3 L$ | $L \succ_3 W$ |
| Panel A | Full sample | | | |
| 1-month | 0.583 | 0.000 | 0.537 | 0.000 |
| 3-month | 0.685 | 0.000 | 0.642 | 0.000 |
| 6-month | 0.673 | 0.000 | 0.684 | 0.000 |
| 9-month | 0.712 | 0.000 | 0.653 | 0.000 |
| 12-month | 0.713 | 0.000 | 0.681 | 0.000 |
| Panel B | Subperiod 1 | | | |
| 1-month | 0.580 | 0.000 | 0.021 | 0.000 |
| 3-month | 0.001 | 0.581 | 0.003 | 0.655 |
| 6-month | 0.039 | 0.000 | 0.147 | 0.000 |
| 9-month | 0.068 | 0.00 | 0.259 | 0.000 |
| 12-month | 0.232 | 0.00 | 0.394 | 0.000 |
| Panel C | Subperiod 1 | | | |
| 1-month | 0.663 | 0.000 | 0.614 | 0.000 |
| 3-month | 0.680 | 0.000 | 0.635 | 0.000 |
| 6-month | 0.663 | 0.000 | 0.627 | 0.000 |
| 9-month | 0.689 | 0.000 | 0.645 | 0.000 |
| 12-month | 0.676 | 0.000 | 0.632 | 0.000 |

Refer to [50] for more details. This table reports $p$-values for the KS test of stochastic dominance for winner and loser portfolios. The full sample period is from January 1, 1989 to December 31, 2001, subperiod 1 is from January 1, 1989 through December 31, 1996, and subperiod 2 is from January 1, 1997 through December 31, 2001. The null hypotheses $H_0^2$ and $H_0^3$ relate to second- and third-order stochastic dominance, respectively.

This table reports $p$-values of the KS test for second-order stochastic dominance ($H_{02}$) and third-order stochastic dominance ($H_{03}$), respectively. Under $H_{02}$, the column "$W \succ_2 L$" shows $p$-values for testing the null hypothesis that winners dominate losers at second order, while the column "$L \succ_2 W$" tests the reverse relation. $p$-values under $H_{03}$ are interpreted similarly. All $p$-values are computed using simulations.

Results for the full sample period (Panel A) show that $p$-values for $W \succ_2 L$ and $W \succ_3 L$ are well above 5% while $p$-values for the opposite hypotheses are nearly zero across all holding periods. These results provide strong evidence of winner dominance over the entire sample period.

Subperiod results (Panels B and C) reveal that winner dominance is more pervasive in the second subperiod (1997-2001). The fact that the momentum strategy is more successful in the latter subperiod suggests that trend-chasing behavior in stock markets has not disappeared in recent times.

The DD test results are shown in Table 8.13. Recall that the DD test $T_j$ ($j = 2, 3$) rejects the null hypothesis $H_0$ if none of the DD statistics $T_j$ ($j = 2, 3$) are significantly positive and at least some of the DD statistics are significantly negative. Since the DD test does not specify what percentage of DD statistics must be significantly negative to reject the null hypothesis, to be conservative, we use a 50% cutoff point. That is, we infer that winners dominate losers if at least 50% of the DD statistics are significantly negative and no DD statistics are significantly positive.

Two main points may be noted from the full sample results (Panel A). First, in terms of second-order stochastic dominance, no DD statistic is significantly positive, while DD statistics for all holding periods are significantly negative. These results strongly indicate that all risk-averse investors would have preferred winners to losers over the entire sample period.

Second, evidence for winner dominance is generally stronger at third order than at second order. This implies that investors who prefer more positive skewness would also have chosen to buy winners and sell losers. This preference may be due to the fact that the distribution of winner returns is right-skewed (Table 8.10).

Subperiod results are consistent with the KS test. In particular, there is only weak evidence of winner dominance in the first subperiod. This result is not entirely surprising. Because the first subperiod was a bullish period for most stock markets, even past losers produced positive returns for each holding period, thus reducing the profitability of the momentum strategy. On the other hand, there is clear evidence of winner dominance in the second subperiod, where at least 63% of DD statistics are significantly negative and none are significantly positive. This result is consistent with the KS test. The success of the momentum strategy in this latter period is due to the continued rise in the price of winners as well as the fall in the price of losers.

Table 8.13: Results of DD Test for Stochastic Dominance

| Holding Period | $\%T_2 < 0$ | $\%T_2 > 0$ | $\%T_3 < 0$ | $\%T_3 > 0$ |
|---|---|---|---|---|
| Panel A | Full sample | | | |
| 1-month | 73 | 0 | 66 | 0 |
| 3-month | 50 | 0 | 81 | 0 |
| 6-month | 85 | 0 | 80 | 0 |
| 9-month | 95 | 0 | 94 | 0 |
| 12-month | 76 | 0 | 95 | 0 |
| Panel B | Subperiod 1 | | | |
| 1-month | 55 | 0 | 45 | 0 |
| 3-month | 0 | 7 | 0 | 0 |
| 6-month | 49 | 32 | 26 | 43 |
| 9-month | 60 | 26 | 45 | 34 |
| 12-month | 44 | 0 | 0 | 0 |
| Panel C | Subperiod 2 | | | |
| 1-month | 83 | 0 | 78 | 0 |
| 3-month | 88 | 0 | 85 | 0 |
| 6-month | 85 | 0 | 81 | 0 |
| 9-month | 95 | 0 | 94 | 0 |
| 12-month | 63 | 0 | 95 | 0 |

Refer to [50] for more details. Results of the DD test of stochastic dominance are reported for winner-loser portfolios. The full sample period is from January 1, 1989 to December 31, 2001. Subperiod 1 is from January 1, 1989 through December 31, 1996 and subperiod 2 is from January 1, 1997 through December 31, 2001. The table reports the percentage of DD statistics $T_j$ ($j = 2, 3$) which are significantly negative or positive at the 5% significance level.

In summary, both tests provide strong evidence of the momentum effect in international stock markets. Although these results are based on second- and third-order stochastic dominance, separate tests using the more general concept of first-order stochastic dominance also indicate winner dominance for the 6- and 9-month holding period. These results cast doubt on the argument that existing asset pricing models do not adequately explain momentum because of omitted risk factors. On the contrary, they suggest that the search for more general asset pricing models to explain momentum may be a futile exercise.

### 8.5.5  Robustness checks

**Results based on five-day skip**  The results reported were based on momentum portfolios formed by skipping one day following the end of the ranking period to avoid bid-ask bounce biases.

Table 8.14: Profitability of Momentum Strategies Using 5-day Skip

| Sample Period | $W - L$ | $\%T_2 < 0$ | $\%T_2 > 0$ | $\%T_3 < 0$ | $\%T_3 > 0$ |
|---|---|---|---|---|---|
| Panel A | Full sample | | | | |
| 1-month | 0.0232 (-2.14) | 74 | 0 | 68 | 0 |
| 3-month | 0.0018 (-0.22) | 41 | 0 | 64 | 0 |
| 6-month | 0.0352 (-4.93) | 89 | 0 | 85 | |
| 9-month | 0.0237 (-3.49) | 97 | 0 | 96 | 0 |
| 12-month | 0.0039 (-0.62) | 69 | 0 | 93 | 0 |
| Panel B | Subperiod 1 | | | | |
| 1-month | 0.0106 (-0.91) | 23 | 0 | 40 | 0 |
| 3-month | -0.0155 (-1.83) | 0 | 30 | 0 | 8 |
| 6-month | 0.0144 (-1.9) | 51 | 34 | 30 | 45 |
| 9-month | 0.0197 (-3.05) | 61 | 27 | 46 | 36 |
| 12-month | 0.0072 (-1.22) | 44 | 0 | 3 | 0 |
| Panel C | Subperiod 2 | | | | |
| 1-month | 0.0433 (-2.23) | 83 | 0 | 79 | 0 |
| 3-month | 0.0295 (-2.01) | 85 | 81 | 0 | |
| 6-month | 0.0687 (-5.59) | 89 | 0 | 86 | 0 |
| 9-month | 0.0302 (-2.31) | 97 | 0 | 96 | 0 |
| 12-month | -0.0014 (-0.11) | 59 | 0 | 89 | 0 |

Refer to [50] for more details. This table reports the profitability of momentum strategies in which momentum portfolios are formed by skipping 5 days after the end of each ranking period. The second column reports average holding period returns of long winner and short loser ($W - L$) momentum portfolios. Average holding period returns are computed as the arithmetic average of all geometric mean returns across overlapping periods.

It may be argued that one day is too short for bid-ask bounce effects to dissipate. Furthermore, a one-day skip may not be long enough to eliminate spurious momentum profits due to nonsynchronous trading of index component stocks.

To address these concerns, we repeat all our tests using a one-week (5 business days) skip period. This should mitigate the problem of non-synchronous trading to the extent that index component stocks trade at least once a week. The results are shown in Table 8.14.

The full sample period is from January 1, 1989 to December 31, 2001. Subperiod 1 is from January 1, 1989 through December 31, 1996 (2087 observations) and subperiod 2 is from January 1, 1997 through December 31, 2001. Results of the DD test of stochastic dominance are shown in the last four columns. The table reports the percentage of DD statistics which are significantly negative or positive at the 5% significance level.

Except for the 1- and 12-month holding periods where average returns are significantly lower with a 5-day skip, results for the other holding periods are quite robust. For example, for the popular 6-month momentum strategy, the average daily return is 0.0352% with 5-day skip, which is very close to the 0.0348% return using 1-day skip (Table 8.11).

The same holds for the 9-month momentum strategy (0.0237% versus 0.0241%). Thus, it is unlikely that all momentum profits can be explained by microstructure biases or nonsynchronous trading.

The statistical significance of momentum profits is confirmed by the DD test, the results of which are very similar to those reported in Table 8.13 based on 1-day skip. As before, there is strong evidence that winners stochastically dominate losers over the full sample and second subperiods.

## Is momentum unique to emerging markets?

Studies show that emerging market stock returns are more autocorrelated than returns to developed markets. Low liquidity may be the main reason for the higher autocorrelation. If momentum profits are solely due to low liquidity, then we cannot claim that they are abnormal.

To examine whether momentum profits are unique to emerging markets, we apply the momentum strategy to developed markets only by excluding the following markets from the sample: Indonesia, Korea, Malaysia, South Africa, Singapore, Spain, Taiwan and Thailand. This leaves us with 16 developed markets. For brevity, we report only the DD test results for this group of countries (Table 8.15).

The second column of Table 8.15 reports the average daily return of the momentum strategy for all 24 markets and for the developed markets only. Consistent with the liquidity story, there is weaker evidence of the momentum effect for developed markets than the overall sample. Nonetheless, the 6- and 9-month mean daily returns for developed markets are still quite large (0.0264% and 0.0290%, respectively) and winners still dominate losers at the 1-, 6- and 9-month horizon. Thus, the momentum effect is not unique to emerging markets. Asset pricing models have difficulty accounting for the momentum phenomenon, which appears to exist globally.

Table 8.15: Stochastic Dominance in Developed Stock Markets

| Holding | Mean return | | SSD | | SSD | |
|---------|-------------|--------------|-----------|-----------|-----------|-----------|
| Period  | All mkts | Dev'd mkts | $\%T_2 < 0$ | $\%T_2 > 0$ | $\%T_3 < 0$ | $\%T_3 > 0$ |
| 1-month  | 0.0357   | 0.0044   | 43 | 0  | 78 | 0  |
|          | (−3.3)   | (−0.62)  |    |    |    |    |
| 3-month  | 0.0040   | 0.0035   | 0  | 25 | 0  | 24 |
|          | (−0.49)  | (−0.7)   |    |    |    |    |
| 6-month  | 0.0348   | 0.0264   | 81 | 0  | 77 | 0  |
|          | (−4.91)  | (−8.03)  |    |    |    |    |
| 9-month  | 0.0241   | 0.0290   | 92 | 0  | 89 | 0  |
|          | (−3.57)  | (−8.8)   |    |    |    |    |
| 12-month | 0.0053   | 0.0162   | 57 | 21 | 44 | 27 |
|          | (−0.84)  | (−5.53)  |    |    |    |    |

Refer to [50] for more details. This table reports performance statistics of the momentum strategy for 16 developed stock markets. The sample period is from January 1, 1989 to December 31, 2001. The portfolio formation procedure is the same as that described in the notes to Table 8.10. Entries under the column "Mean Return" report the arithmetic average of geometric mean returns (in percentages) over the respective overlapping periods (numbers in parentheses are t statistics). The third and fourth column report results of the Davidson-Duclos test of stochastic dominance of winner over loser portfolios at second and third order, respectively.

Preliminary analysis of the data shows that (a) the momentum effect exists globally, (b) the momentum strategy has remained profitable in recent times and (c) for most holding periods, momentum profits could have been earned by simply buying winners without having to short sell losers. Formal tests confirm that winners have stochastically dominated losers at second and third order in recent years. These results are robust to our assumed level of transaction costs. Overall, these results indicate that the momentum effect remains an anomaly for the efficient market hypothesis and standard equilibrium asset pricing models that assume investor risk aversion.

Finally, note that when momentum strategies are implemented on equally weighted portfolios of stocks, they can incur non-trivial transaction costs in terms of direct brokerage costs, bid-ask spreads and price impact, and these may negate gross momentum profits. The effect of transaction costs depends on the design of the momentum strategy, for example, frequency of portfolio rebalancing, number and market size of securities in winner and loser portfolios, etc. The results in this chapter indicate that transaction costs negate momentum profits for the 1-, 3- and 12-month holding periods, but leave the profitability of the 6- and 9-month momentum strategies largely intact, especially in the second subperiod. In addition, momentum studies often underemphasize the point that momentum strategies can be profitable without having to short sell losers. Indeed, shorting past losers often reduces momentum profits.

## 8.6 Exercises

1. Prove Theorem 8.7.

2. Prove Theorem 8.9.

3. Prove Theorem 8.11.

4. Prove Theorem 8.13.

5. Construct an example of two portfolios, say $X$ and $Y$, with majorization order and show in your example that $X$ can be obtained by applying a single Pigou-Dalton transfer on $Y$.

6. Prove Theorem 8.18.

7. From Theorem 8.18, it is known that if $\vec{\beta}_n$ majorizes $\vec{\alpha}_n$, then $\vec{\alpha}_n$ can be obtained from $\vec{\beta}_n$ by applying a finite number of single Pigou-Dalton transfers, and vice versa. Construct an example to illustrate if $\vec{\beta}_n$ majorizes $\vec{\alpha}_n$, then $\vec{\alpha}_n$ can be obtained from $\vec{\beta}_n$ by applying a finite number (more than one) of single Pigou-Dalton transfers.

8. Construct an example to illustrate Theorem 8.19.

9. Prove Theorem 8.21.

10. Prove Theorem 8.23.

11. Construct an example to show that the converse of Theorem 8.23 does not hold.

12. Construct an example to illustrate the ranking of the following assets by second-order stochastic dominance: $\vec{\alpha}_3' \vec{X}_3 = \frac{2}{5}X_1 + \frac{2}{5}X_2 + \frac{1}{5}X_3$ and $\vec{\beta}_3' \vec{X}_3 = \frac{3}{5}X_1 + \frac{1}{5}X_2 + \frac{1}{5}X_3$.

13. Construct an example to illustrate the dominance of two portfolios in which they are not i.i.d.

14. Prove Theorem 8.29.

15. Prove Theorem 8.30.

16. Construct two cumulative distribution functions $F$ and $G$ such that there is no SMSD and no SPSD dominance between $F$ and $G$ but $F \succeq_3^M G$ and $G \succeq_3^P F$.

17. Construct two cumulative distribution functions $F$ and $G$ such that we do not have $F \succeq_3^M G$ or $G \succeq_3^M F$ but we have $G \succeq_3^P F$.

18. Find the ASD, DSD, MSD, PSD relationships between $F$ and $G$:

*The Distributions for Investments F and G*

| Investment F | | Investment G | |
|---|---|---|---|
| Gain | Probability ($f$) | Gain | Probability ($g$) |
| -1,500 | 0.2 | -3,000 | 0.25 |
| 4,500 | 0.8 | 3,000 | 0.75 |

19. Find the ASD, DSD, MSD, PSD relationships between $F$ and $G$:

*The Distributions for Investments F and G*

| Investment F | | Investment G | |
|---|---|---|---|
| Gain | Probability ($f$) | Gain | Probability ($g$) |
| -1,500 | 0.55 | -3,000 | 0.25 |
| 4,500 | 0.45 | 3,000 | 0.75 |

20. Find the ASD, DSD, MSD, PSD relationships between $F$ and $G$:

*The Distributions for Investments F and G*

| Investment F | | Investment G | |
|---|---|---|---|
| Gain | Probability ($f$) | Gain | Probability ($g$) |
| -1,500 | 0.4 | -3,000 | 0.25 |
| 4,500 | 0.6 | 3,000 | 0.75 |

# Chapter 9

# Applications to Risk Management

*We illustrate in this chapter several applications of stochastic dominance in risk management.*

## 9.1   Measures of profit/loss for risk analysis

After the mean variance (MV) criterion was introduced, it has been widely studied how risk averters and risk seekers apply the MV approach or mean standard deviation (MS) to measure risk based on variables of profit. On the other hand, the stochastic dominance (SD) approach to the decision rules for risk averters and risk seekers has also been well investigated.

In this chapter, we discuss the MV and SD theory for decision rules for risk averters and risk seekers based on both variables of profit and loss. The theory is useful in risk management, financial analysis and in many other areas.

Let $X$ be a return or profit random variable with distribution function $F$, and let $u$ be a utility function associated with profit. The loss variable $X^*$ with distribution function $F^*$ and the utility function $u^*$ associated with it are defined as follows.

**Definition 9.1** *Let $X$ be a return or profit random variable with distribution function $F$ defined on $[a, b]$. The* loss *variable $X^*$ with distribution function $F^*$ defined on $[-b, -a]$ is*

$$X^* = -X. \tag{9.1}$$

*Let u be the utility-for-profit function associated with the variable of profit and let u\* be the corresponding utility-of-loss function associated with the variable of loss. Then*

$$u^*(X^*) = u(X).  \qquad (9.2)$$

The variable of loss $X^*$ in (9.1) can also be defined as

$$X^* = TC - X,  \qquad (9.3)$$

where $TC$ is the total cost (assumed to be fixed). As both definitions (9.1) and (9.3) draw the same conclusions based on both SD and MV criteria, one could use (9.1) as well as (9.3). Variables $X$ and $Y$ could be used as the variables of profit or the variables of loss. To avoid any confusion, we denote $X^*$ $(Y^*)$ as the variable of loss with the corresponding distribution function $F^*$ $(G^*)$. ASD and DSD for the variable of loss are defined as follows:

**Definition 9.2** *Given two random variables of loss $X^*$ and $Y^*$ with $F^*$ and $G^*$ as their respective distribution functions defined on $[-b, -a]$, $X^*$ dominates $Y^*$ and $F^*$ dominates $G^*$ in the sense of FASD (SASD, TASD), denoted by*

$$X^* \succeq_1 Y^* \ or \ F^* \succeq_1 G^*, \ (X^* \succeq_2 Y^* \ or \ F^* \succeq_2 G^*, \ X^* \succeq_3 Y^* \ or \ F^* \succeq_3 G^*),$$

*if and only if*

$$F_1^{*A}(x^*) \leq G_1^{*A}(x^*) \ \ (F_2^{*A}(x^*) \leq G_2^{*A}(x^*), \ \ F_3^{A*}(x) \leq G_3^{A*}(x))$$

*for all $x^*$ in $[-b, -a]$.*

**Definition 9.3** *Given two random variables of loss $X^*$ and $Y^*$ with $F^*$ and $G^*$ as their respective distribution functions defined on $[-b, -a]$, $X^*$ dominates $Y^*$ and $F^*$ dominates $G^*$ in the sense of FDSD (SDSD, TDSD), denoted by*

$$X^* \succeq^1 Y^* \ or \ F^* \succeq^1 G^* \ (X^* \succeq^2 Y^* \ or \ F^* \succeq^2 G^*, \ X^* \succeq^3 Y \ or \ F^* \succeq^3 G^*),$$

*if and only if*

$$F_1^{*A}(x^*) \geq G_1^{*A}(x^*) \ \ (F_2^{*A}(x^*) \geq G_2^{*A}(x^*), \ \ F_3^{A*}(x) \geq G_3^{A*}(x))$$

*for all $x^*$ in $[-b, -a]$.*

Strictly ASD and DSD for the variables of loss can be defined similarly. The SD theory is to compare profits by matching certain utility functions with the variable of profit or return. Let $u$ be utility-for-profit function. For $n = 1$, 2, 3, the symbols $\mathcal{U}_n^A$, $\mathcal{U}_n^{SA}$, $\mathcal{U}_n^D$ and $\mathcal{U}_n^{SD}$ denote sets of utility-for-profit functions as defined in Chapter 8. Let $u^*$ be a utility-of-loss function, we define the classes of utility-of-loss functions corresponding to the variables of loss as follows:

**Definition 9.4** *For $n = 1$, 2, 3, the sets $\mathcal{U}_n^{*A}$, $\mathcal{U}_n^{*SA}$, $\mathcal{U}_n^{*D}$ and $\mathcal{U}_n^{*SD}$ are sets of utility-of-loss functions such that:*

$$\mathcal{U}_n^{*A} \ (\mathcal{U}_n^{*SA}) = \{u^* : u^*(X^*) = u(X), \ u \in \mathcal{U}_n^A \ (\mathcal{U}_n^{SA})\}, \ and$$
$$\mathcal{U}_n^{*D} \ (\mathcal{U}_n^{*SD}) = \{u^* : u^*(X^*) = u(X), \ u \in \mathcal{U}_n^D \ (\mathcal{U}_n^{SD})\},$$

*where $X^*$ is defined in (9.1) and $u^*$ is defined in (9.2).*

It is known that the utility-of-loss function is the mirror image (around the vertical axis) of the same investor's utility-for-profit function. Hence, the set of utility-of-loss functions can be defined for the same investors whose utility-for-profit functions belong to $U_n^A$ as $\mathcal{U}_n^{*A}$, and other sets of utility-of-loss functions are defined similarly.

An individual chooses between $F$ and $G$ in accordance with a consistent set of preferences satisfying the NM consistency properties such that $F$ is (strictly) preferred to $G$, or equivalently, $X$ is (strictly) preferred to $Y$, if

$$\begin{aligned} \Delta u &= u(F) - u(G) \geq 0 \, (> 0) \quad \text{and} \\ \Delta u^* &= u^*(F^*) - u^*(G^*) \geq 0 \, (> 0) \end{aligned} \tag{9.4}$$

where

$$\begin{aligned} u(F) &= E[u(X)], \quad u(G) = E[u(Y)], \\ u^*(F^*) &= E[u^*(X^*)], \quad \text{and} \quad u^*(G^*) = E[u^*(Y^*)]. \end{aligned}$$

### 9.1.1 SD criterion for decision-making in risk analysis

As the utility-of-loss function is a mirror image (around the vertical axis) of the same investor's utility-for-profit function, it is easy to show that (a) $u^*$ is a decreasing function whereas $u$ is an increasing function, (b) $u^*$ is convex (concave) if and only if $u$ is concave (convex), (c) the third derivative of $u^*$ is of different sign from that of $u$ and (d) $\mathcal{U}_1^A \equiv \mathcal{U}_1^D$ is the set of increasing functions of the utilities for profit and $\mathcal{U}_1^{*A} \equiv \mathcal{U}_1^{*D}$ is the set of decreasing functions of the utilities of loss.

The following theorem and corollary provide the linkage between $X$ and $X^*$ in the SD theory:

**Theorem 9.5** *For random variables $X$ and $Y$, we have the following:*

1. $X \succeq_i (\succ_i) Y$ *if and only if* $Y^* \succeq^i (\succ^i) X^*$ *for* $i = 1,\ 2$ *or* $3$.

2. $X \succeq_1 (\succ_1) Y$ *if and only if* $X \succeq^1 (\succ^1) Y$.

3. *If $X$ and $Y$ have the same finite mean, then* $X \succeq_2 (\succ_2) Y$ *if and only if* $Y \succeq^2 (\succ^2) X$.

**Corollary 9.6** *For random variables of profit, $X$ and $Y$, we have the following:*

1. $X^* \succeq_i (\succ_i) Y^*$ *if and only if* $Y \succeq^i (\succ^i) X$ *for* $i = 1,\ 2$ *or* $3$,

2. $X^* \succeq_1 (\succ_1) Y^*$ *if and only if* $X^* \succeq^1 (\succ^1) Y^*$, *and*

3. *If $X$ and $Y$ have the same finite mean, then* $X^* \succeq_2 (\succ_2) Y^*$ *if and only if* $Y^* \succeq^2 (\succ^2) X^*$,

*where $X^*$ and $Y^*$ are the variables of loss defined in (9.1).*

The following theorem states the relationship between the SD preferences and the utility preferences for risk averters and risk seekers with respect to the utility functions of loss. See [170].

**Theorem 9.7** *Let $X^*$ and $Y^*$ be random variables of loss with cumulative distribution functions $F^*$ and $G^*$, and $u^*$ be a utility-of-loss function satisfying (9.2), for $m = 1,\ 2$ and $3$, we have*

1. $F^* \succeq_m (\succ_m) G^*$ *if and only if* $u^*(G^*) \geq (>) u^*(F^*)$ *for any $u^*$ in* $\mathcal{U}_m^{*D} (\mathcal{U}_m^{*SD})$, *and*

2. $F^* \succeq^m (\succ^m) G^*$ *if and only if* $u^*(G^*) \geq (>) u^*(F^*)$ *for any $u^*$ in* $\mathcal{U}_m^{*A} (\mathcal{U}_m^{*SA})$,

*where $u^*(F^*)$ and $u^*(G^*)$ are defined in (9.4).*

Theorems 9.5 and 9.7 show that the different classes of utility functions match the corresponding classes of SD totally. These results are important as SD can then be used for utility maximization which, in turn, provides tools for risk averters and risk seekers to make their decisions in risk analysis and investment. We call a person a first-order-ascending-stochastic-dominance (FASD) investor if his/her utility-for-profit function $u$ belongs to $\mathcal{U}_1^A$ or if his/her utility-of-loss function belongs to $\mathcal{U}_1^{*A}$; a first-order-descending-stochastic-dominance (FDSD) investor if his/her utility-for-profit function $u$ belongs to $\mathcal{U}_1^D$ or if his/her utility-of-loss function belongs to $\mathcal{U}_1^{*D}$.

A second-order ascending stochastic dominance (SASD) risk investor, a second-order descending stochastic dominance (SDSD) risk investor, a third-order ascending stochastic dominance (TASD) risk investor and a third-order descending stochastic dominance (TDSD) risk seeker investor can be defined in the same way. Since FASD and FDSD are equivalent, we will call the FASD and FDSD investors FSD investors. For simplicity, from now on we will use $\mathcal{U}_n^A$ ($\mathcal{U}_n^{*A}$) to stand for both $\mathcal{U}_n^A$ ($\mathcal{U}_n^{*A}$) and $\mathcal{U}_n^{SA}$ ($\mathcal{U}_n^{*SA}$) and use $\mathcal{U}_n^D$ ($\mathcal{U}_n^{*D}$) to stand for both $\mathcal{U}_n^D$ ($\mathcal{U}_n^{*D}$) and $\mathcal{U}_n^{SD}$ ($\mathcal{U}_n^{*SD}$).

### 9.1.2 MV criterion for decision-making in risk analysis

The ascending and descending MV rules to cover both risk averters and risk seekers are defined as:

**Definition 9.8** *Given two random variables of profit $X$ and $Y$ with means $\mu_x$ and $\mu_y$ and standard deviations $\sigma_x$ and $\sigma_y$, respectively, then*

1. *$X$ is said to dominate $Y$ (strictly) by the Ascending MS (AMS) rule, denoted by $X\,MS_A\,Y$ if $\mu_x \geq (>)\,\mu_y$ and $\sigma_x \leq (<)\,\sigma_y$; and*

2. *$X$ is said to dominate $Y$ (strictly) by the Descending MS (DMS) rule, denoted by $X\,MS_D\,Y$ if $\mu_x \geq (>)\,\mu_y$ and $\sigma_x \geq (>)\,\sigma_y$.*

**Definition 9.9** *Given two random variables of loss $X^*$ and $Y^*$ with means $\mu_x^*$ and $\mu_y^*$ and standard deviations $\sigma_x^*$ and $\sigma_y^*$, respectively, then*

1. *$X^*$ is said to dominate $Y^*$ (strictly) by the AMS rule, denoted by $X^*\,MS_A^*\,Y^*$ if $\mu_x^* \leq (<)\,\mu_y^*$ and $\sigma_x^* \leq (<)\,\sigma_y^*$; and*

2. *$X^*$ is said to dominate $Y^*$ (strictly) by the DMS rule, denoted by $X^*\,MS_D^*\,Y^*$ if $\mu_x^* \leq (<)\,\mu_y^*$ and $\sigma_x^* \geq (>)\,\sigma_y^*$.*

**Theorem 9.10** *Let $X$ and $Y$ be random variables of profit with means $\mu_x$ and $\mu_y$ and standard deviations $\sigma_x$ and $\sigma_y$, respectively.*

1. *If $X\,MS_A\,Y$ (strictly) and if both $X$ and $Y$ belong to the same location-scale family or the same linear combination of location-scale families, then $E[u(X)] \geq (>) E[u(Y)]$ for the risk-averse investor with the utility-for-profit function $u$ in $\mathcal{U}_2^A$ ($\mathcal{U}_2^{SA}$); and*

2. *If $X\,MS_D\,Y$ (strictly) and if both $X$ and $Y$ belong to the same location-scale family or the same linear combination of location-scale families, then $E[u(X)] \geq (>) E[u(Y)]$ for the risk-seeking investor with the utility-for-profit function $u$ in $\mathcal{U}_2^D$ ($\mathcal{U}_2^{SD}$).*

**Theorem 9.11** *Let $X^*$ and $Y^*$ be random variables of loss with means $\mu_x^*$ and $\mu_y^*$ and standard deviations $\sigma_x^*$ and $\sigma_y^*$, respectively.*

1. *If $X^*\,MS_A^*\,Y^*$ (strictly) and if both $X^*$ and $Y^*$ belong to the same location-scale family or the same linear combination of location-scale families, then $E[u^*(X^*)] \geq (>) E[u^*(Y^*)]$ for the risk-averse investor with the utility-of-loss function $u^*$ in $\mathcal{U}_2^{*A}$ ($\mathcal{U}_2^{*SA}$); and*

2. *If $X^*\,MS_D^*\,Y^*$ (strictly) and if both $X$ and $Y$ belong to the same location-scale family or the same linear combination of location-scale families, then$E[u^*(X^*)] \geq (>) E[u^*(Y^*)]$ for the risk-seeking investor with the utility-of-loss function $u^*$ in $\mathcal{U}_2^{*D}$ ($\mathcal{U}_2^{*SD}$).*

The above theorems will be useful in applying the improved MV criterion in the comparison of profit and loss. In addition, by incorporating the above two theorems into Theorems 9.5 and 9.7, one could link the SD with the MV rules.

We illustrate our approach by analyzing a Production/Operations Management (POM) example. A POM system needs extra capacity to satisfy the expected increased demand. Thus, three mutually exclusive alternative sites have been proposed and the cost $(X^*)$ with their corresponding probabilities $f^*$, $g^*$ and $h^*$ have been estimated and are shown in Table 9.1. Table 9.1 also depicts the ASD integrals of the first three orders for each location.

Table 9.1: The Risk of Three Locations and Their ASD Integrals

| Costs | Probability | | | FASD Integrals | | | SASD Integrals | | | TASD Integrals | | |
|---|---|---|---|---|---|---|---|---|---|---|---|---|
| $x^*$ | $f^*$ | $g^*$ | $h^*$ | $F_1^{*A}$ | $G_1^{*A}$ | $H_1^{*A}$ | $F_2^{*A}$ | $G_2^{*A}$ | $H_2^{*A}$ | $F_3^{*A}$ | $G_3^{*A}$ | $H_3^{*A}$ |
| 1 | 0 | 0.25 | 0.35 | 0 | 0.25 | 0.35 | 0 | 0 | 0 | 0 | 0 | 0 |
| 2 | 0.25 | 0.25 | 0.05 | 0.25 | 0.5 | 0.5 | 0 | 0.25 | 0.35 | 0 | 0.125 | 0.175 |
| 3 | 0.25 | 0 | 0.1 | 0.5 | 0.5 | 0.5 | 0.25 | 0.75 | 0.75 | 0.125 | 0.625 | 0.725 |
| 4 | 0.25 | 0 | 0 | 0.75 | 0.5 | 0.5 | 0.75 | 1.25 | 1.25 | 0.625 | 1.625 | 1.725 |
| 5 | 0.25 | 0.5 | 0.5 | 1 | 1 | 1 | 1.5 | 1.75 | 1.75 | 1.75 | 3.125 | 3.225 |
| 6 | 0 | 0 | 0 | 1 | 1 | 1 | 2.5 | 2.75 | 2.75 | 3.75 | 5.375 | 5.475 |
| Mean | 3.5 | 3.25 | 3.25 | | | | | | | | | |
| Var. | 1.25 | 3.18 | 3.39 | | | | | | | | | |

Refer to [170]. The ASD integral $M_n^{*A}$ is defined in (2) for $n = 1$, 2 and 3; Cost $x^* = 6$ is included in order to measure the effect of $x^* = 5$ on $M_2^{*A}$ and $M_3^{*A}$; $M = F$, $G$ and $H$.

In this example, one could find the preferences of FASD, SASD, TASD risk averters as well as FDSD, SDSD, TDSD risk seekers to both variables of profit and variables of loss. We use $X^*$ to represent the variable of loss or cost and $X$ to represent the variable of profit or return as defined in (9.1) or (9.3). In Tables 9.1 to 9.4, $x^*$ represents different levels of costs with probabilities $f^*$, $g^*$ and $h^*$ and their corresponding CDFs $F^*$, $G^*$ and $H^*$ for the three different locations. The ASD and DSD integrals $M_n^{*A}$ and $M_n^{*D}$, for $n = 1$, 2 and 3 and $M = F$, $G$ and $H$ and the estimates of the ASD integrals are presented in Table 9.1 while the estimates of the DSD integrals for the three locations are depicted in Table 9.2.

Table 9.2: The Risk of Three Locations and Their DSD Integrals

| Costs | FASD Integrals | | | SDSD Integrals | | | TASD Integrals | | |
|---|---|---|---|---|---|---|---|---|---|
| $x^*$ | $F_1^{*D}$ | $G_1^{*D}$ | $H_1^{*D}$ | $F_2^{*D}$ | $G_2^{*D}$ | $H_2^{*D}$ | $F_3^{*D}$ | $G_3^{*D}$ | $H_3^{*D}$ |
| 0 | 1 | 1 | 1 | 3.5 | 3.25 | 3.25 | 6.75 | 6.875 | 6.975 |
| 1 | 1 | 1 | 1 | 2.5 | 2.25 | 2.25 | 3.75 | 4.125 | 4.225 |
| 2 | 1 | 0.75 | 0.65 | 1.5 | 1.5 | 1.6 | 1.75 | 2.25 | 2.3 |
| 3 | 0.75 | 0.5 | 0.6 | 0.75 | 1 | 1 | 0.625 | 1 | 1 |
| 4 | 0.5 | 0.5 | 0.5 | 0.25 | 0.5 | 0.5 | 0.125 | 0.25 | 0.25 |
| 5 | 0.25 | 0.5 | 0.5 | 0 | 0 | 0 | 0 | 0 | 0 |

Refer to [170]. The DSD integral $M_n^{*D}$ is defined in (2) for $n = 1$, 2 and 3; Cost $x^* = 0$ is included in order to measure the effect of $x^* = 1$ on $M_2^{*D}$ and $M_3^{*D}$; $M = F$, $G$ and $H$.

Tables 9.3 and 9.4 show the differences of the ASD integrals and the DSD integrals for each pair of the three locations of the first three orders. In the tables, we define the ASD integral differential $PQ_n^{*A}$ and DSD Integral differential $PQ_n^{*D}$ be

$$PQ_n^{*A} = P_n^{*A} - Q_n^{*A} \quad \text{and} \quad PQ_n^{*D} = P_n^{*D} - Q_n^{*D}, \qquad (9.5)$$

respectively, for $n = 1$, 2 and 3 and for $P, Q = F, G$ or $H$.

*Table 9.3: The ASD Integral Differentials for the Risk of Three Locations*

| $x^*$ | $HF_1^{*A}$ | $GF_1^{*A}$ | $HG_1^{*A}$ | $HF_2^{*A}$ | $GF_2^{*A}$ | $HG_2^{*A}$ | $HF_3^{*A}$ | $GF_3^{*A}$ | $HG_3^{*A}$ |
|---|---|---|---|---|---|---|---|---|---|
| 1 | 0.35 | 0.25 | 0.1 | 0 | 0 | 0 | 0 | 0 | 0 |
| 2 | 0.15 | 0.25 | −0.1 | 0.35 | 0.25 | 0.1 | 0.175 | 0.125 | 0.05 |
| 3 | 0 | 0 | 0 | 0.5 | 0.5 | 0 | 0.6 | 0.5 | 0.1 |
| 4 | −0.25 | −0.25 | 0 | 0.5 | 0.5 | 0 | 1.1 | 1 | 0.1 |
| 5 | 0 | 0 | 0 | 0.25 | 0.25 | 0 | 1.475 | 1.375 | 0.1 |
| 6 | 0 | 0 | 0 | 0.25 | 0.25 | 0 | 1.725 | 1.625 | 0.1 |

Refer to [170]. The ASD Integral Differential $PQ_n^{*A} = P_n^{*A} - Q_n^{*A}$ for $n = 1$, 2 and 3; $P$, $Q = F$, $G$ and $H$.

*Table 9.4: The DSD Integral Differentials for the Risk of Three Locations*

| $x^*$ | $HF_1^{*D}$ | $GF_1^{*D}$ | $HG_1^{*D}$ | $HF_2^{*D}$ | $GF_2^{*D}$ | $HG_2^{*D}$ | $HF_3^{*D}$ | $GF_3^{*D}$ | $HG_3^{*D}$ |
|---|---|---|---|---|---|---|---|---|---|
| 0 | 0 | 0 | 0 | −0.25 | −0.25 | 0 | 0.225 | 0.125 | 0.1 |
| 1 | 0 | 0 | 0 | −0.25 | −0.25 | 0 | 0.475 | 0.375 | 0.1 |
| 2 | −0.35 | −0.25 | −0.1 | 0.1 | 0 | 0.1 | 0.55 | 0.5 | 0.05 |
| 3 | −0.15 | −0.25 | 0.1 | 0.25 | 0.25 | 0 | 0.375 | 0.375 | 0 |
| 4 | 0 | 0 | 0 | 0.25 | 0.25 | 0 | 0.125 | 0.125 | 0 |
| 5 | 0.25 | 0.25 | 0 | 0 | 0 | 0 | 0 | 0 | 0 |

Refer to [170]. The ASD Integral Differential $PQ_n^{*D} = P_n^{*D} - Q_n^{*D}$ for $n = 1$, 2 and 3; $P$, $Q = F$, $G$ and $H$.

From Table 9.3, we find that neither $F^*$, $G^*$ nor $H^*$ dominates one another in the sense of FASD. However, from the table we find that $F^* \succ_n G^* \succ_n H^*$ for $n = 2$ and 3. From Theorem 9.7, we conclude that $H^*$ is preferred to $G^*$ which, in turn, is preferred to $F^*$ by the SDSD or TDSD risk seeker. Similarly, from Table 9.4, we find that neither $F^*$, $G^*$ nor $H^*$ dominates one another in the sense of FDSD. Table 9.4 also shows that $F^*$ does not dominate $G^*$ nor $H^*$ in the sense of SDSD but $H^* \succ^2 G^*$ which implies that $G^*$ is preferred to $H^*$ by the SASD risk averter. Moreover, from the table, we find that $H^* \succ^3 G^* \succ^3 F^*$. Hence $F^*$ is preferred to $G^*$ which, in turn, is preferred to $H^*$ by the TASD risk averter. We note that one could convert the variables of loss to variables of profit and compare their SD relationships.

In addition, using the improved MV criterion, from Table 9.1, we have $G^* MS_A^* H^*$, and $H^* MS_D^* G^* MS_D^* F^*$. This means that $G^*$ dominates $H^*$ by the AMS rule while $H^*$ dominates $G^*$ which, in turn, dominates $F^*$ by the DMS rule. From Theorem 9.11, we conclude that if $G^*$ and $H^*$ belong to the same location-scale family, then $G^*$ is preferred to $H^*$ by the SASD risk averter. Similarly, if $H^*$, $G^*$ and $F^*$ belong to the same location-scale family, then $H^*$ is preferred to $G^*$ which, in turn, is preferred to $F^*$ by the SDSD risk seeker.

The SD theory for both return and loss to the first three orders discussed in this chapter could be used to link the corresponding risk-averse and risk-seeking utility functions to the first three orders. This approach is practical as it provides investors with more tools for empirical analysis, with which they can identify the FSD on return and loss and discern arbitrage opportunities that could increase his/her utility as well as wealth and set up a zero dollar portfolio to make huge profit. In addition, the tools also enable investors to identify the TSD on return and loss so that they can make better choices. The improved MV criterion to decisions in risk analysis for risk-averse and risk-seeking investors is also discussed. These tools could be used to analyze many complex decision problems in business and many other areas.

The combination of SD and MV has superiority to make better choices than by SD or by MV singly, as the MV criterion is easy to compute and gives investors a quick review of the decision with the information of mean and variance only, whereas SD could provide information for the entire distribution of each asset for comparison.

## 9.2  REITs and stocks and fixed-income assets

The collapse of the dot-com mania in 2000 led investors not to stick to the traditional choice between stocks and bonds. Motivated by expectations of falling interest rates, investors switched to real estate markets to maximize their portfolio returns. Investments in real estate are known to trade less frequently and bear high transaction costs. Alternatively, *real estate investment trusts* (REITs) offer investors a better instrument, one that is more liquid and has lower transaction costs compared with traditional real estate investment. In recent years, REITs have developed into a relatively more efficient real estate instrument. Starting in 1992, REITs have grown significantly in both size and number. This is due to the fact that REITs pay stable dividends and are less sensitive to the state of the general economy. REITs provide diversification benefits to mixed-asset portfolios, benefits that appear to come from both the enhanced returns on REITs and their reduced risk.

The statistical analysis of the relationship between real estate returns and the returns on other assets is important to investors, since it provides information to guide portfolio management. In the standard portfolio approach, the return differentials should reflect the risk differentials or other financial characteristics. Since returns on financial assets are often found to display skewness and leptokurtosis, investors' concerns about portfolio return distributions cannot be fully captured by the first two moments. Otherwise, the portfolio's true riskiness will be underestimated. This motivates investors to conduct a statistical analysis to evaluate REITs against stocks and fixed-income assets by considering the effect of the higher moments of the returns. This chapter uses an alternative technique for examining the performance of these assets that accounts for the preferences of risk averters and risk seekers among these assets. In particular, we re-examine market efficiency and the behavior of risk averters and risk seekers via the SD approach by using the whole distribution of returns from these assets.

It is known that in most situations, the Gaussian assumption does not hold, the involved distribution is skewed to either left or right, and fat tails present in the asset return series. Thus, researchers recognize that using traditional MV- or CAPM-based models to analyze investment decisions is not appropriate. To overcome the shortcomings associated with the MV- and CAPM-based models and to investigate the entire distributions of the returns directly, we employ a non-parametric SD approach to analyze the returns of REITs against three stock index returns and two fixed-income investments.

Returns on REITs have been extensively studied in the literature. Many of them examine REITs' efficiency. Studies report that equity REITs and the S&P 500 behave as a random walk and find that the real estate and stock markets are not segmented. Further evidence of random walk behavior and weak-form efficiency in international real estate markets in Europe, Asia and North America has been demonstrated by applying the unit root, variance ratio and run tests. On the other hand, some reports find inefficiency in the price of equity REIT companies whereas some find that efficiency increased for equity REITs.

Several studies test the market efficiency hypothesis for REITs by examining the seasonality and predictability of REITs. Some studies find evidence of seasonality and the January effect in some equity REITs and mortgage REITs while others find size effect in REITs, that is: small firms perform better than large firms.

Reports show that expected excess returns on equity REITs are more predictable than those of small cap stocks and bonds. They decompose excess returns into expected and unexpected excess returns to examine what determines movements in expected excess returns because equity REITs are more predictable than all other assets. On the other hand, studies suggest that REITs and the general stock market are integrated and that there is no predictability in the REIT markets.

Although there is a substantial amount of research on market efficiency, these studies mainly investigate the correlations of dependency over time and/or correlations with other state variables. Very few attempts go beyond the second moments, but there are some exceptions. These mixed findings may arise because different statistical tools were used in these studies, some of which may suffer from mis-specification or distributional problems. In this chapter, we apply an SD approach to analyze the returns of REITs against three stock index returns and two Treasury constant maturities. This approach allows us to examine the first three moments of the return series by focusing on the choice of assets via utility maximization.

## 9.2.1 Data and methodology

We examine daily returns of all REITs, equity REITs, mortgage REITs and US-DS real estate in our study. The index series is based to December 1971 = 100. To simplify, we call this asset group REITs. We compare REITs with three common stock indices: Dow Jones Industrials, the NASDAQ, and the S&P 500; and two fixed-income assets: the 10-year Treasury note and 3-month Treasury bill rate. The sample covers the period from January 1999 through December 2005.

Studies of REIT risk/return to model risk usually apply the CAPM-based model or MV of asset returns. Studies suggest the poor empirical performance of betas to non-normality in return distributions and inadequate specification of investor utility functions. The non-normal aspects of the financial data have been modeled by different distributions and fat tails. Studies propose that portfolios generated with a downside risk (DR) framework are more efficient than those generated with a classic MV and have better risk-return trade-offs. However, a non-stable return distribution is a problem in the application of DR and modern portfolio theory models. In light of the above considerations and evidence, it is clear that if normality does not hold, the MV criterion may produce some misleading results. To circumvent this problem, we use the SD approach.

## 9.2.2   Empirical findings

For comparative purposes, we first apply the MV criterion to obtain a summary of its descriptive statistics of the data being studied in Table 9.5. All assets gain, on average, positive daily returns. The REITs and Treasury bill and Treasury note are statistically significant (greater than zero) but not the stock returns. The daily mean returns on REITs are 0.04% − 0.06%, much higher than the daily mean returns of other asset groups. Consistent with the common intuition, based on daily returns, REITs outperformed the stock indices and Treasury constant maturities for the period under study. However, the unreported pairwise t-tests show that only all REITs and equity REITs are significantly different from the S&P 500 and the two Treasury constant maturities at the 5% level. REITs also exhibit a smaller standard deviation than those of the three stock indices, but they have a larger standard deviation than the Treasury constant maturities. In short, applying the MV criterion, we find that (1) all REITs (except mortgage REITs) dominate the three stock indices but not the Treasury constant maturities, and (2) mortgage REITs dominate the NASDAQ only.

In Table 9.5, we look at the following assets: All REIT (ART), Equity REIT (ERT), Mortgage REIT (MRT), US-DS Real Estate (DRE), Dow Jones Industrials (DJI), NASDAQ (NAS), S&P 500 (SP5), 10-Year T. Note (TB10) and 3-Month T. Bill (TB3).

Table 9.5 shows that the highly significant Jarque-Bera statistics suggest that the return distributions for all assets are non-normal and indicates that all assets have significant skewness and kurtosis. REITs exhibit negative skewness, and mortgage REITs have a very high kurtosis (59.22). The exhibition of significant skewness and kurtosis further supports the non-normality of return distributions. Moreover, on the basis of the findings using the MV criterion, we cannot conclude whether investors' preferences between assets will lead to an increase in wealth or whether their preferences will increase their expected utility. However, the SD approach allows us to address the issue.

Table 9.5: *A Summary of Descriptive Statistics for Various Asset Returns (1999-2005)*

| Asset | Mean | Std. Dev. | Skewness | Kurtosis | J-B |
|-------|------|-----------|----------|----------|-----|
| ART | 0.05973*** | 0.8168 | −0.3653*** | 3.8280*** | 1154.88*** |
| ERT | 0.06148*** | 0.8242 | −0.3427*** | 3.7216*** | 1088.93*** |
| MRT | 0.06181* | 1.414 | 1.2879*** | 59.217*** | 267156*** |
| DRE | 0.03554* | 0.886 | −0.1792*** | 3.2354*** | 805.78*** |
| DJI | 0.01451 | 1.0998 | 0.1974*** | 2.6265*** | 536.40*** |
| NAS | 0.01927 | 1.9509 | 0.3173*** | 3.5756*** | 1002.80*** |
| SP5 | 0.00741 | 1.1473 | 0.2208*** | 2.0591*** | 337.24*** |
| TB10 | 0.01326*** | 0.00218 | 0.4603*** | −0.7526*** | 107.52*** |
| TB3 | 0.008208*** | 0.00477 | 0.3640*** | −1.2980*** | 168.53*** |

Refer to [24]. J-B denotes the Jarque-Bera statistic for testing normality defined by $T[SK^2/6 + (KUR − 3)^2/24]$, which is asymptotically distributed as $\chi^2(2)$. Asterisks ***, **, and * denote statistical significance at the 1%, 5%, and 10% levels, respectively.

Table 9.6: *Results of DD Test for Risk Averters (1999-2005)*

|  | FASD | | | | SASD | | | |
|--|------|------|------|------|------|------|------|------|
|  | T > 0 | T > C | T < 0 | T < C | T > 0 | T > C | T < 0 | T < C |
| ART-DJI | 53 | 18 | 47 | 18 | 0 | 0 | 100 | 27 |
| ART-NAS | 59 | 27 | 41 | 23 | 0 | 0 | 100 | 32 |
| ART-SP5 | 49 | 27 | 51 | 20 | 0 | 0 | 100 | 34 |
| ART-TB10 | 53 | 30 | 47 | 23 | 63 | 33 | 37 | 0 |
| ART-TB3 | 53 | 30 | 47 | 23 | 63 | 33 | 37 | 0 |
| ERT-DJI | 52 | 17 | 48 | 17 | 0 | 0 | 100 | 27 |
| ERT-NAS | 59 | 27 | 41 | 23 | 0 | 0 | 100 | 33 |
| ERT-SP5 | 49 | 24 | 51 | 21 | 0 | 0 | 100 | 34 |
| ERT-TB10 | 52 | 32 | 48 | 23 | 62 | 34 | 38 | 0 |
| ERT-TB3 | 52 | 32 | 48 | 23 | 61 | 34 | 39 | 0 |
| MRT-DJI | 38 | 4 | 62 | 3 | 42 | 4 | 58 | 0 |
| MRT-NAS | 47 | 13 | 53 | 10 | 29 | 1 | 71 | 12 |
| MRT-SP5 | 43 | 5 | 57 | 3 | 41 | 3 | 59 | 0 |
| MRT-TB10 | 41 | 12 | 59 | 10 | 46 | 11 | 54 | 0 |
| MRT-TB3 | 41 | 12 | 59 | 10 | 45 | 11 | 55 | 0 |
| DRE-DJI | 54 | 9 | 46 | 15 | 0 | 0 | 100 | 20 |
| DRE-NAS | 59 | 28 | 41 | 22 | 0 | 0 | 100 | 30 |
| DRE-SP5 | 54 | 21 | 46 | 18 | 2 | 0 | 98 | 25 |
| DRE-TB10 | 53 | 31 | 47 | 26 | 69 | 35 | 31 | 0 |
| DRE-TB3 | 53 | 31 | 47 | 26 | 67 | 35 | 33 | 0 |

Table 9.6: Results of DD Test for Risk Averters
(1999-2005) (continued)

| | TASD | | | |
|---|---|---|---|---|
| | $T > 0$ | $T > C$ | $T < 0$ | $T < C$ |
| ART-DJI | 0 | 100 | 100 | 59 |
| ART-NAS | 0 | 100 | 100 | 74 |
| ART-SP5 | 0 | 100 | 100 | 58 |
| ART-TB10 | 100 | 48 | 0 | 0 |
| ART-TB3 | 100 | 46 | 0 | 0 |
| ERT-DJI | 0 | 0 | 100 | 59 |
| ERT-NAS | 0 | 0 | 100 | 74 |
| ERT-SP5 | 0 | 0 | 100 | 58 |
| ERT-TB10 | 100 | 50 | 0 | 0 |
| ERT-TB3 | 100 | 48 | 0 | 0 |
| MRT-DJI | 57 | 1 | 43 | 0 |
| MRT-NAS | 35 | 1 | 65 | 19 |
| MRT-SP5 | 52 | 1 | 48 | 0 |
| MRT-TB10 | 92 | 15 | 8 | 0 |
| MRT-TB3 | 87 | 15 | 13 | 0 |
| DRE-DJI | 0 | 0 | 100 | 28 |
| DRE-NAS | 0 | 0 | 100 | 64 |
| DRE-SP5 | 1 | 0 | 99 | 45 |
| DRE-TB10 | 100 | 62 | 0 | 0 |
| DRE-TB3 | 100 | 59 | 0 | 0 |

Refer to [24]. The table reports the percentages of positive and negative DD statistics, $T_j^A$ (see Appendix A for $j = 1, 2, 3$) for risk averters, and their significant portions at the 5% significance level, based on the asymptotic critical value of 3.254 of the studentized maximum modulus (SMM) distribution. $T >$ $(<)\, 0\, (c)$ means percentage of values $T_j^A$ statistic $> (<)\, 0$ (critical value). ERT is equity REIT, MRT is mortgage REIT, and DRE is US-DS Real Estate. ART-DJI means pairwise comparison of all REITs with the Dow Jones Industrial index. Other pairs are defined accordingly.

Table 9.6 shows the results of the DD test for risk averters for the entire period. There are four groups showing a pairwise comparison between four types of REITs and other assets.

We take the pair of equity REITs and the S&P 500 as an example. The evidence from Table 9.6 suggests that 21% of $T_1^A$ is significantly negative, and 24% of $T_1^A$ is significantly positive for the risk-averse, implying no FASD between the pair of equity REITs and the S&P 500. We find similar results for all the other pairs, such as equity REITs and Dow Jones Industrials and equity REITs and the NASDAQ.

All $T_2^A$ and $T_3^A$ for the comparison of equity REITs and the S&P 500 are negative along the distributions of returns. In addition, Table 9.6 shows that 34% of $T_2^A$ and 58% of $T_3^A$ are found to be significantly negative at the 5% level. Thus, we conclude that equity REITs dominate the S&P 500 at second and third order under ASD, implying that any risk-averse investor would prefer equity REITs to the S&P 500 for maximizing expected utility. On the other hand, for the comparison of equity REITs and the 3-month Treasury bill (and 10-year Treasury note), we find that 23% of $T_1^A$ is significantly negative and 32% of $T_1^A$ is significantly positive for risk averters. Further inspecting the DD statistics for the second and third order for risk averters, we see that the 3-month Treasury bill (10-year Treasury note) dominates equity REITs, since 34% of $T_2^A$ and 48% (50%) of $T_3^A$ are found to be significantly positive at the 5% level.

Table 9.7: Results of DD Test for Risk Seekers (1999-2005)

| | FDSD | | | | SDSD | | | |
|---|---|---|---|---|---|---|---|---|
| | $T>0$ | $T>C$ | $T<0$ | $T<C$ | $T>0$ | $T>C$ | $T<0$ | $T<C$ |
| ART-DJI | 41 | 18 | 59 | 18 | 38 | 0 | 62 | 25 |
| ART-NAS | 41 | 23 | 59 | 27 | 27 | 0 | 73 | 30 |
| ART-SP5 | 39 | 20 | 61 | 27 | 41 | 0 | 59 | 33 |
| ART-TB10 | 47 | 23 | 53 | 30 | 100 | 40 | 0 | 0 |
| ART-TB3 | 47 | 23 | 53 | 30 | 100 | 44 | 0 | 0 |
| ERT-DJI | 44 | 17 | 56 | 17 | 39 | 0 | 61 | 24 |
| ERT-NAS | 41 | 24 | 59 | 27 | 27 | 0 | 73 | 30 |
| ERT-SP5 | 39 | 21 | 61 | 24 | 41 | 0 | 59 | 33 |
| ERT-TB10 | 48 | 23 | 52 | 32 | 100 | 41 | 0 | 0 |
| ERT-TB3 | 48 | 23 | 52 | 32 | 100 | 45 | 0 | 0 |
| MRT-DJI | 62 | 3 | 38 | 4 | 100 | 3 | 0 | 0 |
| MRT-NAS | 41 | 10 | 59 | 13 | 76 | 0 | 24 | 10 |
| MRT-SP5 | 56 | 3 | 44 | 5 | 100 | 1 | 0 | 0 |
| MRT-TB10 | 59 | 10 | 41 | 12 | 100 | 11 | 0 | 0 |
| MRT-TB3 | 59 | 10 | 41 | 12 | 100 | 12 | 0 | 0 |
| DRE-DJI | 38 | 15 | 62 | 9 | 35 | 0 | 65 | 20 |
| DRE-NAS | 35 | 18 | 65 | 21 | 38 | 0 | 62 | 28 |
| DRE-SP5 | 47 | 26 | 53 | 31 | 38 | 0 | 62 | 28 |
| DRE-TB10 | 47 | 26 | 53 | 31 | 100 | 33 | 0 | 0 |
| DRE-TB3 | 47 | 26 | 53 | 31 | 100 | 34 | 0 | 0 |

*Table 9.7: Results of DD Test for Risk Seekers*
*(1999-2005) continued*

|  | TDSD | | | |
|---|---|---|---|---|
|  | $T > 0$ | $T > C$ | $T < 0$ | $T < C$ |
| ART-DJI | 0 | 0 | 100 | 31 |
| ART-NAS | 0 | 0 | 100 | 47 |
| ART-SP5 | 0 | 0 | 100 | 41 |
| ART-TB10 | 100 | 67 | 0 | 0 |
| ART-TB3 | 100 | 67 | 0 | 0 |
| ERT-DJI | 1 | 0 | 99 | 29 |
| ERT-NAS | 0 | 0 | 100 | 47 |
| ERT-SP5 | 0 | 0 | 100 | 39 |
| ERT-TB10 | 100 | 66 | 0 | 0 |
| ERT-TB3 | 100 | 66 | 0 | 0 |
| MRT-DJI | 100 | 0 | 0 | 0 |
| MRT-NAS | 51 | 0 | 49 | 3 |
| MRT-SP5 | 100 | 0 | 0 | 0 |
| MRT-TB10 | 100 | 33 | 0 | 0 |
| MRT-TB3 | 100 | 36 | 0 | 0 |
| DRE-DJI | 0 | 0 | 100 | 24 |
| DRE-NAS | 0 | 0 | 100 | 34 |
| DRE-SP5 | 0 | 0 | 100 | 34 |
| DRE-TB10 | 100 | 68 | 0 | 0 |
| DRE-TB3 | 100 | 68 | 0 | 0 |

Refer to [24]. The table reports the percentages of positive and negative DD statistics, $T_j^D$ (see Appendix A for $j = 1, 2, 3$) for risk seekers and their significant portions at the 5% significance level, based on the asymptotic critical value of 3.254 of the studentized maximum modulus (SMM) distribution. $T > (<)0(c)$ means percentage of values $T_j^D$ statistic $> (<)$ 0 (critical value). ERT is equity REIT, MRT is mortgage REIT, and DRE is US-DS Real Estate. ART-DJI means pairwise comparison of all REITs with the Dow Jones Industrial index. Other pairs are defined accordingly.

Overall, evidence derived from ascending DD statistics indicates there is no FASD between REITs and other assets, suggesting that investors cannot increase their wealth by switching from one asset to the other and there is no arbitrage opportunity between them. These results are also evidence that we cannot reject market efficiency. However, by considering the statistics from SASD and TASD, we can determine whether investors could increase their expected utility by switching from one asset to another. In our research, it is apparent that risk averters prefer REITs (except mortgage REITs) to stocks, while they prefer Treasury constant maturities over REITs for maximizing their expected utility. This implies that they will increase their expected utility by switching their investments from stocks to real estate and from real estate to fixed-income assets.

Table 9.7 reports the descending DD statistics for risk seekers. As in Table 9.6, taking the comparison of equity REITs and the S&P 500 as an example, we find that 33% of $T_2^D$ and 39% of $T_3^D$ are negative and statistically significant, respectively. Hence, risk-seeking investors will unambiguously prefer the S&P 500 to equity REITs to maximize their expected utility. On the other hand, if we compare equity REITs and the 3-month Treasury bill, risk-seeking investors will prefer equity REITs to the 3-month Treasury bill as is evident from the fact that 45% of $T_2^D$ and 66% of $T_3^D$ are positive and statistically significant, respectively. Different from the evidence for risk averters, evidence from second- and third-order statistics, $T_j^D$, reveals that risk seekers will increase their expected utility by switching from real estate to stocks and from fixed-income assets to real estate.

Is there time-varying behavior for risk averters and risk seekers? Dynamic asset price movements suggest that asset returns are subject to ongoing external shocks in addition to some big events and extraordinary economic/social disturbances. It is of interest to examine whether investors' behavior is influenced by the up and down market trend. To address this issue, we divided the entire sample into two sub-periods. That is, we treat the period from January 1999 to December 2002 as an up market and January 2003 to December 2005 as a down market. This allows us to investigate the behavioral differential conditioned on the financial economic environment.

The results for the two sub-periods are presented in Table 9.8. As we reported earlier, there is no FSD among all assets studied in this paper for each sub-period, implying that an arbitrage opportunity does not exist among these assets in both bull and bear markets. On the other hand, REITs are found to be dominated by Treasury constant maturities under ASD, while REITs dominate Treasury constant maturities under DSD in both sub-periods, indicating that investors' behavior concerning fixed-income assets is not influenced by economic conditions.

Nevertheless, we observe substantial differences among the distributions of other assets during different time periods. For instance, except for mortgage REITs, all other REITs are found to dominate stock indices under ASD in sub-period 1 but not in sub-period 2. For DSD, we find a change in direction of preference from sub-period 1 to sub-period 2. In particular, the S&P 500 dominates equity REITs in sub-period 1; however, it is dominated by equity REITs in sub-period 2 under DSD, implying that investors' behavior concerning stocks could be time-varying and influenced by market conditions.

Table 9.8: Results of DD Test for Sub-Periods

| Sub-period 1 (1999-2002) | | Sub-period 2 (2003-2005) | |
|---|---|---|---|
| Risk Averters | Risk Seekers | Risk Averters | Risk Seekers |
| ART $\succ_2$ DJI | ART $\prec^2$ DJI | ART $\not\succ$ DJI | ART $\succ^2$ DJI |
| ART $\succ_2$ NAS | ART $\prec^2$ NAS | ART $\not\succ$ NAS | ART $\prec^2$ NAS |
| ART $\succ_2$ SP5 | ART $\prec^2$ SP5 | ART $\not\succ$ SP5 | ART $\succ^2$ SP5 |
| ART $\prec_2$ TB10 | ART $\succ^2$ TB10 | ART $\prec_2$ TB10 | ART $\succ^2$ TB10 |
| ART $\prec_2$ TB3 | ART $\succ^2$ TB3 | ART $\prec_2$ TB3 | ART $\succ^2$ TB3 |
| ERT $\succ_2$ DJI | ERT $\prec^2$ DJI | ERT $\not\succ$ DJI | ERT $\succ^2$ DJI |
| ERT $\succ_2$ NAS | ERT $\prec^2$ NAS | ERT $\not\succ$ NAS | ERT $\prec^2$ NAS |
| ERT $\succ_2$ SP5 | ERT $\prec^2$ SP5 | ERT $\not\succ$ SP5 | ERT $\succ^2$ SP5 |
| ERT $\prec_2$ TB10 | ERT $\succ^2$ TB10 | ERT $\prec_2$ TB10 | ERT $\succ^2$ TB10 |
| ERT $\prec_2$ TB3 | ERT $\succ^2$ TB3 | ERT $\prec_2$ TB3 | ERT $\succ^2$ TB3 |
| MRT $\not\succ$ DJI | MRT $\not\succ$ DJI | MRT $\prec_2$ DJI | MRT $\succ^2$ DJI |
| MRT $\succ_2$ NAS | MRT $\prec^2$ NAS | MRT $\not\succ$ NAS | MRT $\not\succ$ NAS |
| MRT $\not\succ$ SP5 | MRT $\not\succ$ SP5 | MRT $\prec_2$ SP5 | MRT $\succ^2$ SP5 |
| MRT $\prec_2$ TB10 | MRT $\succ^2$ TB10 | MRT $\prec_2$ TB10 | MRT $\succ^2$ TB10 |
| MRT $\prec_2$ TB3 | MRT $\succ^2$ TB3 | MRT $\prec_2$ TB3 | MRT $\succ^2$ TB3 |
| DRE $\succ_2$ DJI | DRE $\prec^2$ DJI | DRE $\not\succ$ DJI | DRE $\succ^2$ DJI |
| DRE $\succ_2$ NAS | DRE $\prec^2$ NAS | DRE $\not\succ$ NAS | DRE $\prec^2$ NAS |
| DRE $\succ_2$ SP5 | DRE $\prec^2$ SP5 | DRE $\not\succ$ SP5 | DRE $\succ^2$ SP5 |
| DRE $\prec_2$ TB10 | DRE $\succ^2$ TB10 | DRE $\prec_2$ TB10 | DRE $\succ^2$ TB10 |
| DRE $\prec_2$ TB3 | DRE $\succ^2$ TB3 | DRE $\prec_2$ TB3 | DRE $\succ^2$ TB3 |

Refer to [24]. $Y \succ_j (\prec_j) Z$ means $Y$ dominates (is dominated by) $Z$ under order-$j$ ASD and $Y \succ^j (\prec^j) Z$ means $Y$ dominates (is dominated by) $Z$ under order-$j$ DSD (see Chapter 8), respectively, for $j = 1$, 2 and 3. $Y \not\succ Z$ means no SD between $Y$ and $Z$.

## 9.2.3   Discussion

It is a widely accepted stylized fact that returns on most financial assets exhibit leptokurtosis and sometimes asymmetry and they are not normally distributed. In addition, empirical findings using the MV approach cannot be used to decide whether investors' portfolio preferences will increase wealth or, in the case of risk-averse investors, lead to an increase in expected utility without an increase in wealth. Given the limitation of the MV approach and the lack of a clear solution to the fat-tail distributions, this study is based on the SD approach, which is not distribution-dependent, and can shed light on the expected utility and wealth implications of portfolio preferences by exploiting information obtained from higher-order moments to test their performance.

By investigating the data on REITs and five other assets over the entire sample period of 1999-2005, we find that all REITs (except mortgage REITs) dominate the three stock indices but not the Treasury constant maturities using the MV criterion. We also find no FSD between them, implying that investors cannot increase their wealth by switching from one asset to another.

However, REITs (except mortgage REITs) stochastically dominate returns on the three stock indices but are stochastically dominated by fixed-income securities, the 3-month Treasury bill, and the 10-year Treasury bond at the second and third order for risk averters. We find the reverse case for risk seekers. This means that to maximize their expected utility, all risk-averse investors would prefer to invest in real estate rather than in the stock market, subject to trading costs. However, if we compare real estate to fixed-income assets, they would prefer fixed-income assets to real estate. On the other hand, all risk-seeking investors would prefer to invest in the stock market than in real estate (or in real estate rather than in fixed-income assets) to maximize their expected utility. In addition, we find that investors' behavior concerning fixed-income assets is not influenced by economic conditions, while their behavior concerning stocks is time-varying and influenced by market conditions.

## 9.3 Evaluating hedge funds performance

There is an increasing trend to include alternative investments in managed portfolios because of the benefits obtained from diversification. Hedge funds are one of the most popular alternative investments among institutional investors as well as retail investors. In the last decade, the hedge funds industry has experienced extraordinary growth. Recently, hedge funds focusing on investing in Asia have been established at an increasing pace. The Bank of Bermuda reported 66 new hedge funds established in Asia in 2002 and 90 in 2003. There are more than 520 hedge funds operating in Asia including those in Japan and Australia, with assets under management estimated at more than US$15 billion as of the end of 2004.

Asian hedge funds are expected to have strong growth in the near future given the strong growth in both the real and financial sectors in the Asian economies. This potential along with the general lack of transparency of hedge funds motivates investors to understand Asian hedge funds' performance, specifically to provide a methodology that is useful for investors to filter or rank potential investments based on their past performance.

Hedge funds are commonly believed to generate positive alphas and the returns provided are generally uncorrelated with the traditional asset classes. Hedge fund managers use different strategies to generate such returns. Hedge fund strategies have been classified as directional (or market timing) and non-directional. The directional approach dynamically bets on the expected directions of the markets that fund managers will long or sell-short securities to capture gains from the advance and decline of their counterpart stocks or indices. On the other hand, by exploiting structural anomalies in the financial market, the non-directional approach attempts to extract value from a set of embedded arbitrage opportunities within and across securities.

Comparing the performance of managed funds is not an easy task. Studies found that hedge funds have low correlations with the traditional asset classes like stocks and bonds and attempt to offer protection in falling and/or volatile markets. The MV criterion and the CAPM statistics are commonly used for funds performance evaluation. Many studies using these measures to conclude that hedge funds generate superior results.

However, extensive empirical analyses on hedge fund returns conclude that hedge fund returns are generally more skewed and leptokurtic than normal distribution. Many hedge funds have distributions with fat-tails, and so normality assumptions on the distribution of hedge fund returns are generally not correct. Studies documented that hedge funds have fat tails, resulting in a greater number of extreme events than one would normally anticipate. Hedge funds may offer relatively high means and low variances, such funds give investors both third and fourth moments attribute that are exactly the opposite of those that are desirable.

Measures such as the Sharpe ratio could pose problems due to the option-like returns that hedge funds generate. Studies also pointed out that modern portfolio theory is too simplistic to deal with hedge funds. Sharpe ratios and standard alphas could be misleading in analyzing such investments. This makes the use of traditional performance measures questionable.

On the other hand, volatility may produce misleading results to measure the riskiness of hedge fund returns as some hedge funds having low standard deviation may not possess less risk. The risk could be well harbored in the skewness and kurtosis to make the fund risky. In addition, Sharpe ratios are found to usually overestimate fund performance due to negative skewness and leptokurtic returns.

As traditional MV and CAPM measures could provide erroneous results, many approaches have been offered as alternatives to evaluate hedge fund performance. Multi-factor models could be used to examine hedge fund performance. However, multi-factor models possess low predictive power as exposure to market factors is unlikely to be stable overtime due to dynamic trading strategies being employed by hedge funds. Furthermore, studies find that the aim of hedge funds is to provide superior performance with low volatility in both bull and bear markets as opposed to comparing their relative performance to traditional market indices.

As the dynamic trading strategies employed along with skewed returns distribution could be problematic for evaluating hedge fund performance, academics proposed to use longitudinal analyses to describe temporal features of hedge fund performance. Some academics applied survival analysis to estimate the lifetimes of hedge funds and found that these are affected by factors such as their size, their performance and their redemption period. Some examined the illiquidity exposure of hedge funds and others used a general asset class factor model comprising of option-based and buy-and-hold strategies to benchmark hedge fund performance.

In addition, applying data envelopment analysis (DEA) to evaluate the performance of different classes of hedge funds, it is found that DEA is a useful tool for ranking self appraised and peer group appraised hedge funds. Employing a practical non-linear approach based on first- and cross-moments analyses in up and down markets to assess the risk and performance of Asian hedge funds, studies found that while all funds provide diversification in the sense that they are not perfectly correlated with market index returns, few funds provide downside protection along with upside capture that is assumed to be preferred by investors.

They also found that funds having these preferred attributes provided returns that are not, on average, significantly less than those that do not provide such preferred attributes. Applying SD to the same data sample, we found that selected funds provide both stochastic dominating returns as well as funds that provide upside capture and downside protection, as well as stability in volatile condition at insignificant cost.

## 9.3.1   Data and methodology

The data studied in this chapter are the monthly returns of the 70 Asian hedge funds reported by the Eurekahedge database for the sample period from January 2000 to December 2004. We also include the Eurekahedge Asia ex-Japan (IX1), Eurekahedge Japan (IX2) and Eurekahedge Asian Hedge Fund (IX3) indices and carry out our analysis on two traditional equity market indices, viz. the Morgan Stanley Pacific Index (MSAUCPI) (IX4) and the S&P 500 index (IX5). We include the S&P 500 in our study as it is commonly used as the equity benchmark for comparison, especially when fund managers and investors invest internationally.

In addition, performance of funds can be measured relative to a regional benchmark constructed by Morgan Stanley, like the MSAUCPI. The 3-month U.S. T-bill rate and the Morgan Stanley Capital International (MSCI) global index are used to proxy the risk-free rate and the global market index, respectively, for computing CAPM statistics. For comparison, we first employ the MV criterion and CAPM statistics for funds performance evaluation.

Table 9.9: Names and Symbols of the Asian
Hedge Funds and the Five Market Indices

| Symbol | Name |
|--------|------|
| IX1 | AA EH Asia ex Japan Index |
| IX2 | AA EH Japan Index |
| IX3 | AA EH Index |
| IX4 | MSAUCPI |
| IX5 | S&P 500 |
| AHF01 | ADM Galleus Fund |
| AHF18 | Furinkazan Fund USD |
| AHF39 | LIM Asia Arbitrage Fund |
| AHF47 | Pacific-Asset Alpha Fund |

Refer to [175].

The means and standard deviations vary widely across Asian hedge funds. For example, AHF47 possesses the largest monthly mean return (0.03648) and the largest standard deviation (0.1941) among the 70 hedge funds and the five market indices while both AHF18 and AHF39 exhibit the lowest monthly mean returns (−0.0069) and smallest standard deviation (0.00787), respectively.

Interestingly, using the MV criterion, we have found a fund, AHF47, possessing the largest mean that does not dominate any other fund, including AHF18 and AHF39. Thus, we conclude that, using the MV criterion, a fund with the largest mean return may not be a good investment choice. On the other hand, AHF01 and AHF39 are found to have significant higher means and smaller standard deviations (not significant) than AHF18. Therefore, both AHF01 and AHF39 dominate AHF18 by the MV criterion.

Table 9.10: Summary Statistics of the Asian Hedge Funds and the Five Market Indices

| | Mean | Std. Dev. | Sharpe Ratio | Skewness | Kurtosis | J-B |
|---|---|---|---|---|---|---|
| IX1 | 0.01074 | 0.02474 | 0.3446 | −0.1268 | −0.6269 | 1.1437 |
| IX2 | 0.00779 | 0.01753 | 0.3179 | 0.8404* | 1.2776 | 11.1446** |
| IX3 | 0.00839 | 0.01692 | 0.3647 | −0.1265 | −0.5447 | 0.9019 |
| IX4 | −0.00259 | 0.0495 | −0.097 | −0.0782 | −1.0969 | 3.0693 |
| IX5 | −0.00211 | 0.04708 | −0.0917 | −0.1031 | −0.2931 | 0.3213 |
| Ave. | 0.00913 | 0.04784 | 0.183 | 0.2232 | 1.9470** | 41.3997** |
| Max. | 0.03648 | 0.19413 | 0.8399 | 3.0663** | 16.4066** | 766.96** |
| Min. | −0.0069 | 0.00787 | −0.2948 | −1.6243** | −1.0969 | 0.0718 |
| AHF01 | 0.01267 | 0.01245 | 0.8399 | −0.1083 | −0.4376 | 0.5961 |
| AHF18 | −0.0069 | 0.03092 | −0.2948 | −1.5567** | 7.8761** | 179.32** |
| AHF39 | 0.00601 | 0.00787 | 0.4824 | −0.4813 | 0.1959 | 2.4124 |
| AHF47 | 0.03648 | 0.19413 | 0.1765 | 1.2334** | 1.2206 | 18.9381** |

Refer to [175]. AHF01, AHF18, AHF39 and AHF47 are the "most outstanding funds" in which AHF47 possesses the largest monthly mean return and the largest standard deviation; AHF18 exhibits the lowest monthly mean return and the smallest Sharpe ratio; AHF39 exhibits the smallest standard deviation; AHF01 exhibits the largest Sharpe ratio; AHF39 possesses the highest Treynor and AHF47 obtains the highest Jensen measures. * $p < 0.05$, **$p < 0.01$.

Next, we investigate the CAPM measures for all indices and funds but only report the most important results in Table 9.10. From Table 9.10, AHF01 exhibits the largest Sharpe ratio (0.8399) while AHF18 has the lowest (−0.2948). Furthermore, AHF39 possesses the highest Treynor (0.7199) while AHF47 obtains the highest Jensen (0.0369) measures. A summary of dominance results among the four most outstanding Asian hedge funds and the five indices measured by the MV criterion and all the CAPM statistics are presented in Table 9.11.

From the table, we find that sometimes a fund dominates another fund by a CAPM statistic but the dominance relation can be reversed if measured by other CAPM statistic(s). Thus, we conclude that, in general, different funds are chosen using different CAPM measures.

In addition, our results show that some of the return distributions are non-normal and exhibit both negative skewness and excess kurtosis. As noted by Kat [76] and others, the modern portfolio theory is too simplistic to deal with hedge funds.

*Table 9.11: Pair-Wise Comparison among the Asian Hedge Funds by the MV and CAPM Measures*

|        | IX1  | IX2   | IX3 | IX4     | IX5     |
|--------|------|-------|-----|---------|---------|
| IX1    |      | S,J   | J   | S,T,J,M | S,T,J,M |
| IX2    | T    |       | T   | S,T,J,M | S,T,J,M |
| IX3    | S,T  | S,J,M |     | S,T,J,M | S,T,J,M |
| IX4    |      |       |     |         |         |
| IX5    |      |       |     | S,T,J,M |         |
| AHF01  | M    | M     | M   | M       | M       |
| AHF18  |      |       |     |         |         |
| AHF39  |      |       |     | M       | M       |
| AHF47  |      |       |     |         |         |

|        | AHF01 | AHF18   | AHF39 | AHF47 |
|--------|-------|---------|-------|-------|
| IX1    | J     | S,J,M   | J     | S     |
| IX2    |       | S,T,J,M | J     | S     |
| IX3    |       | S,J,M   | J     | S     |
| IX4    |       | S,J     |       |       |
| IX5    |       | S,J     |       |       |
| AHF01  |       | S,T,M   | S,J   | S,T   |
| AHF18  | N     |         | N     | N     |
| AHF39  | T     | S,T,J,M |       | S,T   |
| AHF47  | J     | S,T,J   | J     |       |

Refer to [171, 174, 175]. M, S, T and J indicate dominance by MV criterion, Sharpe ratio, Treynor index, and Jensen index, respectively. N denotes no dominance by MV, Sharpe ratio, Treynor index and Jensen index.

In the table, the rows indicate whether the fund in the leftmost column dominates any of the funds in the top row while the columns show whether the fund in the top row is being dominated by any of the funds in the leftmost column.

For example, the cells in the first row AHF01 and the second column AHF18 means that AHF01 dominates AHF18 by Sharpe ratio, Treynor index and MV criterion. The five indices IX1-IX5 and the "four most outstanding funds," AHF01, AHF18, AHF39 and AHF47, are defined in Table 9.9.

Furthermore, Gregoriou et al. [60] point out that CAPM measures will usually overestimate and miscalculate hedge fund performance. Therefore, the results drawn by both the MV and CAPM statistics can be misleading.

However, from our analysis using the MV criterion and CAPM statistics, we observe some consistent outcomes. We find, for example, that both AHF01 (fund with the largest Sharpe ratio) and AHF39 (fund with the smallest standard deviation) dominate AHF18 (the fund with the smallest Sharpe ratio) across most of the MV and CAPM statistics.

Thus, one may ask whether the fund with the largest Sharpe ratio and the fund with the smallest standard deviation are the best choices while the fund with the smallest Sharpe ratio is the worst choice. To explore this question and to examine alternative measures to choose funds, we use the SD approach.

**Stochastic dominance results**

DD stated that the null hypothesis of equal distribution could be rejected if any value of the test statistic, $T_j$ ($j = 1, 2, 3$; see Appendix A), is significant. In order to minimize the Type II error and to accommodate the effect of almost SD, we use a conservative 5% cutoff point for the proportion of test statistics in statistical inference. Using a 5% cutoff point, we conclude fund Y dominates fund Z if we find at least 5% of $T_j$ to be significantly negative and no portion of $T_j$ is significantly positive. The reverse holds if the fund Z dominates fund Y.

We depict in Table 9.12 the DD test statistics for the pair-wise comparison of the five market indices and the four most outstanding Asian hedge funds and display in Table 9.13 the summary of their DD dominance results. From the tables, we first find that in general one could not conclude that funds will always outperform indices and vice versa because there exist FSD relationships from funds to indices as well as from indices to funds. In addition, we conclude that risk averters may not always prefer investing in funds than indices and vice versa because there exist SSD relationships from funds to indices as well as from indices to funds.

Table 9.12: Pair-Wise Comparison of Asian Hedge Funds by DD Tests

| | IX1 | IX2 | IX3 | IX4 | IX5 | AHF01 | AHF18 | AHF39 | AHF47 | D |
|---|---|---|---|---|---|---|---|---|---|---|
| IX1 | | ND | ND | FSD | FSD | ND | SSD | ND | SSD | 4 |
| IX2 | ND | | ND | SSD | FSD | ND | FSD | ND | SSD | 4 |
| IX3 | ND | ND | | SSD | SSD | ND | FSD | ND | SSD | 4 |
| IX4 | ND | ND | ND | | ND | ND | ND | ND | SSD | 1 |
| IX5 | ND | ND | ND | ND | | ND | ND | ND | SSD | 1 |
| AHF01 | FSD | SSD | ND | SSD | SSD | | FSD | FSD | SSD | 7 |
| AHF18 | ND | ND | ND | ND | ND | ND | | ND | SSD | 1 |
| AHF39 | SSD | ND | ND | SSD | SSD | ND | FSD | | SSD | 5 |
| AHF47 | ND | ND | ND | ND | ND | ND | ND | ND | | 0 |
| D by | 2 | 1 | 0 | 5 | 5 | 0 | 5 | 1 | 8 | |

Refer to [171, 174, 175]. "D" stands for "Dominates." Results in this table are read based on rows-versus-column basis. For example, the cell in the first row IX1 and the fourth column IX4 tells us that IX1 stochastically dominates IX4 at first order while the cell in the second row IX2 and the first column IX1 informs readers that IX2 does not stochastically dominate IX1. Alternatively, reading along the row IX1, it can be seen that IX1 dominates four other indices/funds while reading down the IX1 column shows that IX1 is dominated by two other indices/funds.

Table 9.13: Summary of the DD Test Statistics

| Index/Fund | Dominates | | | Dominated By | | |
|---|---|---|---|---|---|---|
| | FSD | SSD | TOTAL | FSD | SSD | TOTAL |
| AA EH Asia ex-Japan | 5 | 19 | 24 | 2 | 1 | 3 |
| AA EH Japan Index | 11 | 23 | 34 | 0 | 1 | 1 |
| AA EH Index | 7 | 28 | 35 | 2 | 0 | 2 |
| MSAUCPI | 0 | 2 | 2 | 21 | 19 | 40 |
| S&P 500 | 0 | 2 | 2 | 15 | 12 | 27 |
| AHF47 | 0 | 0 | 0 | 0 | 63 | 63 |
| AHF18 | 0 | 2 | 2 | 17 | 4 | 21 |
| AHF39 | 1 | 46 | 47 | 4 | 0 | 4 |
| AHF01 | 16 | 38 | 54 | 0 | 0 | 0 |

Refer to [171, 174, 175]. The values indicate the number of indices/funds for each index/fund dominates or the number of indices/funds that it is dominated by. Note that in the table the reported number of SSD excludes the number of FSD. As hierarchical relationship exists in SD, FSD implies SSD. Thus, the total number of SSD (inclusive of FSD) is the sum of FSD and SSD (exclusive of FSD). For example, AA EH Asia ex-Japan Index dominates five indices/funds at FSD and dominates 19 indices/funds at SSD (excluding FSD), and thus it dominates 24 indices/funds (including both FSD and SSD) totally. On the other hand, it is dominated by two other indices/funds at FSD, dominated by one index/fund at SSD (excluding FSD), and thus it is dominated by three indices/funds (including both FSD and SSD) totally.

Secondly, we find that in our sample period AA EH Japan Index is the most favorable index and MSAUCPI is the least favorable index as the former dominates 11 (23) other indices/funds at first (second) order but is dominated only by AHF01 at SSD whereas the latter is dominated by 21 (19) indices/funds at first (second) order but dominates only 2 indices/funds at second order. On the other hand, AHF01 is the most favorable fund as it dominates 16 (38) other indices/funds at first (second) order and is not dominated by any other index/fund.

Similarly, from the tables, one could conclude that AHF39 is the second most favorable fund. The two least favorable funds are AHF47 and AHF18. The former is dominated by 63 indices/funds at second order but does not dominate any index/fund while the latter is dominated by 17 (4) indices/funds at first (second) order and dominates two other indices/funds.

Between AHF47 and AHF18, though the former is dominated by more indices/funds, AHF18 is the least favorable as it gets the largest number of indices/funds dominating it at first order, this means that investors will increase their expected wealth by selling AHF18 and longing any fund, say for example, AHF01, dominating it. However, when selling AHF47 and buying any fund dominating it could only increase one's expected utility, not expected wealth. Thus, we conclude that AHF18 is the least favorable one.

Now, we come back to our conclusion drawn using the MV criterion and CAPM statistics where we find that AHF01 and AHF39 are the most favorable funds while AHF18 is the least favorable fund. Using the SD approach, we demonstrate that AHF01 (AHF39) is the (second) most favorable fund whereas AHF18 (AHF47) is the (second) least favorable fund. This finding leads us to conjecture that the SD approach could exploit more information to decide on fund choice than its MV and CAPM counterparts.

Based on this conjecture, we further investigate whether we can acquire additional information by adopting the SD approach to obtain the percentage of significant DD statistics for pairs of these funds in details, namely, AHF47 (largest mean) versus AHF18 (smallest mean), AHF39 (smallest standard deviation) versus AHF47 (largest standard deviation), AHF01 (largest Sharpe ratio) versus AHF18 (smallest Sharpe ratio), AHF01 (the most favorable fund under SD) versus AHF39 (the second most favorable fund under SD).

In addition, we include the comparisons of AHF01-AHF47 and AHF39-AHF18 in our study because these comparisons exhibit FSD relations that seldom occur empirically. The results are reported in Table 9.14.

*Table 9.14: Summary of the DD Test Statistics*

|  | FSD | | SSD | | TSD | |
|---|---|---|---|---|---|---|
| Sample | $\%T_1 > 0$ | $\%T_1 < 0$ | $\%T_2 > 0$ | $\%T_2 < 0$ | $\%T_3 > 0$ | $\%T_3 < 0$ |
| AHF47-AHF18 | 9 | 22 | 11 | 0 | 9 | 0 |
| AHF39-AHF47 | 22 | 11 | 0 | 15 | 0 | 19 |
| AHF39-AHF18 | 0 | 17 | 0 | 27 | 0 | 5 |
| AHF01-AHF18 | 0 | 21 | 0 | 32 | 0 | 12 |
| AHF01-AHF39 | 0 | 27 | 0 | 27 | 0 | 0 |
| AHF01-AHF47 | 21 | 12 | 0 | 16 | 0 | 22 |

Refer to [171, 174, 175]. DD test statistics are computed over a grid of 100 on monthly Asian hedge fund returns. The table reports the percentage of DD statistics which is significantly negative or positive at the 5% significance level, based on the asymptotic critical value of 3.254 of the studentized maximum modulus (SMM) distribution. $T_j$ is the DD statistic for risk averters with $j = 1$, 2 and 3 with $F$ to be the first fund and $G$ to be the second fund stated in the first column.

In Table 9.14, we apply $T_j$ with the preferable fund being the first variable ($F$) and the less preferable fund being the second variable ($G$) in the equation. If the results behave as expected, there will exist $j$ ($j = 1, 2, 3$) such that there will not be any significantly positive $T_j$ but there will exist some significantly negative $T_j$. For example, as investors are risk-averse, we expect the fund with the smallest standard deviation (AHF39) will be preferred to the fund with the largest standard deviation (AHF47). Thus, we place AHF39 as the first variable and AHF47 as the second variable in Equation (A.18). The DD results displayed in Table 9.14 show that there are 22 (11) percentage of $T_1$ to be significantly positive (negative), indicating that AHF39 and AHF47 do not dominate each other at first order. However, we find that all values of $T_2$ ($T_3$) are nonpositive with 15 (19) percent of them significantly negative, implying AHF39 dominates AHF47 at second (third) order SD as expected.

So far, all comparisons behave as expected except the pair AHF47-AHF18. As AHF47 possesses the highest mean while AHF18 attains the smallest mean and it is common sense that all non-satiated investors prefer more to less, we expect AHF47 to be preferred to AHF18. However, their DD results shown in Table 9.14 reveal that all $T_2$ ($T_3$) are nonnegative with 11 (9) percent of $T_2$ ($T_3$) significantly positive, implying that, contradictory to the common belief, the fund, AHF18, with the smallest mean dominates the fund, AHF47, with the largest mean at second (third) order SD. Hence, we confirm that risk averters and risk-averse investors that are decreasing absolute risk averse (DARA), who make portfolio choices on the basis of expected-utility maximization, will unambiguously prefer AHF18 to AHF47 to maximize their expected utilities.

We recall that the MV and CAPM criteria show that AHF18 does not dominate AHF47 whereas AHF47 dominates AHF18 by the Sharpe ratio, Jensen index and Treynor index. As AHF47 possesses an insignificantly larger mean but significantly larger standard deviation than AHF18, one should not be surprised that our SD results reveal that AHF18 dominates AHF47 at second and third order. This result is consistent with Markowitz [105] that investors, especially risk-averse investors, worry more about downside risk than upside profit.

In addition, the results from Table 9.14 show that 9% of $T_1$ is significantly positive in the negative domain whereas 22% of $T_1$ is significantly negative in the positive domain. All this SD information implies that actually AHF47 and AHF18 do not outperform each other. AHF18 is preferable in the negative domain whereas AHF47 is preferable in the positive domain and, overall, risk averters prefer to invest in AHF18 than AHF47. All the information revealed by utilizing SD could not be obtained by adopting the MV or CAPM counterparts.

Note that most of the SD comparisons for assets in the literature behave as in the above comparison: one asset dominates another asset at SSD or TSD. Applying the DD technique, we could obtain more information than the usual SD comparison as we state in the above example: one asset dominates another asset on the downside while the reverse dominance relationship can be found on the upside.

## FSD results and discussions

We next explore FSD relationships in our empirical findings. We find three such: AHF01-AHF18, AHF01-AHF39 and AHF39-AHF18. For illustration, we only discuss AHF01-AHF18 and AHF01-AHF39 in detail. The first pair is interesting as we compare the fund with the largest Sharpe ratio (AHF01) versus the fund with the smallest Sharpe ratio. The second is also interesting as AHF01 is the most preferable fund and AHF39 is the second most preferable fund under SD in our study and thus one may be surprised that the former first order stochastically dominates the latter.

Table 9.13 shows that, for AHF01-AHF18, none of $T_1$ is significantly positive with 21% of it to be significantly negative. Similarly, the table displays that for AHF01-AHF39, none of $T_1$ is significantly positive with 27% of it to be significantly negative. These results imply that AHF01 stochastically dominates both AHF18 and AHF39 at first order and thus investors will increase their expected wealth if they shift their investment from AHF18 and/or AHF39 to AHF01.

Many studies claim that if FSD exists, under certain conditions arbitrage opportunity exist and investors will increase their wealth and expected utilities if they shift from holding the dominated hedge fund to the dominant one. In this chapter, we claim that if FSD exists statistically, arbitrage opportunity may not exist but investors could increase their expected wealth as well as their expected utilities if they shift from holding the dominated hedge fund to the dominant one.

We explain our claim by these examples. It can be shown by plotting real data that the two CDFs do cross and thus the CDF of the dominant fund does not totally lie below that of the dominated one. Thus, the values of $T_1$ are not totally nonpositive: there is a positive portion of $T_1$ in the positive (negative) domain, though these positive values are not significant. This shows that AHF18 (AHF39) does dominate AHF01 in a small portion of the positive (negative) domain or in a small part of bull run (bear market) though this domination is not statistically significant. In other words, AHF01 dominates both AHF18 and AHF39 statistically but not mathematically. Hence, arbitrage opportunity may not exist, but investors can still increase their expected wealth as well as their expected utilities but not their wealth if they shift their investment from AHF18 and/or AHF39 to AHF01.

In addition, even if one fund dominates another fund at first order mathematically, we claim that an arbitrage opportunity may also not exist, because an arbitrage opportunity in FSD exists only if the market is "complete." If the market is not "complete," even if FSD exists, investors may not be able to exploit any arbitrage opportunity there. In addition, if the test detects the FSD of a particular fund over another but the dominance does not last for a long period, these results cannot be used to reject market efficiency and market rationality.

In general, the FSD should not last for a long period of time because market forces cause adjustment to a condition of no FSD if the market is rational and efficient. For example, if fund A dominates fund B at FSD, then all investors would buy fund A and sell fund B. This will continue, driving up the price of fund A relative to fund B, until the market price of fund A relative to fund B is high enough to make the marginal investor indifferent between A and B. In this situation, we infer that the market is still efficient and rational.

In the conventional theories of market efficiency, if one is able to earn an abnormal return for a considerable length of time, the market is considered inefficient. If new information is either quickly made public or anticipated, the opportunity to use the new information to earn an abnormal return is of limited value. On the other hand, in the situation that the first-order dominance holds for a long time and all investors increase their expected wealth by switching their asset choice, we claim that the market is neither efficient nor rational. Another possibility for the existence of FSD to be held for a long period is that investors do no realize such dominance exists. It would be interesting to find out if FSD relationships among some funds would disappear over time. If they do not, then it would be considered a financial anomaly.

## 9.3.2 Discussion

Hedge funds differ from traditional investments in many respects including benchmarks, investment processes, fees and regulatory environment. With its absolute return strategies and non-normality returns distribution, additional investor skill is required to evaluate the quality of hedge funds and how a hedge fund fits into an investor's portfolio. Consistent with other studies, we find that sometimes the traditional MV criterion and CAPM statistics are ambiguous in their evaluation of the Asian hedge funds some of the time. At other times, though some of MV and CAPM measures can identify the dominant funds, they fail to provide detailed information of the dominance relationship nor on the preferences of investors.

This chapter applies a powerful SD test to present a more complete picture for hedge fund performance appraisal and to draw inference on the preference of investors on the funds. Based on a sample of the 70 individual Asian hedge funds from the Eurekahedge database, we find the existence of first-order SD relationship among some hedge funds in the entire sample period, suggesting that all non-satiated investors could increase their expected wealth as well as their expected utilities by investing in the Asian hedge funds to explore these opportunities by shifting their investments from the dominated funds to the dominant funds. We also find the existence of second-order SD relationship among other funds/indices, indicating that the non-satiated and risk-averse investors would maximize their expected utilities, but not their expected wealth by switching from the SSD dominated hedge funds to their corresponding SSD dominant ones. In addition, by applying the DD technique, we also discover that in most SD relationships, one fund dominates another fund in the negative domain while the reverse dominance relationship can be found in the positive domain. Besides the normality assumption in the traditional measures, the difference may also come from the traditional measures definition of an abnormal return as an excess return adjusted to some risk measures, while the SD tests employ the whole distribution of returns. The SD measure is an alternative that is superior to the traditional measures to help investors and fund managers in managing their investment portfolios.

Note that while some hedge funds are dominated by others, other issues may need to be considered before implementing the SD methodology in the selection of hedge funds. One issue is that Asian hedge funds may have different redemption timing, ranging from daily to once a year in rare cases. Hence, investors may prefer funds with high redemption frequency. We also note that while AHF01 dominates AHF47, investors can only redeem AHF01 quarterly, while the redemption timing for AFH47 is once a month. Secondly, entry and exit into hedge funds can be costly. These include search and assessment costs; considerable due diligence is often advised for hedge fund investors. Furthermore, a "hidden" transaction cost may arise in the way hedge fund managers are rewarded using a performance related fee. Studies showed that this fee structure can penalize investors who transact frequently due to the free-rider and claw-back problems.

We conclude that the SD approach is more appropriate as a filter in the hedge fund selection process. Compared with the traditional approaches, the SD approach is more informative, providing greater insights and hence, allowing for better comparison in the performance and risk inherent in a hedge fund's track record relative to that of another.

## 9.4   Evaluating iShare performance

As discussed previously, contemporary finance advocates the use of the MV model and the CAPM statistics, but these tools are not appropriate if return distributions are not normal or investors' utility functions are not quadratic. On the other hand, SD rules offer superior criteria because SD incorporates information on the entire return distribution, rather than the first two moments as with MV and requires no precise assessment as to the specific form of the investor's risk preference or utility function. It also allows us to determine if an arbitrage opportunity exists among the investment alternatives and determine preferences for different types of investors.

These advantages of SD have motivated prior studies to use SD techniques to evaluate the performance of mutual funds. Unfortunately, earlier research was unable to determine the statistical significance of SD. However, as discussed previously, recent advances in SD techniques by Davidson and Duclos and others, allow differences between any two return cumulative density functions to be tested for statistical significance.

An opportunity for applying these innovations emerged with the introduction of country index funds. Standard and Poor's Depository Receipts, (SPDRs or "spiders") track the S&P 500 Index and began trading in January 1993. The acceptance and wide use of SPDRs (ticker symbol: SPY) led to the introduction in March 1996, of 17 exchange traded funds (ETFs) known as World Equity Benchmark Shares (WEBS). WEBS, now known as iShares, are investment companies designed to track the Morgan Stanley Capital International (MSCI) foreign stock market indices. These innovations allow investors to continuously trade shares of several well-diversified portfolios.

Studies find that most of the funds are skew and thus CAPM statistics may not be suitable in analyzing the performance of funds. In addition, studies also find that during the period of extreme market stress, variance and beta are not suitable proxies for risk because of increased skewness in the return distribution.

Note that many studies use stock market indices that are not actually tradable or marketable. The introduction of iShares, securities that mimic international indices, and that are tradable, allows an investigation of realizable return distributions and an examination of the shape of the empirical distribution function.

## 9.4.1   Data and methodology

SPDRs are created and redeemed via "creation units" of 50,000 shares. A listing of the country market indices, their ticker symbols and inception dates for the 17 iShares and the U.S. SPY is presented in Table 9.15.

Subsequently, the following iShares have been added: South Korea EWY (May 2000), Taiwan EWT (June 2000), European Monetary Union EZU (July 2000), Brazil EWZ (July 2000), Pacific ex-Japan EPP (October 2001), South Africa EZA (February 2003) and Emerging Markets EEM (April 2003). Refer to [53] for more information.

*Table 9.15: Ticker Symbols and the Date of Inception for the 17 iShares and the U.S. SPY*

| Country Fund | Symbol | Inception Date |
|---|---|---|
| U.S. SPY | SPY | Jan-93 |
| Australia | EWA | Mar-96 |
| Austria | EWO | Mar-96 |
| Belgium | EWK | Mar-96 |
| Canada | EWC | Mar-96 |
| France | EWQ | Mar-96 |
| Germany | EWG | Mar-96 |
| Hong Kong | EWH | Mar-96 |
| Italy | EWI | Mar-96 |
| Japan | EWJ | Mar-96 |
| Malaysia | EWM | Mar-96 |
| Mexico | EWW | Mar-96 |
| Netherlands | EWN | Mar-96 |
| Singapore | EWS | Mar-96 |
| Spain | EWP | Mar-96 |
| Sweden | EWD | Mar-96 |
| Switzerland | EWL | Mar-96 |
| United Kingdom | EWU | Mar-96 |

Developing iShare to serve as each fund's investment advisor, Morgan Stanley and Barclays Global Fund Advisors use either "replication" or "representative sampling" to construct portfolios designed to mimic a particular country's index. "Replication" is to set up a fund containing essentially all of the securities in the relevant country index in relatively the same proportions as that country's index whereas "representative sampling" is to set up a portfolio possessing a similar investment profile but not all securities in the index are included in the iShare. Changes in portfolio values come from both changes in the share prices in the portfolio and in changes in the exchange rate between U.S. dollars and the currency for a particular country. For consistency, foreign currencies are converted at the same time and at the same rate as used in the determination of each of the MSCI indices.

Although technically iShares are open-end index funds, the "creation units" cause their shares to trade in the secondary market just like ordinary shares. These "creation units" are in-kind deposits of portfolios of securities designed to represent a particular MSCI Index. If price differences between the underlying country index and the associated iShare emerge, arbitrage opportunities exist and large investors will quickly eliminate the price differences. These funds are not actively managed, turnover is virtually nonexistent and operating expenses are low. In addition, only the usual brokerage fee is paid to buy or sell shares, as there are no front-end loads or deferred sales charges.

The daily return data were obtained from the Center for Research in Security Prices (CRSP) for the 17 iShares and SPY for the March 12, 1996 through December 31, 2003 period. For robustness checking, we consider three subperiods with events that are region specific. The first subperiod includes the Asian financial crisis that began in July of 1997 but ended before the Russian devaluation in October 1998. The second subperiod incorporates the U.S. technology bubble boom and the burst of the bubble in March of 2000. Finally, the third subperiod goes from the bursting of the bubble to the end of 2003.

The 3-month U.S. T-bill rate and the Morgan Stanley Capital International (MSCI) index returns proxy the risk-free rate and the global market index, respectively. Mean-variance measures and several statistics derived from the CAPM including beta, Sharpe ratio, Treynor's index and Jensen's alpha (referred to as CAPM statistics) are used along with the stochastic dominance criterion to study the performance of the 18 closed-end funds.

The measures of skewness, kurtosis and the Jarque-Bera statistic indicate that none of the 18 return distributions are normal and the MV criterion and the CAPM statistics are restricted to the first two moments of the data. To overcome the shortcomings associated with the MV and CAPM models and to investigate the performance of the entire distributions of the returns, we apply the Davidson-Duclos [29] (DD) nonparametric stochastic dominance statistics to test for the dominance of any pair of the returns series.

## 9.4.2 Results

We report the descriptive statistics for the returns of the 18 closed-end funds for the entire period in Table 9.16 and skip reporting the results for the three subperiods. From the table, the means and standard deviations vary widely across iShares and over time. A two-sample $t$-test shows that some funds have significantly higher mean returns than others, and the F-statistic indicates some standard deviations are significantly different at the 1% level. For example, the U.S. SPY displays a significantly higher mean and a significantly smaller standard deviation than Malaysia's EWM while Spain EWP exhibits a significantly higher mean but not significantly smaller standard deviation than Japan's EWJ.

*Table 9.16: Summary Statistics, Results of*
*Normality Tests and CAPM Statistics*

| Panel A: March 1996-December 2003 | | | | | | | | | |
|---|---|---|---|---|---|---|---|---|---|
| iShare | Daily Returns | | Normality Test | | | CAPM Statistics | | | |
| | mean | $\sigma$ | skewness | kurtosis | J-B | beta | Sharpe | Treynor | Jensen |
| EWA | 0.0004 | 0.015 | −0.244** | 3.427** | 979.28** | 0.621 | 0.0172 | 0.0004 | 0.0002 |
| EWO | 0.0003 | 0.015 | 0.067 | 3.626** | 1075.96** | 0.332 | 0.0116 | 0.0005 | 0.0001 |
| EWK | 0.0004 | 0.021 | 4.134** | 175.008** | 250.81** | 0.795 | 0.0136 | 0.0004 | 0.0002 |
| EWC | 0.0005 | 0.014 | −0.228** | 2.368** | 475.63** | 0.851 | 0.026 | 0.0004 | 0.0003 |
| EWQ | 0.0004 | 0.015 | −0.06 | 1.716** | 241.79** | 1.131 | 0.0211 | 0.0003 | 0.0002 |
| EWG | 0.0003 | 0.017 | −0.025 | 2.333** | 445.26** | 1.254 | 0.0136 | 0.0002 | 0 |
| EWH | 0.0002 | 0.022 | 0.632** | 7.173** | 4334.90** | 1.275 | 0.0058 | 0.0001 | 0 |
| EWI | 0.0005 | 0.016 | −0.055 | 2.018** | 333.89** | 0.980 | 0.0252 | 0.0004 | 0.0003 |
| EWJ | 0 | 0.017 | 0.581** | 3.978** | 1403.91** | 0.953 | −0.0093 | −0.0002 | −0.0003 |
| EWM | 0 | 0.026 | 0.876** | 7.014** | 4271.62** | 0.839 | −0.0017 | 0 | −0.0001 |
| EWW | 0.0006 | 0.022 | 0.138* | 7.896** | 5100.81** | 1.273 | 0.0217 | 0.0004 | 0.0003 |
| EWN | 0.0003 | 0.016 | −0.071 | 1.791** | 263.92** | 1.104 | 0.01 | 0.0001 | 0 |
| EWS | 0 | 0.023 | 0.443** | 4.477** | 1702.23** | 1.136 | −0.0065 | −0.0001 | −0.0003 |
| EWP | 0.0006 | 0.016 | 0.103 | 1.743** | 251.77** | 1.014 | 0.0312 | 0.0005 | 0.0004 |
| EWD | 0.0005 | 0.020 | 0.087 | 2.162** | 384.66** | 1.303 | 0.0182 | 0.0003 | 0.0002 |
| EWL | 0.0003 | 0.016 | 0.032 | 1.555** | 198.01** | 0.841 | 0.0105 | 0.0002 | 0 |
| EWU | 0.0004 | 0.014 | 0.045 | 1.529** | 191.77** | 0.959 | 0.0179 | 0.0003 | 0.0002 |
| SPY | 0.0004 | 0.013 | 0.010 | 2.283** | 426.20** | 1.128 | 0.0232 | 0.0003 | 0.0002 |
| MSCI | 0.0002 | 0.001 | −0.111 | 2.603** | 558.01** | | | | |

This chapter only reports the entire period from March 1996 to December 2003. Readers may refer to [53] for the results of the three subperiods: March 1996–June 1998, July 1998–February 2002 and March 2002–December 2003. The risk-free asset is a 3-month T-bill in the U.S. and the market return is from the MSCI World Index. J-B is Jarque-Bera statistics. The CAPM statistics use as the risk-free asset the 3-month U.S. T-bill rate and the market portfolio is the MSCI World Index. * means significant at 5% level, and ** means significant at 1% level.

Nonetheless, the results show that the return distributions are non-normal and display both skewness and kurtosis and, hence, the distributions violate the normality requirements of the traditional CAPM measures.

In addition, the table highlights the ambiguity that is present both between and within the traditional CAPM measures. For example, the Sharpe ratios exhibit wide variation. At the extremes are Spain (EWP) with a value of 0.0312 and Japan (EWJ) at –0.0093.

Surprisingly, although their difference is large, the Sharpe ratio test shows no statistically significant difference between these funds. Applying the pair-wise Sharpe ratio test statistic shows that none of the Sharpe ratios are significantly different.

However, when we apply the multivariate Sharpe ratio test statistic to the iShares, the equality of the Sharpe ratios among all iShares is rejected significantly. In addition, the evaluation issue is exacerbated in that the Treynor and Jensen measures suggest different rankings for the 18 closed-end funds.

The three subperiod results in our unreported tables allow the examination of the performance measures during different economic conditions. Again, we observe substantial differences among the distributions during different time periods.

Skewness appears to be reduced over time as the number exhibiting significant skewness at the 1% level decreases from 11, to 8, to 3 (including MSCI) over the three subperiods. Changes in kurtosis and the JB statistic are much less with a maximum of 3 kurtosis measures being not significant in the first subperiod.

The column on the far right shows the number of funds each individual fund dominates and the last row shows the number of funds each individual fund is dominated.

Superior to both MV and CAPM criteria, SD procedures allow us to determine whether one iShare stochastically dominates another at different orders based on the entire empirical return distribution. Table 9.17 exhibits the results of the Davidson-Duclos SD tests for the entire period.

Readers may refer to the comments below the table on how to read the table. From the table, we find that Spain (EWP) dominates five other funds and SPY dominates most of the other funds at 14. On the other hand, Malaysia (EWM) is dominated by 13 other funds while four funds are not dominated by any other fund.

We skip reporting the dominance relationship among each iShare for each of the sub-periods. Nevertheless, a summary of the number of dominated iShares and number of dominant iShares of the entire period as well as each of the sub-periods is displayed in Table 9.18.

*Table 9.17: Pairwise Results of the DD Tests Between iShares*
*March 1996 - December 2003*

|   | A | O | K | C | Q | G | H | I | J | M | W | N | S | P | D | L | U | Y | D |
|---|---|---|---|---|---|---|---|---|---|---|---|---|---|---|---|---|---|---|---|
| A |   | N | N | N | N | N | S | N | N | S | S | N | S | N | S | N | N | N | 5 |
| O | N |   | N | N | N | N | S | N | N | S | S | N | S | N | S | N | N | N | 5 |
| K | N | N |   | N | N | N | N | N | N | N | N | N | N | N | N | N | N | N | 0 |
| C | N | N | N |   | N | S | S | N | S | S | S | N | S | N | S | N | N | N | 7 |
| Q | N | N | N | N |   | N | S | N | N | S | S | N | S | N | S | N | N | N | 5 |
| G | N | N | N | N | N |   | S | N | N | S | S | N | S | N | S | N | N | N | 5 |
| H | N | N | N | N | N | N |   | N | N | N | N | N | N | N | N | N | N | N | 0 |
| I | N | N | N | N | N | N | S |   | N | S | S | N | S | N | S | N | N | N | 5 |
| J | N | N | N | N | N | N | S | N |   | S | S | N | S | N | S | N | N | N | 5 |
| M | N | N | N | N | N | N | N | N | N |   | N | N | N | N | N | N | N | N | 0 |
| W | N | N | N | N | N | N | N | N | N | N |   | N | N | N | N | N | N | N | 0 |
| N | N | N | N | N | N | N | S | N | N | S | N |   | S | N | S | N | N | N | 4 |
| S | N | N | N | N | N | N | N | N | N | N | N | N |   | N | N | N | N | N | 0 |
| P | N | N | N | N | N | N | S | N | N | S | S | N | S |   | S | N | N | N | 5 |
| D | N | N | N | N | N | N | N | N | N | S | N | N | N | N |   | N | N | N | 1 |
| L | N | N | N | N | N | N | S | N | N | S | S | N | S | N | S |   | N | N | 5 |
| U | N | N | N | N | N | S | S | N | S | S | S | N | S | N | S | N |   | N | 7 |
| Y | S | N | N | N | S | S | S | S | S | S | S | S | S | S | S | S | S |   | 14 |
| D | 1 | 0 | 0 | 0 | 1 | 3 | 12 | 1 | 3 | 13 | 11 | 1 | 12 | 1 | 12 | 1 | 1 | 0 |   |

Refer to [53] for more information. In the first column and the first row, "Y" refers to "SPY" and "X" refers to "EWX" where X = A, O, K, etc. "N" refers to "ND" or "no dominance" whereas "S" refers to SSD. The results in this table are read based on row versus column. For example, the first row EWA and the second column EWO means that EWA does not stochastically dominate EWO while the second row EWO and the first column EWA means that EWO does not stochastically dominate EWA. Alternatively, reading along the row SPY it can be seen that SPY dominates 14 other funds while reading down the SPY column shows that SPY is not dominated by any other fund.

Interestingly, we find one instance of first-order stochastic dominance in the March 1996-June 1998 period before the Asian financial crisis that Spain, EWP, exhibited FSD over Japan, EWJ. Thus, investors could have increased both their expected wealth and their expected utility by switching from Japan to Spain. This is an interesting finding, as most prior studies find no FSD. However, we must conclude it is time-specific as the relation does not appear in any other period.

Table 9.18: Country Fund, Ticker Symbol
and a Summary of SD Results

| Country | Symbol | Entire Period | | Mar 96-Jun 98 | | Jul 98-Feb 02 | | Mar 02-Dec 03 | |
|---|---|---|---|---|---|---|---|---|---|
| | | D | D By | D | D By | D | D By | D | D By |
| Australia | EWA | 5 | 1 | 3 | 4 | 4 | 0 | 6 | 0 |
| Austria | EWO | 5 | 0 | 4 | 0 | 1 | 1 | 9 | 0 |
| Belgium | EWK | 0 | 0 | 5 | 0 | 4 | 0 | 0 | 0 |
| Canada | EWC | 7 | 0 | 6 | 0 | 4 | 0 | 4 | 0 |
| France | EWQ | 5 | 1 | 5 | 0 | 4 | 0 | 1 | 2 |
| Germany | EWG | 5 | 3 | 6 | 0 | 1 | 1 | 1 | 6 |
| H.K. | EWH | 0 | 12 | 0 | 13 | 1 | 11 | 1 | 8 |
| Italy | EWI | 5 | 1 | 3 | 1 | 4 | 0 | 5 | 0 |
| Japan | EWJ | 5 | 3 | 2 | 11* | 3 | 0 | 3 | 2 |
| Malaysia | EWM | 0 | 13 | 0 | 15 | 0 | 16 | 3 | 0 |
| Mexico | EWW | 0 | 11 | 1 | 10 | 0 | 11 | 1 | 4 |
| Netherlands | EWN | 4 | 1 | 6 | 0 | 4 | 0 | 1 | 5 |
| S'pore | EWS | 0 | 12 | 0 | 14 | 1 | 10 | 0 | 9 |
| Spain | EWP | 5 | 1 | 5* | 0 | 4 | 1 | 2 | 2 |
| Sweden | EWD | 1 | 12 | 5 | 0 | 1 | 0 | 0 | 15 |
| S'land | EWL | 5 | 1 | 5 | 0 | 4 | 1 | 1 | 1 |
| U.K. | EWU | 7 | 1 | 5 | 0 | 4 | 0 | 6 | 0 |
| U.S. | SPY | 14 | 0 | 7 | 0 | 8 | 0 | 10 | 0 |

Refer to [53] for more information. "D" and "D by" mean "Dominated" and "Dominated by," respectively. The values indicate the number of funds for each fund dominates and the number of funds that it is dominated by. For example, for the entire period the Australian fund dominates five other funds and is dominated by one other fund. * indicates significance of FSD between EWP and EWJ.

Our results show that risk-averse investors will increase their expected utility by switching from these 14 other closed-end funds to SPY. The existence of second-order stochastic dominance does not imply any arbitrage opportunity, and neither implies the failure of market efficiency nor market rationality. Thus, we conclude that although SPY does not significantly outperform most other funds from a wealth perspective, risk-averse investors prefer SPY as they will increase their expected utility by switching from 14 other funds to SPY. At the other extreme, all risk-averse investors holding EWM, EWK, EWH, EWW and EWS will increase their expected utility by switching to some other funds.

We further analyze SPY (U.S.) and EWM (Malaysia) and exhibit the results in Table 9.19 because, although the Sharpe Ratio indicates SPY and EWM are not significantly different, the DD tests show that SPY dominates the most other funds (14) while EWM is dominated by the most other funds (13); and the SSD shows a statistically significant difference between these two funds.

Table 9.19: Results of the DD Tests
Comparing SPY with EWM

|  | FSD | SSD | TSD |
|---|---|---|---|
| Panel A: Mar 1996-Dec 2003 | | | |
| %DD+ | 37% | 0 | 0 |
| %DD− | 27% | 37% | 75% |
| Panel B: Mar 1996-June 1998 | | | |
| %DD+ | 22% | 0 | 0 |
| %DD− | 24% | 50% | 79% |
| Panel C: July 1998-Feb 2002 | | | |
| %DD+ | 30% | 0 | 0 |
| %DD− | 20% | 28% | 43% |
| Panel D: Mar 2002-Dec 2003 | | | |
| %DD+ | 0 | 0 | 0 |
| %DD− | 0 | 0 | 0 |

Refer to [53] for more information. The risk-free asset is a 3-month T-bill in US and market return is from the MSCI World Index. * indicates significant at 5% level, and ** significant at 1% level.

When subperiods are considered, the first two subperiods are similar to the entire period, but the third subperiod (March 2002-December 2003) is different. A closer examination of Table 9.22, Panels B through D, reveals that SPY possesses a significantly higher mean in the first sub-period, but EWM possesses a significantly higher mean in the second sub-period and a higher but insignificant mean in the third sub-period. On the other hand, SPY possesses a significantly smaller standard deviation in the first and second sub-periods, but an insignificantly smaller standard deviation in the third sub-period. In addition, SPY obtains a higher Sharpe Ratio in sub-periods 1 and 3 but not in sub-period 2.

Table 9.19 shows the SD results between SPY and EWM do not change sign, and basically all three sub-periods draw a similar conclusion, that EWM dominates SPY in the upside returns while SPY dominates to EWM in the downside returns in the sense of FSD. The dominance of SPY over EWM in SSD still remains the same as that in the entire period. The difference is that the SD of SPY is significant in the first two sub-periods but becomes insignificant in the third sub-period. In this period, the DD tests indicate SPY does not dominate EWM nor does EWM dominate SPY.

We further analyze the second set of funds: EWP (Spain) and EWJ (Japan) and display the results in Table 9.15 since the DD test indicates FSD of EWP over EWJ in the first subperiod. The FSD finding is important as most past studies find no evidence of FSD. For the entire period, the unreported figure shows that the CDFs and the first-order DD statistics $(T_1)$ show that EWP first order dominates EWJ in the negative return region with marginal significance (2%), but EWJ dominates EWP (but not significant) in the positive region. As we use a conservative 5% cutoff point for the proportion of t-statistics to minimize type II error of finding dominance, and to avoid Almost SD [90], 2% dominance implies that we cannot conclude FSD of EWP over EWJ over the entire period.

However, in the first sub-period, Table 9.14 reveals that there are 11% (41% and 70%) FSD (SSD and TSD, respectively) of EWP over EWJ and the unreported figure confirms that these dominances are in the negative region. The FSD (though marginal) of EWP over EWJ implies that all investors with increasing utility will prefer EWP to EWJ. There is an arbitrage opportunity between EWP and EWJ such that all investors will increase both their expected wealth and their expected utility if they shift their investments from EWJ to EWP. However, Table 9.14 and the unreported figures show that FSD, SSD and TSD disappear in the following two sub-periods. This could be due to its exploitation after investors realize this arbitrage opportunity.

In addition we examine SD in up-markets and down-markets by applying a regime shifting technique. We use the MSCI index to classify the up-market and down-market regimes and to estimate the likelihood of being in an up-market or down-market on each day by applying the Hamiltonian regime switching approach. By using the regime switching technique, we find days with low likelihood of being a down market prevail in our first two sub-periods, while those with high likelihood of being down market are more pronounced within the third sub-period. The unreported figure shows the long upward trend in the MSCI until the bursting of the technology bubble in early 2000, followed by the substantial decline until the end of the period. The unreported figure shows the probability of a down market over the entire period. We see that early in the period, the data are dominated by a lower probability of a down market, while later in the period there is a high probability of a down market. Because the results of the up- and down-market regime changes are similar to the original analyses of the sub-periods, we do not report the results here.

### 9.4.3   Discussion

This section compares the performance of 18 country market indices represented by iShares. Our examination considers their returns from their inception through 2003, and we find that, empirically, iShare returns exhibit both skewness and kurtosis and are not normally distributed. We find that the traditional CAPM measures are ambiguous in their evaluation of the iShares, present both between and within measures. We apply SD procedures to determine whether statistically significant stochastic dominance occurs among 18 marketable iShares. We find that over the entire 1996–2003 period certain iShares dominate others. Conversely, some do not dominate any other iShares, but they themselves are not dominated by all iShares. We find that U.S. dominates most of the other funds while Malaysia is dominated by most of the other funds. Spain and Japan show the greatest difference in Sharpe ratio, but do not appear different from an SD perspective.

The results show that even though there appear to be large differences among the Sharpe ratios, the Sharpe ratio test indicates none of the differences are statistically significant. Thus, the ratios are indistinguishable on the first two moments, possibly due to the existence of skewness and kurtosis. Furthermore, the Treynor and Jensen measures provide conflicting rankings. This may be caused by the use of betas that may be biased due to volatile markets and non-normal return distributions. Variations in these measures over the subperiods support this view.

By using the DD tests of significance on the entire period as well as on the three subperiods, we identify the existence for FSD, SSD and TSD and the levels of significance. Although the results vary over time, SD appears to be more robust than the CAPM in the ranking of the iShares.

## 9.5   Exercises

1. For the illustration shown in Table 9.1, convert the variables of profits to the variables of loss.

(a) Apply the the improved MV criterion to find the preferences of these assets for risk averters and risk seekers, and

(b) Apply stochastic dominance criterion to find the preferences of these assets for risk averters and risk seekers.

2. The following table shows the experiment of gain one month later for an investor who invests $10,000 either in stock A or in stock B:

*Experiment 1*

| Stock A | | Stock B | |
|---|---|---|---|
| Gain (in 1000) | Probability | Gain (in 1000) | Probability |
| 0.5 | 0.3 | −0.5 | 0.1 |
| 2 | 0.3 | 0 | 0.1 |
| 5 | 0.4 | 0.5 | 0.1 |
| | | 1 | 0.2 |
| | | 2 | 0.1 |
| | | 5 | 0.4 |

Refer to [95].

(a) Define variable of profit in this experiment and compute the ASD and DSD integral differentials for the profits of investing in Stocks A and B in Experiment 1. Draw a conclusion for the preferences of different risk averters and risk seekers for investing in Stocks A and B.

(b) What are the preferences of different risk averters and risk seekers in investing in Stocks A and B if you use the AMS rule or the DMS rule based on the losses?

(c) Define variable of loss in this experiment and compute the ASD and DSD integral differentials for the losses of investing in Stocks A and B in Experiment 1. Draw a conclusion for the preferences of different risk averters and risk seekers for investing in Stocks A and B.

(d) What are the preferences of different risk averters and risk seekers in investing Stocks A and B if you use AMS rule or DMS rule based on the losses?

3. The following table shows the experiment of gain one month later for an investor who invests $10,000 either in stock A or in stock B:

*Experiment 2*

| Stock A | | Stock B | |
|---|---|---|---|
| Gain (in 1000) | Probability | Gain (in 1000) | Probability |
| −1.6 | 0.25 | −1 | 0.25 |
| −0.2 | 0.25 | −0.8 | 0.25 |
| 1.2 | 0.25 | 0.8 | 0.25 |
| 1.6 | 0.25 | 2 | 0.25 |

Refer to [95].

Re-do (a) to (d) in Question 2 for Experiment 2.

4. The following table shows the experiment of gain one month later for an investor who invests $10,000 either in stock A or in stock B:

*Experiment 3*

| Stock A | | Stock B | |
|---|---|---|---|
| Gain (in 1000) | Probability | Gain (in 1000) | Probability |
| −1.6 | 0.25 | −1 | 0.25 |
| −0.2 | 0.25 | −0.8 | 0.4 |
| 1.2 | 0.25 | 0.8 | 0.3 |
| 1.6 | 0.25 | 2 | 0.05 |

Refer to [170].

Re-do (a) to (d) in Question 1 for Experiment 3.

5. Refer to the data in Section 9.2, and apply the MV criterion and the CAPM statistics to find the preferences of the returns from following series: All REIT (ART), Equity REIT (ERT), Mortgage REIT (MRT), US-DS real estate (DRE), Dow Jones Industrials (DJI), NASDAQ (NAS), S&P 500 (SP5), 10-year T. note (TB10), 3-month T. bill (TB3). Comment on your findings.

6. Refer to the REIT data in Section 9.2. Compute the DD statistic discussed in Appendix to study the preferences of risk averters for each pair of the series as stated in Section 9.2 and comment on your findings.

7. Refer to the REIT data in Section 9.2. Compute the modified DD test as discussed in Appendix to study the preferences of risk seekers for each pair of the series as stated in Section 9.2 and comment on your findings.

8. Refer to the REIT data in Section 9.2. Compute both the DD test and the modified DD test as discussed in Appendix to study the preferences of investors with S-shaped utility functions for each pair of the series as stated in Section 9.2 and comment on your findings.

9. Refer to the REIT data in Section 9.2. Compute both the DD test and the modified DD test as discussed in Appendix to study the preferences of investors with reverse S-shaped utility functions for each pair of the series as stated in Section 9.2 and comment on your findings.

10. Apply the SD test developed by Linton, Maasoumi and Whang [97] to study the preferences of risk averters for each pair of the series of the REIT data in Section 9.2.

11. Apply the modified LMW test as discussed in Appendix to study the preferences of risk seekers for each pair of the series of the REIT data in Section 9.2.

12. Apply the SD test developed in [97] and the modified LMW test as discussed in Appendix to study the preferences of investors with S-shaped and reverse S-shaped utility functions for each pair of the series of the REIT data in Section 9.2.

13. Refer to the "hedge funds" data in Section 9.3. Apply the MV criterion to find the preferences of the hedge fund returns. Comment on your findings.

14. Refer to the "hedge funds" data in Section 9.3. Compute the DD statistic discussed in Appendix to study the preferences of risk averters for each pair of the hedge fund returns. Comment on your findings.

15. Refer to the "hedge funds" data in Section 9.3. Compute the modified DD test as discussed in Appendix to study the preferences of risk seekers for each pair of the hedge fund returns. Comment on your findings.

16. Refer to the "hedge funds" data in Section 9.3. Compute both the DD test and the modified DD test as discussed in Appendix to study the preferences of investors with S-shaped utility functions for each pair of the hedge fund returns. Comment on your findings.

17. Refer to the "hedge funds" data in Section 9.3.  Compute both the DD test and the modified DD test as discussed in Appendix to study the preferences of investors with reverse S-shaped utility functions for each pair of the hedge fund returns.  Comment on your findings.

18. Apply the SD test developed by Linton, Maasoumi and Whang [97] to study the preferences of risk averters for each pair of the hedge fund returns for the "hedge funds" data in Section 9.3.

19. Apply the modified LMW test as discussed in Appendix to study the preferences of risk seekers for each pair of the hedge fund returns for the "hedge funds" data in Section 9.3.

20. Apply the SD test developed by Linton, Maasoumi and Whang [97] and the modified LMW test as discussed in Appendix to study the preferences of investors with S-shaped and reverse S-shaped utility functions for each pair of the hedge fund returns for the "hedge funds" data in Section 9.3.

21. Refer to the data "iShare" in Section 9.4.  Apply the MV criterion and the CAPM statistics to find the preferences of the iShare returns.  Comment on your findings.

22. Refer to the "iShare" data in Section 9.4.  Compute the DD statistic discussed in Appendix to study the preferences of risk averters for each pair of the iShare returns.  Comment on your findings.

23. Refer to the "iShare" data in Section 9.4.  Compute the modified DD test as discussed in Appendix to study the preferences of risk seekers for each pair of the iShare returns.  Comment on your findings.

24. Refer to the "iShare" data in Section 9.4.  Compute both the DD test and the modified DD test as discussed in Appendix to study the preferences of investors with S-shaped utility functions for each pair of the iShare returns.  Comment on your findings.

25. Refer to the "iShare" data in Section 9.4.  Compute both the DD test and the modified DD test as discussed in Appendix to study the preferences of investors with reverse S-shaped utility functions for each pair of the iShare returns.  Comment on your findings.

26. Apply the SD test developed by Linton, Maasoumi and Whang [97] to study the preferences of risk averters for each pair of the iShare returns for the "iShare" data in Section 9.4.

27. Apply the modified LMW test as discussed in Appendix to study the preferences of risk seekers for each pair of the iShare returns for the "iShare" data in Section 9.4.

28. Apply the SD test developed by Linton, Maasoumi and Whang [97] and the modified LMW test as discussed in Appendix to study the preferences of investors with S-shaped and reverse S-shaped utility functions for each pair of the iShare returns for the "iShare" data in Section 9.4.

# Chapter 10

# Applications to Economics

*This chapter presents several applications of stochastic dominance to economics.*

## 10.1  Indifference curves/location-scale family

In this first section, we discuss the shapes of indifference curves for risk averters, risk seekers and risk-neutral investors for generalized utility functions. Next, we will discuss the relationships among the FSD, SSD and MS efficient sets. In addition, we will discuss the relationship between risk aversion and the indifference curve slope.

The location-scale family $D_X$ generated by a random variable $X$, with zero mean and variance one, is defined as

$$\mathcal{D}_X = \{Y : Y = \mu + \sigma \cdot X, \, \mu \in \mathbb{R}, \, \sigma \in \mathbb{R}^+\}. \tag{10.1}$$

### 10.1.1  Portfolio and expected utility

Consider a portfolio of one risky asset and one risk-free asset and let $\lambda_0$ be the wealth proportion invested in the risk-free asset. Then, the return of the portfolio is given by

$$X = (1 - \lambda_0)R + \lambda_0 R_0,$$

where $R$ and $R_0$ are the returns of risky and risk-free asset, respectively, then

$$X - R_0 = (1 - \lambda_0)(R - R_0).$$

Let $Z$ be the random variable obtained from $R$ using the normalizing transformation $Z = (R-\mu)/\sigma$, where $\mu$ and $\sigma$ are the mean and standard deviation of $R$. Then, the expected utility can be written as

$$V(\mu - R_0, \sigma) \;=\; Eu\left(\frac{X - R_0}{1 - \lambda_0}\right) = Eu[(\mu - R_0) + \sigma z]$$

$$=\; \int_a^b u((\mu - R_0) + \sigma z)dF(z),$$

where $-\infty \le b < a \le +\infty$ and $F$ is the distribution function of $Z$.

For generality, in the following discussion, let $\mu_o$ be the reference level and $a = -\infty, b = +\infty$. Then

$$V(\mu - \mu_o, \sigma) \equiv \int_{-\infty}^{\infty} u((\mu - \mu_o) + \sigma z)dF(z).$$

Without loss of generality, we assume $X = Z$ be the seed variable with mean zero ($\mu_o = 0$) and unit variance, then the mean-variance function in equation (10.2) becomes the expected utility $V(\sigma, \mu)$, for the utility $u$ on the random variable $Y$ and can be expressed as:

$$V(\sigma, \mu) \equiv Eu(Y) = \int_a^b u(\mu + \sigma x)\, d\,F(x) \qquad (10.2)$$

where $[a, b]$ is the support of $X$, $F$ is the distribution function of $X$, and the mean and variance of $Y$ are $\mu$ and $\sigma^2$, respectively. Note that the requirement of the zero mean and unit variance for $X$ is not necessary. However, without loss of generality, we can make this assumption as we will always be able to find such a seed random variable in the location-scale family.

### Indifference curve

For any constant $\alpha$, the indifference curve drawn on the $(\sigma, \mu)$ plane such that $V(\sigma, \mu)$ is a constant is

$$C_\alpha = \{(\sigma, \mu) : V(\sigma, \mu) = \alpha\}.$$

On the indifference curve, we have:

$$V_\mu(\sigma, \mu)\, d\mu + V_\sigma(\sigma, \mu)\, d\sigma = 0$$

or

$$S(\sigma, \mu) = \frac{d\mu}{d\sigma} = -\frac{V_\sigma(\sigma, \mu)}{V_\mu(\sigma, \mu)} \qquad (10.3)$$

where

$$V_\mu(\sigma, \mu) = \frac{\partial V(\sigma, \mu)}{\partial \mu} = \int_a^b u^{(1)}(\mu + \sigma x)\, d\, F(x)$$

$$V_\sigma(\sigma, \mu) = \frac{\partial V(\sigma, \mu)}{\partial \sigma} = \int_a^b u^{(1)}(\mu + \sigma x)\, x\, d\, F(x).$$

In addition, in order to study the curvature of the indifference curve, from (10.3), we have:

$$S_\sigma(\sigma, \mu) = \frac{d^2\mu}{d\sigma^2} = \frac{V_{\mu\sigma} V_\sigma - V_{\sigma\sigma} V_\mu}{(V_\mu)^2} \tag{10.4}$$

where

$$V_{\mu\sigma}(\sigma, \mu) = \int_a^b u^{(2)}(\mu + \sigma x) x\, dF(x)$$

$$V_{\sigma\sigma}(\sigma, \mu) = -\int_a^b u^{(2)}(\mu + \sigma x) x^2\, dF(x).$$

Let

$$r(y) = -\frac{u''(y)}{u'(y)} > 0$$

be the Pratt-Arrow measure of absolute aversion, and

$$g(y) = \frac{r'(y)}{u(y)}$$

be a function that measures the speed of increase in absolute risk aversion. The following proposition establishes the relationship between risk aversion and the shape of indifference curve.

**Proposition 10.1** *We have $\frac{\partial S(\sigma, \mu)}{\partial \mu} \le (=, \le) 0$ for all $\mu$ and for all $\sigma \ge 0$ if and only if $u(\mu + \sigma x)$ displays decreasing (constant, increasing) absolute risk aversion.*

Risk aversion ($r > 0$) implies $\frac{d^2\mu}{d\sigma^2} > 0$. As the quantity $r(y)$ is actually $-(\log u')'$, $r > 0$ means that $\log u'$ is decreasing. In other words, $u'$ is the exponential of a decreasing function. Decreasing or constant absolute risk aversion implies that the indifference curve slope respectively declines or stays constant with an increase in $\mu$, given $\sigma > 0$, that is:

$$S_\mu < (=) 0 \quad \text{for} \quad g < (=) 0 \quad \text{and} \quad \sigma > 0.$$

The indifference curve slope increases with an increase in $\sigma$, given $\mu$, if absolute risk aversion is non-increasing.

**Proposition 10.2** *For non-decreasing continuous and piece-wise differentiable utility function $u(x)$ and $Y = \mu + \sigma \cdot X$ defined in (10.1), we have $V_\mu(\sigma, \mu) \geq 0$.*

**Proposition 10.3** *For utility $u(x)$ satisfying $u^{(1)}(x) \geq 0$ and and $Y = \mu + \sigma \cdot X$ defined in (10.1), we have*

1. *If $u^{(2)}(x) \leq 0$, then $V_\sigma(\sigma, \mu) \leq 0$; and*

2. *If $u^{(2)}(x) \geq 0$, then $V_\sigma(\sigma, \mu) \geq 0$.*

Combining the above, we have

**Property 10.4** *If the distribution function of the return or investment with mean $\mu$ and variance $\sigma^2$ belongs to a location-scale family, then*

1. *For any $u \in \mathcal{U}_1$,*
$$V_\mu(\sigma, \mu) > 0;$$

2. *For any $u$ in $\mathcal{U}$ such that $\mathcal{U}_2^A \subseteq \mathcal{U} \subseteq \mathcal{U}_2^{EA}$ and $u^{(1)}$ exists,*
$$V_\sigma(\sigma, \mu) \leq 0;$$

3. *For any $u$ in $\mathcal{U}$ such that $\mathcal{U}_2^{SA} \subseteq \mathcal{U} \subseteq \mathcal{U}_2^{ESA}$ and $u^{(1)}$ exists,*
$$V_\sigma(\sigma, \mu) < 0;$$

4. *For any $u$ in $\mathcal{U}$ such that $\mathcal{U}_2^D \subseteq \mathcal{U} \subseteq \mathcal{U}_2^{ED}$ and $u^{(1)}$ exists,*
$$V_\sigma(\sigma, \mu) \geq 0.$$

5. *For any $u$ in $\mathcal{U}$ such that $\mathcal{U}_2^{SD} \subseteq \mathcal{U} \subseteq \mathcal{U}_2^{ESD}$ and $u^{(1)}$ exists,*
$$V_\sigma(\sigma, \mu) > 0.$$

It is important to study the convexity of the indifference curve $C_\alpha$ with the restriction of $V(\sigma, \mu) \equiv \alpha$. Under the constraint of $(\sigma, \mu) \in C_\alpha$, the following proposition for $\frac{\partial S(\sigma, \mu)}{\partial \sigma}$ can be obtained:

**Proposition 10.5** *If the distribution function of the return with mean $\mu$ and variance $\sigma^2$ belongs to a location-scale family and for any utility function $u$, if $u^{(1)} > 0$, then the indifference curve $C_\alpha$ can be parameterized as $\mu = \mu(\sigma)$ with slope*
$$S(\sigma, \mu) = -\frac{V_\sigma(\sigma, \mu)}{V_\mu(\sigma, \mu)}.$$

*In addition,*

1. If $u^{(2)} \leq 0$, then the indifference curve $\mu = \mu(\sigma)$ is an increasing and convex function of $\sigma$,

2. If $u^{(2)} = 0$, then the indifference curve $\mu = \mu(\sigma)$ is a horizontal function of $\sigma$, and

3. If $u^{(2)} \geq 0$, then the indifference curve $\mu = \mu(\sigma)$ is a decreasing and concave function of $\sigma$.

One could rewrite Proposition 10.5 into the following property:

**Property 10.6** *If the distribution function of the return or investment with mean $\mu$ and variance $\sigma^2$ belongs to a location-scale family, then*

1. *For any risk averter with utility function $u$, his/her indifference curve $\mu = \mu(\sigma)$ is an increasing and convex function of $\sigma$,*

2. *For any risk neutral investor with utility function $u$, his/her indifference curve $\mu = \mu(\sigma)$ is a horizontal function of $\sigma$, and*

3. *For any risk seeker with utility function $u$, his/her indifference curve $\mu = \mu(\sigma)$ is a decreasing and concave function of $\sigma$.*

The following property could then be obtained:

**Property 10.7** *For the investments $Y$ and $X$ with means $\mu_Y$ and $\mu_X$ and variances $\sigma_Y^2$ and $\sigma_X^2$ such that $Y = p + qX$, in the $\sigma - \mu$ indifference curves diagram,*

1. *If $(\mu_Y, \sigma_Y)$ is in the north of $(\mu_X, \sigma_X)$, then $Y$ is preferred to $X$ for any FSD risk investor;*

2. *If $(\mu_Y, \sigma_Y)$ is in the north-west of $(\mu_X, \sigma_X)$, then $Y$ is preferred to $X$ for any SASD risk averter;*

3. *If $(\mu_Y, \sigma_Y)$ is in the north-east of $(\mu_X, \sigma_X)$, then $Y$ is preferred to $X$ for any SDSD risk lover; and*

4. *If $(\mu_Y, \sigma_Y)$ is in the north-west, north or north-east of $(\mu_X, \sigma_X)$, then $Y$ is preferred to $X$ for any risk neutral investor.*

Note that the location-scale family includes a wide family of symmetric distributions (Student's $t$) ranging from Cauchy to normal inclusively such that

$$f(r\,;\,p) \propto \frac{1}{\sigma}\left\{1 + \frac{(r-\mu)^2}{k\sigma^2}\right\}^{-p}$$

where $k = 2p - 1$ and $p \geq 2$. This extends the family of normal distributions as studied previously in the mean-variance criterion.

The following is another interesting result in the comparison for different risk averters and risk seekers:

**Property 10.8** *If the distribution function of the return or investment with mean $\mu$ and variance $\sigma^2$ belongs to a location-scale family, in the loci $\{(\sigma, \mu)\}$ of the $\sigma - \mu$ diagram, then*

1. $S_1(\sigma, \mu) \geq S_2(\sigma, \mu)$ *for all* $(\sigma, \mu)$ *if and only if* $u_1(y)$ *is more risk averse than* $u_2(y)$ *for all* $y$.

2. $S_1(\sigma, \mu) \leq S_2(\sigma, \mu)$ *for all* $(\sigma, \mu)$ *if and only if* $u_1(y)$ *is more risk seeking than* $u_2(y)$ *for all* $y$.

## 10.1.2   A dilemma in using the mean-variance criterion

Consider the following example:

| Investment $X$ | | Investment $Y$ | |
|---|---|---|---|
| $X$ | $P(X)$ | $Y$ | $P(Y)$ |
| 1 | 0.8 | 10 | 0.99 |
| 100 | 0.2 | 1000 | 0.01 |
| Mean | 20.80 | | 19.9 |
| Standard Deviation | 39.60 | | 98.5 |

In this example, we have $\mu_X > \mu_Y$ and $\sigma_X < \sigma_Y$. First, according to the mean-variance criterion, $X$ dominates $Y$ at least for risk averters. In addition, if we plot the $X$ and $Y$ in the $\sigma - \mu$ diagram, we find that $X$ lies "further north-west" of $Y$. The indifference curves for risk neutral investors are horizontal and hence one may draw a conclusion that: $X$ is preferred to $Y$ for any risk neutral investor or risk averter.

Now, let's consider the utility function

$$u_0 = \log_{10}(x)$$

to illustrate this example. For this utility function, however, we find that

$$E_X(u_0) = 0.4 < E_Y(u_0) = 1.02 \quad \text{and} \quad u_0 \in \mathcal{U}_2^{SA}.$$

This contradicts the above conclusion and hence one may wonder about the validity of the findings in this chapter.

However, the above example cannot conclude the invalidity of the above theorems as this example violates the condition in the theorems. In the theorems, the mean and the standard deviation can be varied but the distribution functions of the return or investment must belong to the local-scale family. In this example, the distribution functions of $X$ and $Y$ do not belong to the local-scale family and hence the theorems cannot be applied.

## 10.2 LS family for $n$ random seed sources

This section discusses an LS family with general $n$ random seed sources. The extensions are carried out in two different directions. First, we allow for the possibility that the returns on the risky assets could be driven by more than one seed random variables, and we do not impose any distributional assumption on the seed random variables. Second, investors' preferences do not necessarily conform to an expected utility class.

The research has taken into consideration the perspectives of both economics and behavioral science regarding modern portfolio choice theory and asset pricing theory. On the one hand, the impact of multivariate seed variables on asset returns, in theory, provides more realistic and general framework for studying the randomness of asset returns. The returns on risky projects driven by a finite number of risky factors are not only a theoretical concept but are also commonly used in practice. For example, the relationship among the economic activities of the firm and the market returns on the debt and the equity of the firm are of interest to financial economists. Thus, there has been a renewed interest in the empirical relations between market return to equity and basic characteristics of the firm, such as the size, leverage, earnings yield, dividend-yield, book-to-market ratios and leverage of the firm. In addition, empirical evidence is in favor of a multi-factor rather than single-factor asset pricing model.

On the other hand, there exist substantial experimental and empirical evidence in decision theory, all leading to the rejection of the expected utility functions in representing investors' behaviors in the presence of risk. This last set of observations leads us to consider general non-expected utility functions.

For the purpose of this section, we shall focus on the class of betweenness utility functions. The betweenness utility function is obtained by replacing the independence axiom of von Neumann and Morgenstern's expected utility representation with the so-called betweenness axiom. The betweenness axiom has been found to be well supported through experimental evidence, and provides predictions that are in line with Allais' paradox. The usefulness of the betweenness utility functions for resolving the well-known empirical puzzles in finance has been overwhelming.

To understand the importance of the LS family, we need to go back to the classical mean-variance analysis and mutual fund separation theorem. It is well known that if investors rank risky portfolios through their mean and variance, two-fund separation holds, and the separating portfolios will be located on Markowitz's efficient frontier. In the presence of a risk-free asset, investors would optimally hold a combination of the risk-free asset and a common risky portfolio. A well-known question: How robust is the mutual-fund separation phenomenon for rational investors whose behaviors conform to some normality axioms such as those underlying von Neumann and Morgenstern's expected utility functions?

Seeking answers to this question has been an enduring task for academics in economics for more than 40 years. The research on this subject can be roughly divided into two branches, each following its own school of thought. The first branch of research focuses on investors' behavior assumptions. The second branch aims at identifying the distributional assumptions on asset returns that are sufficient for mutual fund separation for expected-utility investors. This section falls into this second school.

Academics have developed distributional conditions on asset returns to ensure that two-fund separation holds with the underlying separating portfolios common to all risk-averse expected-utility investors. Ross [137] showed that two-fund separation holds if and only if asset returns are driven by two common factors with residual returns (to the factors) having zero mean conditional on the linear span formed by the factors. This insight into two-fund separation is then further extended to general observations on $k$-fund separation.

In fact, the seed variable may follow any distribution. Though the location-scale expected utility functions defined over the LS family are summarized through two parameters, the location-scale expected utility functions, in general, differ from the classical mean-variance criterion. This is because the underlying expected utility functions defined over LS family can still be well-defined even when the seed random variable has no finite mean and variance.

In light of the above established findings, this section discusses the geometric and topological properties of the LS expected utility functions and non-expected utility functions defined over the LS family with $n$ random seed variables. Our results show that the indifference curves are convex upward for risk averters, and concave downward for risk-lovers, while keeping in mind that we are dealing with a wider $n$-dimensional LS family of distributions for general location-scale expected and non-expected utility functions.

We also discuss several well-defined partial orders and dominance relations defined over the LS family. These include the first- and second-order stochastic dominances (FSD, SSD), the mean-variance (MV) rule, and a newly defined location-scale dominance (LSD).

## 10.2.1 Location-scale expected utility

We assume that the returns of risky projects are driven by a finite number, say $n$, risky factors that are summarized by an $R^n$-valued random vector $X = [X_1, \ldots, X_n]$. Let $X_i$ be the $i$th factor, and let $X_{-i}$ be the vector of the factors excluding the $i$th factor. For notational simplicity, we may write $X = [X_i, X_{-i}]$ for all $i$. We assume that $E[X_i \mid X_{-i}] = 0$ for all $i$. The random vector $X$ satisfying these conditions is known to be a vector of random seeds. It is noted that the conditions for a zero conditional mean for the random seeds are satisfied when the random factors have zero mean and are independently distributed. So all observations and results derived later on in this paper are valid under the stronger assumption of independently distributed random factors.

Again, for any given vector, $X$, of random seeds, we let

$$\mathcal{D}_X = \{\mu + \sigma \cdot X : \mu \in \mathbb{R}, \ \sigma \in \mathbb{R}^n_+\} \qquad (10.5)$$

to denote the LS family induced by $X$. Here, $x \cdot y$ stands for the inner product defined on $R^n$, and $R^n_+$ represents the nonnegative cone of $R^n$. Later, we shall use $R^n_{++}$ to represent the positive cone with all entries to be strictly positive. Elements in $D$ can be interpreted as payoffs or returns associated with each of the risky projects. Here, all scaling factors $\sigma_i$ in $\sigma$ are restricted to be nonnegative. We write $\sigma \leq \sigma'$ whenever $\sigma_i \leq \sigma'_i$ for all $i$.

Investors are thus assumed to express their preferences over all random payoffs in $D$. Let $(\sigma, \mu) \to V(\sigma, \mu)$ be a location-scale utility function that represents investors' preference on $D$. A location-scale utility function $V$ is said to be located in Meyer's LS expected utility class if there exists a monotonic transformation of $V$, still denoted by $V$, and a well-defined utility index $u$ so that

$$V(\sigma, \mu) = \int_{\mathbb{R}^n} u(\mu + \sigma \cdot x) \, d\, F(x) \qquad (10.6)$$

for all $(\sigma, \mu) \in R^n_+ \times R$.

Here, $F$ is the distribution function of the $X$. Unless otherwise specified, we shall assume that the utility index $u \in C^1(\mathbb{R})$ is monotonic increasing and continuously differentiable, and $F$ satisfies Feller's property so that the LS expected utility function $V(\sigma, \mu)$ is well-defined and is continuously differentiable in $(\sigma, \mu)$.

Our first observation is that the monotonicity of the utility index $u$ implies and is implied by the monotonicity of the utility function $V(\sigma, \mu)$ with respect to the location variable $\mu$. Particularly, for any smooth utility index, $u$, with

$$V_\mu(\sigma, \mu) = \int_{\mathbb{R}^n} u'(\mu + \sigma \cdot x) \, d\, F(x).$$

Thus, we have

$$V_\mu(\sigma, \mu) \geq 0 \Leftrightarrow u'(\cdot) \geq 0.$$

The marginal expected utility with respect to each of the scaling factors that is summarized by the $n$-dimensional gradient function

$$V_\sigma(\sigma, \mu) \equiv \left[ \frac{\partial V(\sigma, \mu)}{\partial \sigma_i} \right]_{n \times 1}$$

can be computed as

$$V_\sigma(\sigma, \mu) = \left[ \int_{\mathbb{R}^n} u'(\mu + \sigma \cdot x) x_i \, d\, F(x) \right]_{n \times 1}.$$

The marginal expected utility may take either + or - signs, depending on the curvature/convexity of the utility index $u(\cdot)$. With $u'(\cdot) \geq 0$, we can easily prove the validity of the following relationships respectively for risk averse, risk loving and risk neutral investors:

$$x \mapsto u(x) \text{ is concave} \Rightarrow V_\sigma \leq 0;$$
$$x \mapsto u(x) \text{ is convex} \Rightarrow V_\sigma \geq 0;$$
$$x \mapsto u(x) \text{ is linear} \Rightarrow V_\sigma \equiv 0.$$

We only need to prove the validity of the first relationship as follows and the rest can be obtained similarly: The concavity of the utility index implies that, for all $x = (x_i, x_{-i}) \in R^n$, it must hold true that

$$u'(\mu + \sigma \cdot x) x_i \leq u'(\mu + \sigma_{-i} \cdot x_{-i}) x_i$$

and that

$$V_{\sigma_i}(\sigma, \mu) = 0. \qquad (10.7)$$

The converse to the above relationships are, in general, not valid. But for distribution function $F$ to have a finite second moment and to satisfy Feller's property, the validity of the converse relationships can be proved under fairly general conditions. For example, if we assume that there exists an $i$ such that $X_i$ has its support located within a bounded open interval $(a_i, b_i)$, and if the utility function is twice continuously differentiable, then we can readily prove the following:

$$V_\sigma \le 0 \;\; \Rightarrow \;\; u'' \le 0;$$
$$V_\sigma \ge 0 \;\; \Rightarrow \;\; u'' \ge 0;$$
$$V_\sigma = 0 \;\; \Rightarrow \;\; u'' \equiv 0.$$

Again, we only need to prove the validity of the first relationship as follows: Let $F_i$ be the marginal distribution function for $X_i$. Under Feller's condition, the marginal expected utility function $(\sigma, \mu) \to V_\sigma(\sigma, \mu) \le 0$ is continuous. So, we may set $\sigma_{-i} = \emptyset$ for $\sigma$ and for $V_{\sigma_i}(\sigma, \mu)$ so that, for all $\mu$ and $\sigma_i > 0$, we obtain

$$V_{\sigma_i}(\sigma_i, \mu) = \int_{a_i}^{b_i} u'(\mu + \sigma_i x) x \, dF_i(x) \le 0.$$

Since, by assumption, $E[X_i] = \int_{a_i}^{b_i} x \, dF_i(x) = 0$, and since $u(\cdot)$ is continuously differentiable on $R$, which have bounded first-order derivatives over $(a_i, b_i)$, we have

$$\lim_{x \to a_i} u'(\mu + \sigma_i x) \int_{a_i}^x y \, dF_i(y) = 0,$$
$$\lim_{x \to b_i} u'(\mu + \sigma_i x) \int_{a_i}^x y \, dF_i(y) = 0.$$

Applying integration by parts, we obtain

$$V_{\sigma_i}(\sigma_i, \mu) = -\sigma_i \int_{a_i}^{b_i} u''(\mu + \sigma_i x) \left( \int_{a_i}^x y \, dF_i(y) \right) dx.$$

This yields

$$\int_{a_i}^{b_i} u''(\mu + \sigma_i x) \left( \int_{a_i}^x y \, dF_i(y) \right) dx \ge 0, \forall \mu, \sigma_i > 0.$$

With $\int_{a_i}^{b_i} \int_{a_i}^{x} y dF_i(y) \, dx = -E\left[X_i^2\right] < 0$, by Feller's condition, we may set $\sigma_i \to 0_+$ to the above inequality to obtain $u''(x) \leq 0, \forall x \in R$.

The assumption on the existence of bounded support for the "only if" part of Meyer's Property 2 can, in fact, be further relaxed. The arguments prevail if there exists a random source, $X_i$, with finite second moment so that, for all $\mu$ and $\sigma_i > 0$, the following limits exist:

$$\lim_{x \to \infty} x \int_{-\infty}^{x} y dF_i(y) = 0 \qquad (10.8)$$

$$\lim_{x \to \pm\infty} u'(\mu + \sigma_i x) \int_{-\infty}^{x} y dF_i(y) = 0.$$

The second condition is valid if the utility index $u(\cdot)$ has bounded first-order derivatives. The first condition is to ensure that the improper integral

$$\int_{-\infty}^{\infty} \int_{-\infty}^{x} y dF_i(y) \, dx$$

is well-defined and takes a negative value. We have,

$$\int_{-\infty}^{\infty} \int_{-\infty}^{x} y dF_i(y) \, dx = -E\left[X_i^2\right]. \qquad (10.9)$$

It is easy to verify that the condition $\lim_{x \to +\infty} x \int_{-\infty}^{x} y dF_i(y) = 0$ is satisfied when $X_i$ is normally distributed with zero mean.

The above observations on the monotonicity of the LS expected utility functions defined over the $n$-dimensional LS family are expressed formally in

**Proposition 10.9** *Consider the expected utility function, $V(\sigma, \mu)$, on an $n$-dimensional LS family $D$. Letting $u \in C^1(\mathbb{R})$, we have*

1. $u' \geq 0 \Leftrightarrow V_\mu \geq 0$.

2. *If $u' \geq 0$, then it must hold true that*

$$x \mapsto u(x) \text{ is concave} \quad \Rightarrow \quad V_\sigma \leq 0;$$
$$x \mapsto u(x) \text{ is convex} \quad \Rightarrow \quad V_\sigma \geq 0;$$
$$x \mapsto u(x) \text{ is linear} \quad \Rightarrow \quad V_\sigma \equiv 0.$$

3. *If $u \in C^2(\mathbb{R})$ with $u' \geq 0$, and if there exists $i$ so that condition (10.8) is satisfied, then it must hold true that*

$$V_\sigma \leq 0 \Rightarrow u'' \leq 0;$$
$$V_\sigma \geq 0 \Rightarrow u'' \geq 0;$$
$$V_\sigma = 0 \Rightarrow u'' \equiv 0.$$

**Proposition 10.10** *Consider the expected utility function, $V(\sigma, \mu)$, on an n-dimensional LS family $D$. Letting $u \in C^1(\mathbb{R})$, we have*

$$(\sigma, \mu) \to V(\sigma, \mu) \text{ is concave} \Leftrightarrow u \text{ is concave}.$$

In this proposition, we intentionally drop the differentiability condition of the utility function. Literature used to include the condition that $u$ is twice continuously differentiable but one could easily construct examples to show that the concavity of $(\sigma, \mu) \to V(\sigma, \mu)$ does not necessarily imply that $u$ is twice continuously differentiable. This is true even if $V(\sigma, \mu) \in C^\infty (\mathbb{R}^n_{++} \times \mathbb{R})$ is infinitely, continuously differentiable.

### 10.2.2 Indifference curves

We further explore the topological properties for the indifference curves induced by an LS expected utility function $V$. For an arbitrary constant $a$, let

$$C_a \equiv \{(\sigma, \mu) \in \mathbb{R}^n_+ \times \mathbb{R} : V(\sigma, \mu) = a\} \tag{10.10}$$

be the indifference curve at utility level $a$. We can readily obtain the following observation with respect to the shapes of the indifference curves: The indifference curve $C_a$ is upward-sloping if $u$ is concave and downward-sloping if $u$ is convex as shown in the following proposition on the shapes of the indifference curves respectively for risk-averse, risk-loving and risk-neutral investors:

**Proposition 10.11** *Let $u \in C^1(\mathbb{R})$ be increasing and continuously differentiable. Then*

1. *The indifference curve $C_a$ is convex upward if $u$ is concave;*

2. *It is concave downward if $u$ is convex; and*

3. *It is horizontal if $u$ is a straight line.*

Note that the statements made in Proposition 10.11 about the shape and curvature of the indifference curves can be restated analytically in terms of the gradient and Hessian matrix of the indifference curve $\mu(\sigma)$, $\sigma \in R^n_+$. These, of course, require the standard regularity conditions on the utility function. For instance, by the implicit function theorem, the gradient vector $\mu_\sigma \equiv \left[\frac{\partial \mu}{\partial \sigma_j}\right]_{n \times 1}$ along the indifference curve is given by

$$\mu_\sigma = -\frac{V_\sigma (\sigma, \mu)}{V_\mu (\sigma, \mu)}, \forall (\sigma, \mu) \in C_a \tag{10.11}$$

which is nonnegative (nonpositive) when $u(\cdot)$ is concave (convex). We may further compute the Hessian matrix $\mu_{\sigma\sigma} \equiv \left[\frac{\partial^2 \mu}{\partial \sigma_k \partial \sigma_j}\right]_{n \times n}$ for the $\mu(\cdot)$-function. This, of course, requires the utility index to be twice continuously differentiable. For all $(\sigma, \mu) \in C_a$, we have:

$$\mu_{\sigma\sigma} = -\frac{[\mu_\sigma, I_n] \, H(\sigma, \mu) \, [\mu_\sigma, I_n]^T}{V_\mu(\sigma, \mu)} \tag{10.12}$$

in which $H(\sigma, \mu)$ is the $(n+1) \times (n+1)$ Hessian matrix for $V(\sigma, \mu)$, and $I_n$ is the $n \times n$ unit matrix. From this expression, we see that concavity (convexity) of the utility index $u(\cdot)$ implies negative (positive) semi-definiteness of the Hessian matrix $H(\sigma, \mu)$. With $V_\mu > 0$, the latter, in turn, implies $\mu_{\sigma\sigma}$ to be positive (negative) semi-definite.

By virtue of the above observations, we obtain the following analytic version of Proposition 10.11:

**Corollary 10.12** *Let* $u \in C^2(\mathbb{R})$ *with* $u' > 0$. *Along the indifference curve* $\mu(\sigma), \sigma \in R_+^n$, *it must hold true that*

$$u'' \leq 0 \quad \Rightarrow \quad \mu_\sigma \geq 0, \mu_{\sigma\sigma} \geq 0;$$
$$u'' \geq 0 \quad \Rightarrow \quad \mu_\sigma \leq 0, \ \mu_{\sigma\sigma} \leq 0;$$
$$u'' \equiv 0 \quad \Rightarrow \quad \mu_\sigma = 0, \ \mu_{\sigma\sigma} \equiv 0.$$

## 10.2.3 Expected versus non-expected LS utility functions

This section introduces a class of LS utility functions that are not necessarily located in the expected utility class. To motivate our effort for considering a general class of non-expected utility functions, we raise and discuss the following so-called "inverse problem" with respect to Meyer's location scale (LS) expected utility functions: for any arbitrarily given utility function $V(\sigma, \mu)$ defined over the LS family $D$, which may satisfy all desirable topological properties (such as monotonicity and concavity), we wonder whether $V(\sigma, \mu)$ admits an expected utility representation or not.

Upon a negative answer to the inverse problem as illustrated below, we introduce a class of non-expected utility functions over the LS family admitting all desirable properties that are possessed by the standard LS expected utility functions. We extend here the betweenness utility functions to random variables belonging to the LS family.

## An inverse problem

The inverse problem raised above can be formulated as the following mathematical problem:

**Problem 10.13** *For any given utility function $V(\sigma, \mu) \in C(\mathbb{R}^n \times \mathbb{R})$ on the LS family D, is there a utility index $u \in C(\mathbb{R})$ and a monotonic increasing function $\varphi \in C(\mathbb{R})$ such that*

$$\varphi(V(\sigma, \mu)) = \int_{\mathbb{R}^n} u(\mu + \sigma \cdot x) \, dF(x) \qquad (10.13)$$

*for all $(\sigma, \mu) \in R_+^n \times R$?*

Here, we take into account the ordinal property of the expected utility representation. It is well known that for all arbitrary monotonic increasing functions $f$, $E[u(x)]$ and $f(E[u(x)])$ represent the same preference ordering. In light of the LS utility function, $V(\sigma, \mu) \in C(\mathbb{R}^n \times \mathbb{R})$ admits an expected utility representation if there exists a monotonic transformation of $V(\sigma, \mu)$ so that $\varphi(V(\sigma, \mu))$ admits an expected utility representation.

The following observation can be readily proved in working toward an answer to this inverse problem:

**Proposition 10.14** *The inverse problem has a solution if and only if there exists a monotonic increasing function $\varphi \in C(\mathbb{R})$ such that*

$$V(\sigma, \mu) = \varphi^{-1}\left(\int_{\mathbb{R}^n} \varphi(V(\emptyset, \mu + \sigma \cdot x)) \, dF(x)\right) \qquad (10.14)$$

*for all $(\sigma, \mu) \in R_+^n \times R$; in particular, if a solution exists, the utility index is given by $u(x) = \varphi(V(\emptyset, x))$.*

Assuming further that $(\sigma, \mu) \to V(\sigma, \mu)$ is continuously differentiable, from the above proposition, we can readily identify

$$V_\sigma(\emptyset, x) = \emptyset, \forall x \in \mathbb{R}, \qquad (10.15)$$

as a necessary condition for the existence of a solution to the inverse problem. In fact, let $\varphi \in C^1(\mathbb{R})$ be a solution to Equation (10.14). We may compute the utility gradient with respect to the scaling variables $\sigma$ on both sides of Equation (10.14), and set $\sigma \to \emptyset$ to obtain

$$\varphi'(V(\emptyset, \mu)) V_\sigma(\emptyset, \mu) = \emptyset, \forall \mu \in \mathbb{R}.$$

Since, by assumption, $\varphi' > 0$, we conclude that $V_\sigma(\emptyset, \mu) = \emptyset, \forall \mu \in R$.

The necessary condition $V_\sigma(\emptyset, x) = \emptyset \ \forall x \in R$ for the existence of a solution to the inverse problem is, in general, too weak to constitute a sufficient condition for the existence of a solution. So, in general, we might expect a negative answer to the inverse problem raised above; that is, it would not admit an LS expected utility representation for all $(\sigma, \mu)$-preferences. The next section studies a class of non-expected utility functions defined over the LS family.

## Location-scale non-expected utility

In light of the above example for a negative answer to the inverse problem for LS expected utility representation, we consider a general class of non-expected utility functions defined over the LS family. Although these utility functions may not necessarily admit some expected utility representations, the underlying behavior assumptions are well understood in decision theory and economics. The treatment below is based on the betweenness utility functions, though much of the analysis can be readily extended to the broad class of Gateaux differentiable utility functions.

**Definition 10.15** *A utility function $u$ is said to be in the betweenness class if there exists a betweenness function $H : R \times R \to R$, which is increasing in its first argument, and is decreasing in its second argument, and $H(x,x) \equiv 0$ for all $x \in R$, such that, for all $X$, $u(X)$ is determined implicitly by setting $E[H(X, u(X))] = 0$. The corresponding LS betweenness utility function $V : R_+^n \times R \to R$ on the LS family $D_X \equiv \{\mu + \sigma \cdot X : \mu \in R, \sigma \in R_+^n\}$ is, accordingly, defined by setting $V(\sigma, \mu) = U(\mu + \sigma \cdot X)$ as a unique solution to*

$$\int_{\mathbb{R}^n} H(\mu + \sigma \cdot x, V(\sigma, \mu)) \, dF(x) = 0 \qquad (10.16)$$

*for all $(\sigma, \mu)$.*

The betweenness utility function is known to be obtained by weakening the key independence axiom underlying the expected utility representation with the so-called betweenness axiom. The betweenness utility function is said to display risk aversion if, for all $X$, $u(X) \leq U(EX)$, or, equivalently, $E[H(X, u(EX))] \leq 0$. It is well known that the betweenness utility function displays risk aversion if and only if the betweenness function is concave in its first argument.

The following result summarizes the properties of the LS betweenness utility function:

**Proposition 10.16** *Let $H \in C^1 (\mathbb{R} \times \mathbb{R})$ be a betweenness function. We have*

1. $\mu \to V(\sigma, \mu)$ *increasing; moreover,*

2. *If $H$ is concave in its first argument, then $\sigma \to V(\sigma, \mu)$ must be monotonic decreasing, and $(\sigma, \mu) \to V(\sigma, \mu)$ must be quasi-concave; and*

3. *If $H$ is jointly concave in both arguments, then $(\sigma, \mu) \to V(\sigma, \mu)$ must be concave in both arguments.*

Similar to LS expected utility functions, the monotonicity of a betweenness utility function with respect to $\mu$ and $\sigma$ implies the monotonicity of the corresponding indifference curves; and the concavity of the utility function $(\sigma, \mu) \to V(\sigma, \mu)$ implies the quasi-concavity of the utility function, while the latter is equivalent to the convexity of the indifference curve $C_a$. Keeping in mind the equivalence between the concavity of $x \to H(x, v)$ and the risk aversion of the betweenness utility function, the relevance of the risk aversion and its implications for the shape of the indifference curve for this betweenness LS class can be readily established. Similar observations can be made when the betweenness utility functions display risk-loving or risk-neutrality, keeping in mind that the betweenness utility function displays risk-loving (risk-neutrality) if the betweenness function $H$ is convex (linear) in its first argument. We may thus state without proof the following property:

**Corollary 10.17** *Let $H \in C^1 (\mathbb{R} \times \mathbb{R})$ be a betweenness function. We have*

1. *The indifference curve $C_a$ is convex upward if the corresponding betweenness utility function displays risk aversion;*

2. *The indifference curve $C_a$ is concave downward if the corresponding betweenness utility function displays risk-loving; and*

3. *The indifference curve $C_a$ is horizontal if the corresponding betweenness utility function displays risk-neutrality.*

As an aside, the expected utility functions form a subclass to the class of betweenness utility functions. In fact, the standard expected utility function certainty equivalent induced by utility index $u$ is obtained by setting $H(x, y) = u(x) - u(y)$.

## 10.2.4    Dominance relationships over the LS family

This section discusses several useful stochastic dominance relationships as partial orders defined over the LS family. These include FSD and SSD, in addition to a newly defined location-scale dominance (LSD) relationship defined over the LS family. These dominance relationships are known to admit no utility representations. Their properties over the LS family can be, nevertheless, readily studied. Note that the LSD defined in our paper differs from the MV criterion used in the literature, more information on which can be found in Definition 10.18 below.

### Location-scale dominance

To link stochastic dominance efficient sets with the mean-variance efficient set, we introduce the following LS dominance relationship. Recall that stochastic dominance sets are defined as follows. For any pair of real-valued random variables $Y$ and $Y'$ with distribution functions, $F_Y, F_{Y'}$, respectively, we write $(Y, Y') \in D_{FSD}$ and $(Y, Y') \in D_{SSD}$ if the corresponding dominance relationships do not exist between the two random variables. $D_{FSD}$ and $D_{SSD}$ are respectively known as FSD- and SSD-efficient sets.

**Definition 10.18** *Let $X$ be an $R^n$-valued random variable with zero mean and conditional mean $E[X_i \mid X_{-i}] = 0$ for all $i$. Let $D$ be an LS family generated from $X$. For all $Y = \mu + \sigma \cdot X$ and $Y' = \mu' + \sigma' \cdot X$, we say that $Y$ dominates $Y'$ according to the LS rule if $\mu \geq \mu'$ and $\sigma \leq \sigma'$. We write $Y \succeq_{LS} Y'$ whenever $Y$ dominates $Y'$ according to the LS rule. Otherwise, we write $(Y, Y') \in D_{LSD}$ if $Y$ and $Y'$ do not dominate each other in the sense of LSD. The set $D_{LSD}$ is referred to as the LS-efficient set.*

For $n = 1$, when the random seed $X$ has zero mean and a finite second moment, the LS rule defined on $D$ is equivalent to Markowitz's MV rule defined over the family [104]. The equivalence breaks down when $X$ does not have a finite second moment, for which the variance of $X$ does not exist; yet, the LS expected utility functions are still well-defined for all bounded continuous utility indexes.

For random payoffs belonging to the high-dimensional $(n > 1)$ LS family, the equivalence between the LS rule and the MV criterion breaks down even when the seeds random variable $X$ have finite second moments. In fact, with

$$\sigma[Y] = (\sigma^{\mathsf{T}} \Sigma_X \sigma)^{1/2} \text{ and}$$
$$\sigma[Y'] = ((\sigma')^{\mathsf{T}} \Sigma_X \sigma')^{1/2}$$

where $\Sigma_X$ is the positive variance-covariance matrix for the vector, $X$, of random seeds, we have: $\sigma \geq \sigma'$ implies but is not implied by $\sigma[Y] \geq \sigma[Y']$. Accordingly, for LS expected utility functions, monotonicity in $\sigma$ does not necessarily imply monotonicity in $\sigma[Y]$. More generally, as a direct consequence of Proposition 10.9, one could easily obtain the following observation on LSD defined over an LS family:

**Proposition 10.19** *For $n = 1$, let $Y$ and $Y'$ belong to the same LS family $D_X$. Suppose $X$ has zero mean and finite second moment. Then $Y$ dominates $Y'$ according to the MV rule if and only if $Y \succ_{LS} Y'$. Moreover, for $n > 1$, for all $Y$ and $Y'$ belonging to the same LS family $D_X$, we have*

$$Y \succeq_{LS} Y' \Rightarrow E[u(Y)] \geq E[u(Y')]$$

*for all increasing and concave utility indexes $u \in C^1(\mathbb{R})$.*

## FSD, SSD and LSD

The relationships among the three forms of dominance relationships, namely, FSD, SSD and LSD, defined over an $n$-dimensional LS family can be readily studied. The following proposition summarizes the findings:

**Proposition 10.20** *Let $D$ be an LS family induced by an $n$-dimensional vector, $X$, of seed random variables with bounded supports. We have:*

1. $D_{SSD} \subset D_{FSD}$

2. $D_{SSD} \subset D_{LSD}$

3. $D_{LSD} - D_{FSD} \neq \emptyset$

4. $D_{FSD} - D_{LSD} \neq \emptyset$

So, we see that both notions of FSD and LSD relations are stronger than that of SSD. Part 3 of Proposition 10.20 suggests that there is no specific logical relationship between FSD and LSD. The LSD neither implies nor is implied by the FSD.

## 10.3    Elasticity of risk aversion and trade

### 10.3.1    International trade and uncertainty

In the last decade exchange rates of the major industrial countries have shown substantial volatility. Exchange rate uncertainty became a concern of international firms and, therefore, affected and is affecting international trade and foreign investments, although empirical findings are mixed; empirical studies regarding the relationship between exchange rate risk and international trade flows do not necessarily confirm the intuition, that higher exchange rate volatilities lead to a reduction in international trade. In this section, we give an explanation why a positive link between exchange rate risk and exports is possible from a portfolio theoretical point of view. We apply the mean-standard deviation approach for a scale and location family of probability distributions as to examine an exporting firm that is subjected to revenue risk without hedging opportunities.

In order to study the decision problem of a risk-averse competitive exporting firm under exchange rate risk, we use a basic model from the literature. The firm produces the quantity $Q$ of a final good at increasing marginal cost: $C'(Q) > 0, C''(Q) > 0$. The foreign exchange rate $\tilde{e}$ is random. The commodity price $P$, denominated in foreign currency is given. The objective is to maximize the expected value of a von Neumann-Morgenstern utility function of profit $u(\Pi)$, with $u' > 0$ and $u'' < 0$. $\tilde{\Pi} = \tilde{e}PQ - C(Q)$ denotes risky profit of the exporting firm. Hence, the export decision problem reads:

$$\max_Q \; Eu(\tilde{e}PQ - C(Q)),$$

where $E$ denotes the expectation operator. It is well known that under some conditions, the expected utility decision problem can be transformed into the mean ($\mu$)-standard deviation ($\sigma$) framework. That is to say,

(i) There exists a function $V(\mu, \sigma)$ such that

$$V(\mu, \sigma) = Eu(\tilde{\Pi}) = \int_{-\infty}^{\infty} u(\Pi) \, f_{\tilde{\Pi}}(\Pi; \mu, \sigma) \, d\Pi,$$

where $\mu$ denotes the mean and $\sigma$ the standard deviation of risky profit for the pdf $f_{\tilde{\Pi}}$; and

(ii) The function $V$ satisfies the following properties, where $V_x = \partial V/\partial x$ is the partial derivative: $V_\mu > 0$, $V_{\mu\mu} \leq 0$, $V_\sigma < 0$, $\sigma > 0$ and $V_\sigma(\mu, 0) = 0$; the partial derivatives $V_{\sigma\sigma}$ and $V_{\mu\sigma}$ exist and $V$ is a strictly concave function. The indifference curves are upward sloping and concave in the $(\mu, \sigma)$-space.

## 10.3.2 LS parameter condition and elasticity

Let us start by defining the so-called location and scale (LS) parameter condition of a probability distribution. This framework applies for our model of the exporting firm, since the random profit of the firm $\tilde{\Pi}$ is a positive linear transformation of the random foreign exchange rate $\tilde{e}$.

**Definition 10.21 (Seed random variable)** *Let $\tilde{\eta}$ be the seed random variable with zero mean and unit standard deviation. The nondegenerate random foreign exchange rate $\tilde{e}$ is defined to be*

$$\tilde{e} = \mu_{\tilde{e}} + \sigma_{\tilde{e}}\,\tilde{\eta}, \ \ with\ \sigma_{\tilde{e}} > 0.$$

**Definition 10.22** *By using Definition 10.21, let $\tilde{e}_i = \mu_i + \sigma_i\tilde{\eta}$, $\sigma_i > 0$, for $i = 1, 2$.*

1. *$E(\tilde{e}_i) = \mu_i$ and $Var(\tilde{e}_i) = \sigma_i^2$ for $i = 1, 2$.*

2. *The probability distributions of $\tilde{c}_1$ and $\tilde{e}_2$ are area preserving under location-scale transformations such that $\mathrm{Prob}(\tilde{e}_1 \leq e_1) = \mathrm{Prob}(\tilde{e}_2 \leq e_2)$ if $e_2 = \mu_2 + \frac{\sigma_2}{\sigma_1}(e_1 - \mu_1)$.*

3. *We say that $\tilde{e}_2$ is more risky than $\tilde{e}_1$ if $\mu_2 = \mu_1 > 0$ and $\sigma_2 > \sigma_1 > 0$. (Increase in risk.)*

4. *We say that $\tilde{e}_2$ is more expected than $\tilde{e}_1$ if $\mu_2 > \mu_1 > 0$ and $\sigma_2 = \sigma_1 > 0$. (Increase in mean.)*

Note that the pdf of $\tilde{e}$ is a function of $\mu_{\tilde{e}}$, $\sigma_{\tilde{e}}$ and the pdf of $\tilde{\eta}$. A change in expected value and/or standard deviation of the foreign exchange rate can now be introduced within a comparative static analysis of export production.

We now present the elasticity of risk aversion in the mean-standard deviation approach. This concept allows for a distinct investigation of risk and expectation effects on the export decision of the firm.

Let $S = -V_\sigma/V_\mu$, with $\sigma > 0$. Then $S$ is positive and denotes the marginal rate of substitution between expectation $\mu$ and risk $\sigma$, i.e., the positive slope of the indifference curve. Therefore, $S$ can be interpreted as a measure of risk aversion within the mean-standard deviation framework.

**Definition 10.23 (Standard deviation elasticity)** *The elasticity of risk aversion with respect to the standard deviation of the firm's risky profit is stated by*

$$\varepsilon_\sigma = -\frac{\partial \ln S}{\partial \ln \sigma}, \ \sigma > 0.$$

Note that $\varepsilon_\sigma$ indicates the percentage change in risk aversion over the percentage change in profit standard deviation, the profit mean being fixed.

**Definition 10.24 (Mean elasticity)** *The elasticity of risk aversion with respect to the mean of the firm's risky profit is given by*

$$\varepsilon_\mu = \frac{\partial \ln S}{\partial \ln \mu}, \ \mu > 0.$$

Note that $\varepsilon_\mu$ indicates the percentage change in risk aversion over the percentage change in profit mean, the profit standard deviation being fixed.

In the following we examine the relationship between trade and a change in the expected value of the foreign exchange rate and its standard deviation, respectively. The relationships are investigated by using the introduced elasticity measures.

## 10.3.3   Risk and mean effects on international trade

We model an increase in exchange rate risk by augmenting the standard deviation $\sigma_{\tilde{e}}$, holding the mean $\mu_{\tilde{e}}$ constant.

**Proposition 10.25 (Trade and risk)** *Suppose the exchange rate becomes more risky, i.e., the standard deviation of the foreign exchange rate increases. Then, the firm's export decreases (remains constant, increases) if and only if the standard deviation elasticity of risk aversion is less than (equal to, greater than) unity.*

Now we model an increase in exchange rate expectation by augmenting the mean $\mu_{\tilde{e}}$, holding the standard deviation $\sigma_{\tilde{e}}$ constant.

**Proposition 10.26 (Trade and expectation)** *Suppose a higher exchange rate becomes more expected, i.e., the mean of the foreign exchange rate increases. If the mean elasticity of risk aversion is less than or equal to unity, then the firm's export increases.*

Whether or not there is a risk effect on international trade depends upon the magnitude of the profit standard deviation elasticity of risk aversion. With unit elastic risk aversion there is no risk effect. That is, the firm's optimum export production remains unchanged although the exchange rate becomes more risky. If risk aversion is (in)elastic then the firm will (diminish) extend its export production. Hence, the elasticity measure provides a distinct answer to the question of how a change in exchange rate risk affects international trade. With this result in mind contradicting empirical findings are not unlikely when the elasticity is highly unstable over time.

The intuition that a devaluation of the home currency stimulates export production of the firm is not true in general. A sufficient condition which supports this intuition is that we have unit elastic or inelastic risk aversion with respect to the profit mean. Furthermore, there exist a critical elasticity level, $1/R$, which implies that there is no mean effect at all on international trade. Sufficient elastic risk aversion, i.e., $\varepsilon_\mu > 1/R$ induces the firm to lessen export production although the expected value of the foreign exchange rate increases. Again, contradicting empirical results that build on different samples should not be that surprising.

Note that $R(S)$ can be interpreted as a measure of relative (absolute) risk aversion within the mean-standard deviation approach.

## 10.4 Income inequality

We close our series of three chapters devoted to applications of stochastic dominance with the area of welfare economics. Unlike other illustrations of applications, we simply point out the reasons why SD has become a useful statistical tool in the studies of income inequality. The focus is on the striking analogy between financial risk measures (Chapter 4) and inequality measures, and on the role stochastic dominance plays in these two areas.

In social welfare, it is of interest to compare income distributions in different geographic regions or at different times. The main interest is on income inequality. The concept of inequality (in a specific context of income) is well understood in a common sense. In quantitative economics, it is desirable to be able to quantify inequality. Specifically, we wish to "measure" the inequality in an income distribution $F_X$ of a population $X$, i.e., to assign a numerical value $I(X)$ or $I(F_X)$ to $F_X$. This of course reminds us of numerical risk measures in financial economics! Note that, the study of inequality measures preceded that of risk measures. As we will see, the approach to financial risk measures is exactly the same as that of inequality measures in welfare economics, although this is not stated as such in the literature!

Consider two populations $X, Y$ with distribution functions $F, G$, respectively. Here, $X, Y$ are (finite) populations of individual incomes. We wish to compare $F, G$ in terms of inequality, i.e., whether there is less inequality in $F$ than in $G$. From a social welfare viewpoint, if there is less inequality in $F$ than in $G$ then $F$ is better than $G$. It all boils down to define appropriate orders on random variables, or more specifically on their distribution functions. This is a familiar situation in statistics, for example, to compare two estimators of a given population parameter, such as the population mean, we compare them in terms of the concept of error (and more generally, in terms of other desirable properties, such as consistency) by using, for example one type of error measure such as the mean squared error. This numerical measure provides a total ordering of estimators. In parameter estimation problems, we talk about loss functions in the sense of error. For example, the mean squared error is defined by using the square loss function. The mean squared error is the expected loss with respect to this choice of loss function. In other words, we quantify the risk of an estimator by the expected loss, and use this concept of numerical risk measure to compare estimators.

However, there is a delicate difference, both in the context of financial risk assessment and inequality measurement, namely we are not estimating some given population parameter, but rather trying to come up with some reasonable numerical value to characterize a distribution function. In other words, the problem is to define a population parameter to represent numerically the concept we have in mind which is the risk of losing money in an investment, or the inequality in an income distribution. This is the problem of quantitative modeling of a linguistic concept. Of course, even for finite physical populations of individuals, after identifying an appropriate population parameter, we will rely on statistical inference for obtaining it. We are not comparing estimators of that population parameter (of course, this is done within the standard procedures of statistics) but rather comparing various distribution functions using that specified population parameter, such as the variance $V(X)$. In this example, it is the variance (of the loss variable) which represents the risk of an investment, or a measure of inequality in an income distribution.

Now consider the variance of an income distribution as an inequality index, just as in financial risk analysis. The question is: Does the variance really capture the concept of inequality? Since $V(\lambda X) = \lambda^2 V(X)$, it violates an obvious requirement of inequality, namely scale independence. The scale independence property is intuitive: the inequality measure should be invariant with respect to uniform proportional changes in individuals' income. The defect of this inequality measure is similar to the critics of the variance, or of the value-at-risk as risk measures. We do have in mind desirable properties of an inequality measure. Cowell [27] went on to propose an "axiomatic approach" to inequality measures, by requiring that an inequality measure should satisfy a reasonable list of properties. This is similar to the concept of coherent risk measures. These properties are

1. Scale independence

2. Pigou-Dalton transfer principle

3. Principle of population

4. Symmetry

5. Decomposability

A class of inequality measures satisfying the above "axioms" consists of generalized entropy inequality measures, such as the well-known Gini coefficient. The analogy with risk measures is the class of Choquet integral risk measures with respect to 2-alternating capacities.

Just like risk measures, ranking of income distributions according to a reasonable inequality measure depends on that choice of the inequality measure, even the inequality measure is required to be defined solely in terms of the distributions, a property similar to "law invariance" of risk measures. An ideal approach is not ranking distributions according to some choice of a population parameter, but according to the distributions themselves. But how should we rank distributions in the context of income inequality? Well, in welfare economics, the counterpart of utility functions is social welfare functions (!) and, similarly, leads to stochastic dominance orders. While stochastic dominance rules avoid subjective choices of numerical inequality measures (a property referred to as robustness), they are partial orders, and as such, it is not always possible to rank distributions. In practice, we should start with stochastic dominance rules. If they fail, then we call upon inequality measures. Thus, inequality measures should be consistent with respect to stochastic dominance rules, just like the case of risk measures.

The interpretation of stochastic dominance order in income inequality is this. The value $F_X(x) = P(X \leq x)$ is the proportion of individuals, in the population $X$, with incomes below the income level $x$. Thus, for two populations $X$ and $Y$, if $F_X(.) \leq F_Y(.)$, i.e., $X \preceq_1 Y$, then for all income levels $x$, $F_X(x) \leq F_Y(x)$ meaning that $X$ is "better" than $Y$ as far as inequality is concerned. It is found in the studies of income inequality that all three stochastic dominance orders are appropriate in ranking income distributions with various social welfare functions.

As a final note, here is the relation between stochastic dominance and the usual ranking of income inequality by the Lorenz curve. First, for a population $X$ with $0 < EX < \infty$, the Lorenz curve of $X$ (or of its distribution function $F_X$) is

$$L_X : [0,1] \to R \text{ where } L_X(t) = (1/EX) \int_0^t F_X^{-1}(x)dx.$$

The Lorenz order expresses the fact that "$X$ is at least as unequal as $Y$" when $L_X(.) \leq L_Y(.)$.

A more general way to rank distributions is the generalized Lorenz curve, defined by Shorrocks [147] as

$$GL_X(t) = \int_0^t F_X^{-1}(x)dx.$$

It can be shown that generalized Lorenz order (i.e., $X$ dominates $Y$ when $GL_X(.) \leq GL_Y(.)$) is precisely $\succeq_2$. Note that the Gini coefficient of $X$ is $1 - 2\int_0^1 L_X(x)dx$.

## 10.5   Exercises

1. Prove Proposition 10.2.

2. Prove Proposition 10.3.

3. Prove Property 10.6.

4. Prove Equation 10.7.

5. Prove Equation 10.9.

6. Prove Proposition 10.10.

7. Prove Proposition 10.11.

8. Prove Proposition 10.14.

9. In Section 10.2.3, the necessary condition $V_\sigma\left(\emptyset, x\right) = \emptyset, \forall x \in \mathbb{R}$ for the existence of a solution to the inverse problem is, in general, too weak to constitute a sufficient condition for the existence of a solution. Construct an example to get a negative answer to the inverse problem.

10. Prove Proposition 10.16.

11. For all arbitrary random variables $X$ and $Y$, prove that

$$X \succeq_1 Y \Leftrightarrow E\left[u\left(X\right)\right] \geq E\left[u\left(Y\right)\right]$$

for all bounded and increasing utility indices $u \in C^1\left(\mathbb{R}\right)$.

12. A c.d.f. $F\left(\cdot\right)$ is said to satisfy the *asymptotic condition* if

$$1 - F\left(x\right) = o\left(\frac{1}{x}\right) \quad \text{and} \quad F\left(x\right) = o\left(\frac{1}{x}\right)$$

as $x \to +\infty$ and $-\infty$, respectively. A c.d.f. $F\left(\cdot\right)$ is said to satisfy the *integrability conditions* if the improper integrals

$$\int_{-\infty}^{0} F\left(x\right) dx \geq 0 \quad \text{and} \quad \int_{0}^{\infty} \left[1 - F\left(x\right)\right] dx \geq 0$$

exist and take finite values. Suppose $X$ and $Y$ with c.d.f.'s satisfy both the asymptotic and the integrability conditions. Prove that

$$X \succeq_2 Y \Leftrightarrow E\left[u\left(X\right)\right] \geq E\left[u\left(Y\right)\right]$$

for all increasing and concave utility indices $u \in C^2\left(\mathbb{R}\right)$ with bounded first-order derivatives.

13. From Section 10.2.4, we know that, for LS expected utility func tions, monotonicity in $\sigma$ does not necessarily imply monotonicity in $\sigma\left[Y\right]$. Construct an example for this observation.

14. Prove Proposition 10.20.

15. Prove Proposition 10.25.

16. Prove Proposition 10.26.

# Appendix

# Stochastic Dominance Tests

## A.1   CAPM statistics

The Sharpe-Lintner Capital Asset Pricing Model (CAPM) is a parsimonious general equilibrium model in which the excess return, $R$, on a security is formulated by:

$$R = \alpha + \beta R_m + e \qquad (A.1)$$

where $R_m$ is the excess return on market portfolio, and $e$ is the random error.

The Sharpe ratio is defined as

$$\text{Sharpe ratio} = \frac{\hat{\mu}_R}{\hat{\sigma}_R} \qquad (A.2)$$

which is the ratio of the estimated excess expected return, $\hat{\mu}_R$, of an investment to its estimated return volatility or estimated standard deviation, $\hat{\sigma}_R$.

The Treynor index is defined as

$$\text{Treynor index} = \frac{\hat{\mu}_R}{\hat{\beta}} \qquad (A.3)$$

which is the ratio of the estimated excess expected return, $\hat{\mu}_R$, of an investment to its estimated beta, $\hat{\beta}$, where $\beta$ is defined in (A.1).

The Jensen index is defined as

$$\text{Jensen index} = \hat{\alpha} = \hat{R} - \hat{\beta}\hat{R}_m \qquad (A.4)$$

where $R$ is the excess return on a security, $R_m$ is the excess return on market portfolio, and $\beta$ is the beta defined in (A.1).

In addition, the estimate $\hat{\beta}$ of $\beta$ is commonly used in comparing the performance of a prospect, where $\beta$ is defined in (A.1).

## A.2   Testing equality of multiple Sharpe ratios

Recall that the Sharpe ratio is defined as the ratio of the excess expected return to the standard deviation of returns where the excess expected return is usually computed relative to the risk-free rate, $r_f$. Consider $k$ ($k \geq 2$) portfolios of expected excess returns, $X_{1t}, \ldots, X_{kt}$, at time $t$ with means, $\mu_1, \ldots, \mu_k$, variance, $\sigma_1^2, \ldots, \sigma_k^2$, and covariance, $\sigma_{ij}$, of portfolio $i$ and $j$. For simplicity, we assume the excess returns to be serially i.i.d. and not subject to change over time.

Note that the excess returns being studied in the Sharpe ratio analysis are usually assumed to be serially i.i.d. normal-distributed. However, in reality, the assumptions of normality may be violated. For example, empirical studies have demonstrated that the normality assumption in the distribution of a security or portfolio return is violated such that the distribution is "fat-tailed." Some studies suggest a family of stable Paretian distributions between normal and Cauchy distributions for stock returns, some suggest a mixture of normal distributions for the stock returns, while others suggest that a mixture of non-normal stable distributions would be a better representation of the distribution of returns. Thus, it is of practical interest to consider the situation in which the excess returns are not normally distributed.

In addition, note that for many economic or financial data, $\{X_{it}\}$ may not be i.i.d. For example, it is well known that stock returns could be heteroskedastic in their variances. To circumvent this problem, it is common to include a generalized autoregressive conditional heteroskedasticity innovation to allow both autoregressive and moving average components in the heteroskedastic variance of the returns to display a high degree of persistence. However, in this situation, one could transform them to be i.i.d. For example, if $\{X_{it}\}$ follows a GARCH model such that

$$X_{it} = \mu_i + \varepsilon_{it} \tag{A.5}$$

for $i = 1, \ldots, k$ and $t = 1, \ldots, T$; where $\varepsilon_{it}$ follows a generalized autoregressive conditional heteroskedasticity (GARCH) $(p, q)$ model such that

$$
\begin{aligned}
\varepsilon_{it} &= z_{it}\sqrt{h_{it}} \\
z_{it} &\sim \ iid \ N(0,1) \\
h_{it} &= w_i + \sum_{j=1}^{p} a_{i,j}\varepsilon_{i,t-j}^2 + \sum_{k=1}^{q} b_{i,k}h_{i,t-k}.
\end{aligned}
$$

Then, one could easily estimate $\{h_{it}\}$ by $\{\hat{h}_{it}\}$, transfer $\{X_{it}\}$ in (A.5) into $\{X_{it}^*\}$ by setting

$$
X_{it}^* = \frac{X_{it}}{\hat{h}_{it}}
$$

and thereafter $\{X_{it}^*\}$ become i.i.d. for each $i$. Thus, one could further relax the assumption of serial independence and homoskedasticity by transforming non-i.i.d. data to be i.i.d. Hence, for simplicity and tractability, we assume $\{X_{it}\}$ to be i.i.d. for each $i$ here.

Let $\theta = (\mu_1, \ldots, \mu_k, \sigma_1^2, \ldots, \sigma_k^2)'$ be a $2k \times 1$ vector of unknown parameters, the vector of the Sharpe ratios of the $k$ portfolios is then a $k \times 1$ vector such that

$$
\begin{aligned}
u(\theta) &= (r_1, \ldots, r_k)' \\
&= \left( \frac{\mu_1}{\sigma_1}, \ldots, \frac{\mu_k}{\sigma_k} \right)'. 
\end{aligned}
\tag{A.6}
$$

The MSR statistic test the hypothesis, $H_0$, of the equality of multiple Sharpe ratios defined in (A.6) such that

$$
H_0 : r_1 = \cdots = r_k.
\tag{A.7}
$$

Let $\hat{\theta} = (\bar{x}_1, \ldots, \bar{x}_k, s_1^2, \ldots, s_k^2)'$ be the vector of the usual sample means and sample variances calculated from $n$ observations. The parameter $u(\theta)$ in (A.6) can then be estimated by

$$
u(\hat{\theta}) = \left( \frac{\bar{x}_1}{s_1}, \ldots, \frac{\bar{x}_k}{s_k} \right)'.
\tag{A.8}
$$

According to the standard asymptotic distribution theory, we have

$$
\sqrt{n}[\hat{\theta} - \theta] \xrightarrow{D} N(0, \Sigma)
\tag{A.9}
$$

where $\Sigma = \begin{pmatrix} \Sigma_1 & \mathbf{0} \\ \mathbf{0} & \Sigma_2 \end{pmatrix}$ is a $2k \times 2k$ matrix such that

$$\Sigma_1 = \begin{pmatrix} \sigma_1^2 & \cdots & \sigma_{1k} \\ \vdots & \ddots & \vdots \\ \sigma_{k1} & \cdots & \sigma_k^2 \end{pmatrix},$$

$$\Sigma_2 = \begin{pmatrix} 2\sigma_1^4 & \cdots & 2\sigma_{1k}^2 \\ \vdots & \ddots & \vdots \\ 2\sigma_{k1}^2 & \cdots & 2\sigma_k^4 \end{pmatrix}$$

and $\mathbf{0}$ is a $k \times k$ matrix of zeros.

Applying the multivariate version of the $\delta$-method, we have

$$\sqrt{n}[u(\hat{\theta}) - u(\theta)] \xrightarrow{D} N\left[0, \left(\frac{\partial u}{\partial \theta}\right) \Sigma \left(\frac{\partial u}{\partial \theta}\right)'\right] \tag{A.10}$$

where $(\partial u/\partial \theta) = [D_1 \vdots D_2]$ is a $k \times 2k$ matrix with $D_1$ and $D_2$ to be $k \times k$ diagonal matrices whose diagonal elements are $(1/\sigma_1, \ldots, 1/\sigma_k)$ and $[-\mu_1/(2\sigma_1^3), \ldots, -\mu_k/(2\sigma_k^3)]$, respectively. It can be shown that

$$\left(\frac{\partial u}{\partial \theta}\right) \Sigma \left(\frac{\partial u}{\partial \theta}\right)' = D_1 \Sigma_1 D_1' + D_2 \Sigma_2 D_2' \tag{A.11}$$

where

$$D_1 \Sigma_1 D_1' = \begin{bmatrix} 1 & \rho_{12} & \cdots & \rho_{1k} \\ \rho_{21} & 1 & \cdots & \rho_{2k} \\ \vdots & \vdots & \ddots & \vdots \\ \rho_{k1} & \rho_{k2} & \cdots & 1 \end{bmatrix},$$

$$D_2 \Sigma_2 D_2' = \frac{1}{2} \begin{bmatrix} r_1^2 & r_1 r_2 \rho_{12}^2 & \cdots & r_1 r_k \rho_{1k}^2 \\ r_1 r_2 \rho_{12}^2 & r_2^2 & \cdots & r_2 r_k \rho_{2k}^2 \\ \vdots & \vdots & \ddots & \vdots \\ r_1 r_k \rho_{k1}^2 & r_2 r_k \rho_{k2}^2 & \cdots & r_k^2 \end{bmatrix}$$

and $\rho_{ij} = \sigma_{ij}/(\sigma_i \sigma_j)$ is the correlation of the excess returns of portfolios $i$ and $j$. Thereafter, (A.10) can be expressed as

$$\sqrt{n}[u(\hat{\theta}) - u(\theta)] \xrightarrow{D} N(0, \Omega) \tag{A.12}$$

where

$$\Omega = \frac{1}{2} \begin{bmatrix} 2 + r_1^2 & 2\rho_{12} + r_1 r_2 \rho_{12}^2 & \cdots & 2\rho_{1k} + r_1 r_k \rho_{1k}^2 \\ 2\rho_{12} + r_1 r_2 \rho_{12}^2 & 2 + r_2^2 & \cdots & 2\rho_{2k} + r_2 r_k \rho_{2k}^2 \\ \vdots & \vdots & \ddots & \vdots \\ 2\rho_{k1} + r_1 r_k \rho_{k1}^2 & 2\rho_{k2} + r_2 r_k \rho_{k2}^2 & \cdots & 2 + r_k^2 \end{bmatrix}.$$

Note that the asymptotic variance of $r_i$ in (A.12) is $1 + r_i^2/2$. When $k = 2$,

$$Var(r_1 - r_2) = \frac{1}{n}[2 - 2\rho_{12} + \frac{1}{2}(r_1^2 + r_2^2 - 2r_1 r_2 \rho_{12}^2)] \qquad (A.13)$$

which is the same as the pair-wise test statistic for the equality of two Sharpe ratios.

## A.3  Hypothesis testing

Having derived the asymptotic distribution of the vector of Sharpe ratios such that $u(\theta) = (r_1, \ldots, r_k)'$, we are going to test the hypothesis $H_0$ stated in (A.7). We use the standard multivariate method known as the repeated measures design for comparing treatments to test $H_0$. We first define the $(k - 1) \times k$ constant matrix $C$ such that

$$C = \begin{pmatrix} 1 & -1 & 0 & \cdots & 0 \\ 0 & 1 & -1 & \cdots & 0 \\ \vdots & & \ddots & \ddots & \vdots \\ 0 & \cdots & \cdots & 1 & -1 \end{pmatrix}. \qquad (A.14)$$

The hypothesis $H_0$ in (A.7) is then equivalent to

$$H_0 : C\,u(\theta) = 0 \quad \text{versus} \quad H_1 : C\,u(\theta) \neq 0. \qquad (A.15)$$

Thereafter, we incorporate Hotelling's approach to obtain the multivariate Sharpe ratio statistic, $T^2$, such that

$$T^2 k[C\,u(\hat{\theta})]'(C\,\hat{\Omega}\,C'^{-1}[C\,u(\hat{\theta})] \qquad (A.16)$$

can be applied to test this hypothesis. For the $\alpha$ level, we reject the hypothesis $H_0$ in (A.15) if

$$T^2 > \frac{(n-1)(k-1)}{(n-k+1)} F_{k-1,n-k+1}(\alpha) \qquad (A.17)$$

and conclude in $H_1$ that $C u(\theta) \neq 0$ where $F_{k-1,n-k+1}(\alpha)$ is the upper $(100\alpha)$th percentile of an $F$-distribution with $k-1$ and $n-k+1$ degrees of freedom and $\hat{\Omega}$ is the estimate of $\Omega$ in (A.12) such that the unknown parameters $r_i$ and $\rho_{ij}$ are replaced by their sample estimates.

Note that the testing procedure proposed here is a multivariate testing procedure. The correlations between all of the $k$ Sharpe ratios have been taken into consideration. The pair-wise test is, in fact, a special case ($k = 2$) of our proposed multivariate test. More importantly, if we apply the pair-wise statistic to test the equality of all possible ($k(k-1)/2$) pairs of Sharpe ratios simultaneously, the family-wise type I error will then be greatly inflated. The proposed multivariate test circumvents this problem with the correct type I error asymptotically, since our test takes care of the correlations among all the Sharpe ratios as well.

## A.4   Davidson-Duclos (DD) test

### A.4.1   Stochastic dominance tests for risk averters

Let $\{f_i\}$ ($i = 1, 2, \ldots, N_f$) and $\{g_i\}$ ($i = 1, 2, \ldots, N_g$) be observations drawn from the independent random variables $Y$ and $Z$, respectively, with distribution functions $F$ and $G$, respectively. For a grid of pre-selected points $\{x_k, \; k = 1, \ldots, K\}$, the $j$th order ascending DD test statistics, $T_j^A(x)$ ($j = 1, 2$ and $3$), is:

$$T_j^A(x) = \frac{\hat{F}_j^A(x) - \hat{G}_j^A(x)}{\sqrt{\hat{V}_j^A(x)}}, \tag{A.18}$$

where

$$\hat{V}_j^A(x) = \hat{V}_{F_j}^A(x) + \hat{V}_{G_j}^A(x),$$

$$\hat{H}_j^A(x) = \frac{1}{N_h(j-1)!} \sum_{i=1}^{N_h} (x - h_i)_+^{j-1},$$

$$\hat{V}_{H_j}^A(x) = \frac{1}{N_h} \left[ \frac{1}{N_h((j-1)!)^2} \sum_{i=1}^{N_h} (x - h_i)_+^{2(j-1)} - \hat{H}_j^A(x)^2 \right],$$

$$H = F, G; \; h = f, g.$$

It seems difficult to test the null hypothesis for the full support of the distributions. Thus, we have to test the null hypothesis for a pre-designated finite number of values $\{x_k, \ k = 1, \ldots, K\}$. Specifically, the following hypotheses are tested:

$H_0$ : $F_j^A(x_k) = G_j^A(x_k)$ for all $x_k$,

$H_A$ : $F_j^A(x_k) \neq G_j^A(x_k)$ for some $x_k$,

$H_{A1}$ : $F_j^A(x_k) \leq G_j^A(x_k)$ for all $x_k$, $F_j^A(x_k) < G_j^A(x_k)$ for some $x_k$, and

$H_{A2}$ : $F_j^A(x_k) \geq G_j^A(x_k)$ for all $x_k$, $F_j^A(x_k) > G_j^A(x_k)$ for some $x_k$.

Note that in the above hypotheses, $H_A$ is set to be exclusive of both $H_{A1}$ and $H_{A2}$; this means that if the test accepts $H_{A1}$ or $H_{A2}$, it will not be classified as $H_A$. Under the null hypothesis, DD show that $T_j^A$ is asymptotically distributed as the studentized maximum modulus (SMM) distribution. To implement the DD test, the $T_j^A$ at each grid point is computed and the null hypothesis, $H_0$, is rejected if $T_j^A$ is significant at any grid point. The SMM distribution with $K$ and infinite degrees of freedom at $\alpha\%$, denoted by $M_{\infty,\alpha}^K$, is used to control the probability of rejecting the overall null hypotheses. The following decision rules are adopted based on the $1 - \alpha$ percentile of $M_{\infty,\alpha}^K$:

1. If $\left|T_j^A\right| < M_{\infty,\alpha}^K$ for $k = 1, \ldots, K$, accept $H_0$;

2. If $T_j^A < M_{\infty,\alpha}^K$ for all $k$ and $-T_j^A > M_{\infty,\alpha}^K$ for some $k$, accept $H_{A1}$;

3. If $-T_j^A < M_{\infty,\alpha}^K$ for all $k$ and $T_j^A > M_{\infty,\alpha}^K$ for some $k$, accept $H_{A2}$;

4. If $T_j^A > M_{\infty,\alpha}^K$ for some $k$ and $-T_j^A > M_{\infty,\alpha}^K$ for some $k$, accept $H_A$.

The DD test compares distributions at a finite number of grid points. Various studies examine the choice of grid points. It is found that an appropriate choice of $K$ for reasonably large samples ranges from 6 to 15. Too few grids will miss information about the distributions between any two consecutive grids, and too many grids will violate the independence assumption required by the SMM distribution. To make the comparisons comprehensive without violating the independence assumption, we suggest to make 10 major partitions with 10 minor partitions within any two consecutive major partitions in each comparison and show the statistical inference based on the SMM distribution for $K = 10$ and infinite degrees of freedom. This allows the consistency of both the magnitude and sign of the DD statistics between any two consecutive major partitions to be examined.

## A.4.2   Stochastic dominance tests for risk seekers

There are several SD tests for risk seekers. One could modify the DD test for risk averters to be the descending DD test statistic. Let $\{f_i\}$ $(i = 1, 2, \ldots, N_f)$ and $\{g_i\}$ $(i = 1, 2, \ldots, N_g)$ be observations drawn from the independent random variables $Y$ and $Z$, respectively, with distribution functions $F$ and $G$, respectively. For a grid of pre-selected points $\{x_k, k = 1, \ldots, K\}$, the $j$th order descending DD test statistic, $T_j^D(x)$ $(j = 1, 2$ and $3)$, is

$$T_j^D(x) = \frac{\hat{F}_j^D(x) - \hat{G}_j^D(x)}{\sqrt{\hat{V}_j^D(x)}}, \tag{A.19}$$

where

$$\hat{V}_j^D(x) = \hat{V}_{F_j}^D(x) + \hat{V}_{G_j}^D(x) - 2\hat{V}_{FG_j}^D(x),$$

$$\hat{H}_j^D(x) = \frac{1}{N_h(j-1)!} \sum_{i=1}^{N_h}(h_i - x)_+^{j-1},$$

$$\hat{V}_{H_j}^D(x) = \frac{1}{N_h}\left[\frac{1}{N_h((j-1)!)^2}\sum_{i=1}^{N_h}(h_i - x)_+^{2(j-1)} - \hat{H}_j^D(x)^2\right],$$

$$H = F, G; \; h = f, g;$$

For $k = 1, \ldots, K$, the following hypotheses are tested for risk seekers:

$H_0$ : $F_j^D(x_k) = G_j^D(x_k)$ for all $x_k$;

$H_D$ : $F_j^D(x_k) \neq G_j^D(x_k)$ for some $x_k$;

$H_{D1}$ : $F_j^D(x_k) \geq G_j^D(x_k)$ for all $x_k$, $F_j^D(x_k) > G_j^D(x_k)$ for some $x_k$; and

$H_{D2}$ : $F_j^D(x_k) \leq G_j^D(x_k)$ for all $x_k$, $F_j^D(x_k) < G_j^D(x_k)$ for some $x_k$.

We note that in the above hypotheses, $H_D$ is set to be exclusive of both $H_{D1}$ and $H_{D2}$; this means that if the test accepts $H_{D1}$ or $H_{D2}$, it will not be classified as $H_D$. Under the null hypothesis, DD show that $T_j^D$ is asymptotically distributed as the studentized maximum modulus (SMM) distribution. To implement the DD test, the $T_j^D$ at each grid point is computed and the null hypothesis, $H_0$, is rejected if $T_j^D$ is significant at any grid point. The SMM distribution with $k$ and infinite degrees of freedom at $\alpha\%$, denoted by $M_{\infty,\alpha}^K$, is used to control the probability of rejecting the overall null hypotheses.

The following decision rules are adopted based on the $1 - \alpha$ percentile of $M_{\infty,\alpha}^K$:

1. If $\left| T_j^D \right| < M_{\infty,\alpha}^K$ for $k = 1, \ldots, K$, accept $H_0$;

2. If $T_j^D < M_{\infty,\alpha}^K$ for all $k$ and $-T_j^D > M_{\infty,\alpha}^K$ for some $k$, accept $H_{D1}$;

3. If $-T_j^D < M_{\infty,\alpha}^K$ for all $k$ and $T_j^D > M_{\infty,\alpha}^K$ for some $k$, accept $H_{D2}$;

4. If $T_j^D > M_{\infty,\alpha}^K$ for some $k$ and $-T_j^D > M_{\infty,\alpha}^K$ for some $k$, accept $H_D$.

As in the case of the test for risk averters, accepting either $H_0$ or $H_D$ implies the non-existence of any SD relationship between $F$ and $G$, the non-existence of any arbitrage opportunity between these two markets, and neither of these markets is preferred to the other. If $H_{D1}$ ($H_{D2}$) of order one is accepted, asset $F(G)$ stochastically dominates $G(F)$ at the first order. In this situation, an arbitrage opportunity exists and non-satiated investors will be better off if they switch their investments from the dominated market to the dominant one. On the other hand, if $H_{D1}$ or $H_{D2}$ is accepted at order two or three, a particular market stochastically dominates the other at the second or third order. In this situation, an arbitrage opportunity does not exist and switching from one market to another will only increase the risk seekers' expected utilities, but not their wealth.

## A.5  Barrett and Donald (BD) test

### A.5.1  Stochastic dominance tests for risk averters

Barrett and Donald [10] developed a Kolmogorov-Smirnov-type test for SD of any pre-specified order for two independent samples of possibly unequal sample sizes. The BD test evaluates the following two sets of null and alternative hypotheses:

$$
\begin{aligned}
H_{A0} &: \quad F_j^A(x) \leq G_j^A(x) \text{ for all } x \quad \text{and} \\
H_{A1} &: \quad F_j^A(x) > G_j^A(x) \text{ for some } x. \\
H'_{A0} &: \quad G_j^A(x) \leq F_j^A(x) \text{ for all } x \quad \text{and} \\
H'_{A1} &: \quad G_j^A(x) > F_j^A(x) \text{ for some } x.
\end{aligned}
$$

The null hypothesis that $Y$ dominates (but does not strictly dominate) $Z$ is stated in $H_{A0}$ and the null hypothesis that $Z$ dominates (but does not strictly dominate) $Y$ is stated in $H'_{A0}$. The BD test statistic for risk averters is:

$$\hat{K}_j^A = \left(\frac{N^2}{2N}\right)^{1/2} \sup_x \left[\hat{F}_j^A(x) - \hat{G}_j^A(x)\right]. \tag{A.20}$$

To compute the critical value for BD statistic, Barrett and Donald [10] provide the following theorem:

**Theorem A.1** *Define the random variable,*

$$K_j^{F,G} = \sup_x \left[\sqrt{\lambda}F_j^A(x, B_F \circ F) - \sqrt{1-\lambda}G_j^A(x, B_G \circ G)\right]$$

*where $\lambda \in (0,1)$ and $B_F \circ F$ and $B_G \circ G$ are independent Brownian bridge processes. The limiting distributions of the test statistics under the null hypothesis are characterized as $\sqrt{N}(\hat{F}_j^A - F) \to B_F \circ F$ and $\sqrt{N}(\hat{G}_j^A - G) \to B_G \circ G$. If $H_0$ is true, $\lim_{n\to\infty} P(\text{reject } H_0) \leq P(K_j^{F,G} > c_j) = \alpha(c_j)$. If $H_0$ is false, $\lim_{n\to\infty} P(\text{reject } H_0) = 1$.*

If $H_{A0}$ is rejected, $Y$ is concluded not to dominate $Z$ for risk averters. Similarly, if $H'_{A0}$ is rejected, $Z$ is concluded not to dominate $Y$ for risk averters.

Theorem A.1 shows how to compute the critical values corresponding to any desired Type I error, such as $\alpha(c_j) = 0.05$. It indicates that if the null hypothesis is true, the Type I error rate will not exceed $\alpha(c_j)$ asymptotically. For $j \geq 2$, it is analytically intractable to derive the critical values of the test statistic because the limiting distribution of $K_j^{F,G}$ depends on the underlying cumulative distribution functions. The $p$-values can be simulated by using arbitrarily fine grids for calculating the suprema of $F_j^A$ and $G_j^A$.

## A.5.2    Stochastic dominance tests for risk seekers

Barrett and Donald's Kolmogorov-Smirnov-type test could be extended to the test for risk seekers for testing the following two sets of null and alternative hypotheses:

$$
\begin{aligned}
H_{D0} &: \quad F_j^D(x) \leq G_j^D(x) \text{ for all } x \quad \text{and} \\
H_{D1} &: \quad F_j^D(x) > G_j^D(x) \text{ for some } x. \\
H'_{D0} &: \quad G_j^D(x) \leq F_j^D(x) \text{ for all } x \quad \text{and} \\
H'_{D1} &: \quad G_j^D(x) > F_j^D(x) \text{ for some } x.
\end{aligned}
$$

The null hypothesis that $Y$ dominates (but does not strictly dominate) $Z$ is stated in $H_{D0}$ and the null hypothesis that $Z$ dominates (but does not strictly dominate) $Y$ is stated in $H'_{D0}$. The BD test statistic for risk seekers is:

$$\hat{K}_j^D = \left(\frac{N^2}{2N}\right)^{1/2} \sup_x \left[\hat{F}_j^D(x) - \hat{G}_j^D(x)\right]. \tag{A.21}$$

If $H_{A0}$ is rejected, $Y$ is concluded not to dominate $Z$ for risk averters. Similarly, if $H'_{D0}$ is rejected, $Z$ is concluded not to dominate $Y$ for risk seekers.

## A.6  Linton, Maasoumi and Whang test

### A.6.1  Stochastic dominance tests for risk averters

Linton et al. [97] propose a procedure (LMW test) for estimating the critical values of the extended Kolmogorov-Smirnov tests of SD based on sub-sampling. Their SD tests are consistent and powerful against some $N^{-1/2}$ local alternatives. Another advantage of this method is that it allows for general dependence amongst the prospects, and for non-i.i.d. observations. The hypotheses set-up are the same as for the BD test. Their test statistic for risk averters is

$$T_j^A = \min \sup {}_x \sqrt{N \left[\hat{F}_j^A(x) - \hat{G}_j^A(x)\right]}. \tag{A.22}$$

They follow the DD test for the computation of $\hat{H}_j^A$ and compute approximations to the suprema in $\hat{H}_j^A$ based on taking maxima over some smaller grid of points, $x_1, \ldots, x_k$.

### A.6.2  Stochastic dominance tests for risk seekers

The LMW test could be extended to a test for risk seekers with the corresponding hypotheses as follows:

$$H_{D0} : F_j^D(x) \leq G_j^D(x) \text{ for all } x \quad \text{and}$$
$$H_{D1} : F_j^D(x) > G_j^D(x) \text{ for some } x.$$
$$H'_{D0} : G_j^D(x) \leq F_j^D(x) \text{ for all } x, \quad \text{and}$$
$$H'_{D1} : G_j^D(x) > F_j^D(x) \text{ for some } x.$$

Their test statistic is

$$T_j^D = \min \sup_x \sqrt{N \left[ \hat{F}_j^D(x) - \hat{G}_j^D(x) \right]}. \qquad (A.23)$$

Now we turn to construct statistics to test equality or dominance of $F_j^w$ and $G_j^w$, where $w = M$ or $P$. That is, to construct statistics to test equality or dominance of $F$ and $G$ for the following:

1. $H_j^A(x)$ for $x \leq 0$

2. $H_j^D(x)$ for $x > 0$

3. $H_j^d(x)$ for $x \leq 0$

4. $H_j^a(x)$ for $x > 0$

To test the equality or dominance of $F$ and $G$ under the situation of (a), we could simply use the statistic $T_j^A$ and the corresponding procedure for the range of $x \leq 0$. Similarly, to test for equality or dominance of $F$ and $G$ under the situation of (b), we could simply use the statistic $T_j^D$ and the corresponding procedure for the range of $x > 0$.

## A.7   Stochastic dominance tests for MSD and PSD

One could develop an MSD and PSD test.[1] In this book, we recommend a very easy way: We only discuss how to get an MSD and PSD test by using Davidson-Duclos (DD) statistics. Readers could use the same approach to get an MSD and PSD test by using Barrett and Donald (BD) statistics and Linton, Maasoumi and Whang (LMW) statistics.

One could simply examine $T_j^A(x)$ in (A.18) over both positive and negative domains of the empirical return distributions to reveal risk averters' preferences. A similar examination of $T_j^D(x)$ in (A.19) reveals risk seekers' preferences over both positive and negative domains. Thus, examining both $T_j^A(x)$ over the positive domain and $T_j^D(x)$ over the negative domain allows identification of the risk preferences of investors with $j$th order S-shaped utility functions. These investors exhibit $j$th order

---

[1]Readers may refer to Z. Bai, H. Liu and W-K Wong (November 16, 2007) Test Statistics for Prospect and Markowitz Stochastic Dominances with Applications, Social Science Research Network Working Paper Series, *http://ssrn.com/abstract=1030728* for more information.

risk aversion over the positive domain and risk seeking over the negative domain. On the other hand, examining both $T_j^P(x)$ over the positive domain and $T_j^A(x)$ over the negative domain allows us to identify investors with $j$th order reverse S-shaped utility functions. These investors exhibit $j$th order risk seeking over the positive domain and risk aversion over the negative domain. Thus, after modifying DD to study the behavior of risk seekers, combining the DD test for risk aversion and the modifying DD test risk seeking allows an identification of S-shaped and reverse S-shaped utility functions.

We discuss the above for testing PSD first. Let $\{f_i\}$ $(i = 1, 2, \ldots, N_f)$ and $\{g_i\}$ $(i = 1, 2, \ldots, N_g)$ be observations drawn from the independent random variables $Y$ and $Z$, respectively, with distribution functions $F$ and $G$, respectively. The integrals $F_j^A$, $G_j^A$, $F_j^D$ and $G_j^D$ for $F$ and $G$ are defined in (8.2) for $j = 1$, 2 and 3. If one applies $T_i^A(x)$ $(i = 1, 2, 3)$ in (A.18) over the positive domain, one could draw one of the following conclusions:

- Do not reject $H_{A1}$ in Section A.4.1 that $F_j^A(x_k) \leq G_j^A(x_k)$ for all $x_k$, and do not reject $F_j^A(x_k) < G_j^A(x_k)$ for some $x_k$ on the positive domain and conclude that $Y \succeq_i^A Z$ on the positive domain.

- Do not reject $H_{A2}$ in Section A.4.1 that $G_j^A(x_k) \leq F_j^A(x_k)$ for all $x_k$, and do not reject $G_j^A(x_k) < F_j^A(x_k)$ for some $x_k$ on the positive domain and conclude that $Y \succeq_i^A Z$ on the positive domain.

- Do not reject $H_0$ in Section A.4.1 that $F_j^A(x_k) = G_j^A(x_k)$ for all $x_k$ on the positive domain and conclude that $F(x) = G(x)$ for any $x$ over the positive domain.

- Do not reject $H_A$ in Section A.4.1 that $F_j^A(x_k) \neq G_j^A(x_k)$ for some $x_k$ on the positive domain and conclude that no ASD between $Y$ and $Z$ on the positive domain.

Similarly, if one applies $T_j^P(x)$ in (A.19) over the negative domain, one could conclude any of the following:

- Do not reject $H_{D1}$ in Section A.4.2 that $Y \succeq_j^D Z$ on the negative domain.

- Do not reject $H_{D2}$ in Section A.4.2 that $Y \succeq_j^D Z$ $(j = 1, 2, 3)$ on the negative domain.

- Do not reject $H_0$ in Section A.4.2 that $F(x) = G(x)$ for any $x$ over the negative domain.

- Do not reject $H_D$ in Section A.4.2 that no DSD between $Y$ and $Z$ on the negative domain.

Combining the above, we have the following cases for testing PSD:

1. If $Y \succeq_i^A Z$ on the positive domain and $Y \succeq_j^D Z$ on the negative domain with $i = j$, then $Y \succeq_i^P Z$.

2. If $Y \succeq_i^A Z$ on the positive domain and $Y \succeq_j^D Z$ on the negative domain with $i < j$, then $Y \succeq_j^P Z$ because $Y \succeq_i^A Z$ implies $Y \succeq_j^A Z$.

3. If $Y \succeq_i^A Z$ on the positive domain and $Y \succeq_j^D Z$ on the negative domain with $i > j$, then $Y \succeq_i^P Z$ because $Y \succeq_j^D Z$ implies $Y \succeq_i^D Z$.

4. If $F(x) = G(x)$ for any $x$ over the positive domain and $Y \succeq_j^D Z$ on the negative domain, then $Y \succeq_j^P Z$.

5. If $Y \succeq_i^A Z$ on the positive domain and $F(x) = G(x)$ for any $x$ over the negative domain, then $Y \succeq_i^P Z$.

6. If $Y \succeq_i^A Z$ on the positive domain and $Z \succeq_j^D Y$ on the negative domain, then there is no PSD between $Y$ and $Z$.

7. If there is no ASD between $Y$ and $Z$ on the positive domain (no matter whether $Y \succeq_j^D Z$, $Z \succeq_j^D Y$, $F(x) = G(x)$ for any $x$, or there is no DSD between $Y$ and $Z$ on the negative domain), then there is no PSD between $Y$ and $Z$.

8. If there is no DSD between $Y$ and $Z$ on the negative domain (no matter whether $Y \succeq_i^A Z$, $Z \succeq_i^A Y$, $F(x) = G(x)$ for any $x$, or there is no ASD between $Y$ and $Z$ on the positive domain), then there is no PSD between $Y$ and $Z$.

The case for MSD could be obtained similarly. We briefly discuss as follows: If one applies $T_i^A(x)$ $(i = 1, 2, 3)$ in (A.18) over the negative domain, one could draw one of the following conclusions:

- Do not reject $H_{A1}$ in Section A.4.1 on the negative domain and conclude that $Y \succeq_i^A Z$ on the negative domain.

- Do not reject $H_{A2}$ in Section A.4.1 on the negative domain and conclude that $Y \succeq_i^A Z$ on the negative domain.

- Do not reject $H_0$ in Section A.4.1 on the negative domain and conclude that $F(x) = G(x)$ for any $x$ over the negative domain.

- Do not reject $H_A$ in Section A.4.1 on the negative domain and conclude that there is no ASD between $Y$ and $Z$ on the negative domain.

If one applies $T_j^D(x)$ in (A.19) over the positive domain, one could conclude any of the following:

- Do not reject $H_{D1}$ in Section A.4.2 that $Y \succeq_j^D Z$ on the positive domain.

- Do not reject $H_{D2}$ in Section A.4.2 that $Y \succeq_j^D Z$ $(j = 1, 2, 3)$ on the positive domain.

- Do not reject $H_0$ in Section A.4.2 that $F(x) = G(x)$ for any $x$ over the positive domain.

- Do not reject $H_D$ in Section A.4.2 that there is no DSD between $Y$ and $Z$ on the positive domain.

Combining the above, we have the following cases for testing MSD:

1. If $Y \succeq_i^A Z$ on the negative domain and $Y \succeq_j^D Z$ on the positive domain with $i = j$, then $Y \succeq_i^M Z$.

2. If $Y \succeq_i^A Z$ on the negative domain and $Y \succeq_j^D Z$ on the positive domain with $i < j$, then $Y \succeq_j^M Z$ because $Y \succeq_i^A Z$ implies $Y \succeq_j^A Z$.

3. If $Y \succeq_i^A Z$ on the negative domain and $Y \succeq_j^D Z$ on the positive domain with $i > j$, then $Y \succeq_i^M Z$ because $Y \succeq_j^D Z$ implies $Y \succeq_i^D Z$.

4. If $F(x) = G(x)$ for any $x$ over the negative domain and $Y \succeq_j^D Z$ on the positive domain, then $Y \succeq_j^M Z$.

5. If $Y \succeq_i^A Z$ on the negative domain and $F(x) = G(x)$ for any $x$ over the positive domain, then $Y \succeq_i^M Z$.

6. If $Y \succeq_i^A Z$ on the negative domain and $Z \succeq_j^D Y$ on the positive domain, then there is no PSD between $Y$ and $Z$.

7. If there is no ASD between $Y$ and $Z$ on the negative domain (no matter whether $Y \succeq_j^D Z$, $Z \succeq_j^D Y$, $F(x) = G(x)$ for any $x$, or there is no DSD between $Y$ and $Z$ on the positive domain), then there is no PSD between $Y$ and $Z$.

8. If there is no DSD between $Y$ and $Z$ on the positive domain (no matter whether $Y \succeq_i^A Z$, $Z \succeq_i^A Y$, $F(x) = G(x)$ for any $x$, or there is no ASD between $Y$ and $Z$ on the negative domain), then there is no PSD between $Y$ and $Z$.

# Bibliography

[1] C. Acerbi (2002) Spectral measures of risk: a coherent representation of subjective risk aversion, *J. of Banking and Finance* (7) 1505–1518.

[2] J. Aczel (2006) *Lectures on Functional Equations and Their Applications*, Dover Publ., New York.

[3] T. Amemiya (1985), *Advanced Econometrics*, Cambridge University Press.

[4] G. Andersen (1996) Nonparametric tests of stochastic dominance in income distributions, *Econometrica* 64(5) 1183–1193.

[5] G. J. Anderson (2004) Toward an empirical analysis of polarization, *J. Econometrics* (122) 1–26.

[6] J. S. Armstrong (2001) *Principles of Forecasting: A Handbook for Researchers and Practitioners*, Springer Verlag, New York.

[7] K. J. Arrow (1974) *Essays in the Theory of Risk Bearing*, North Holland.

[8] P. Artzner, F. Delbaen, J. M. Eber, and D. Heath (1999) Coherent measures of risk, *Mathematical Finance* (9)(3) 203–228.

[9] A. B. Atkinson and F. Bourguignon (1982) The comparison of multi-dimensional distributions of economic status, *The Review of Economic Studies* 49(2) 183–201.

[10] G. Barrett and S. Donald (2003) Consistent tests for stochastic dominance, *Econometrica* (71) 71–104.

[11] V. S. Bawa, J. N. Bodurtha, M. R. Rao, and H. L. Suri (1985) On determination of stochastic dominance optimal sets, *Journal of Finance* (XL)(2) 417–431.

[12] V. L. Bernard and H. N. Seyhun (1997) Does post-earnings-announcement drift in stock prices reflect a market inefficiency? A stochastic dominance approach, *Review Quan. Finance and Accounting* (9) 17–34.

[13] G. Bian and W. K. Wong (1997) An Alternative Approach to Estimate Regression Coefficients, *The Journal of Applied Statistical Science* (6)(1) 21–44.

[14] P. Bickel and E. L. Lehmann (1969) Unbiased estimation in convex families, *Ann. Math. Statist.* (40) 1523–1535.

[15] P. Billingsley (1968) *Convergence of Probability Measures*, John Wiley & Sons, New York.

[16] P. Billingsley (1995) *Probability and Measure*, John Wiley & Sons, New York.

[17] T. Bollerslev (1986) Generalized autoregressive conditional heteroskedasticity, *Journal of Econometrics* (31) 307–327.

[18] D. Bosq and H. T. Nguyen (1996) *A Course in Stochastic Processes: Stochastic Models and Statistical Inference*, Kluwer Academic.

[19] U. Broll, J. E. Wahl, and W. K. Wong (2006) Elasticity of risk aversion and international trade, *Economics Letters* (92)(1) 126–130.

[20] R. Carroll et al. (2006) *Measurement Error in Nonlinear Models*, Chapman & Hall/CRC.

[21] K. Chan, A. Hameed, and W. Tong (2000) Profitability of momentum strategies in the international equity markets, *J. Finance. and Quan. Analysis* 35(2) 153–172.

[22] R. L. Cheu, H. T. Nguyen, T. Magoc, and K. Kreinovich (2009) Logit discrete choice model: a new distribution-free justification, *Soft Computing* (13) 133–137.

[23] C. Chevalley (1999) *Theory of Lie Groups*, Princeton University Press, Princeton, New Jersey.

[24] T. C. Chiang, H. H. Lean, and W. K. Wong (2009) Do REITs outperform stocks and fixed-income assets? New evidence from mean-variance and stochastic dominance approaches, *Journal of Risk and Financial Management* (forthcoming).

[25] J. Chipman (1960) The foundations of utility, *Econometrica* (28) 193–224.

[26] G. Choquet (1953/54) Theory of capacities, *Ann. Inst. Fourier* (V), 131–295.

[27] F. A. Cowell (1985) Measures of distributional change: an axiomatic appraoch, *Review Econ. Studies* (52) 27–41.

[28] F. A. Cowell (1995) *Measuring Inequality, 2nd Ed.*, Harvester Wheatsheaf, Hemel Hempstead.

[29] R. Davidson and J.-Y. Duclos (2000) Statistical inference for stochastic dominance and for the measurement of poverty and inequality, *Econometrica* (68) 1435–1464.

[30] E. De Giorgi and T. Hens (2006) Making prospect theory fit for finance, *Financial Markets and Portfolio Management* (20)(3) 339–360.

[31] G. Debreu and R. D. Luce (1960) Review of individual choice behavior, *Amer. Econ. Review* (50) 186–188.

[32] C. Dellachcrie (1971) Quelques commentaires sur le prolongement de capacite's, *Lecture Notes in Mathematics*, Springer-Verlag, Berlin (191) 77–81.

[33] D. Denneberg (1989) Distorted probabilities and insurance premium, *Proceedings of the 14th SOR*, Ulm Athenaum, Frankfurt.

[34] D. Denneberg (1994) *Non-Additive Measure and Integral*, Kluwer Academic.

[35] M. Denuit, J. Dhaene, M. Goovarts, R. Kaas, and R. Laeven (2006) Risk measurement with utility principles, *Statistics and Decisions* (24) 1–25.

[36] J. Dhaene et al. (2006) Risk measures and comonotonicity: a review, *Stochastic Models* (22) 573–606.

[37] R. M. Dudley (1989) *Real Analysis and Probability*, Wadsworth and Brooks/Cole, Belmont.

[38] J.-M. Dufour, L. Khalaf, J.-T. Bernard, and I. Genest (2004) Simulation-based finite-sample tests for heteroskedasticity and ARCH effects, *Journal of Econometrics* (122)(2) 317–347.

[39] W. Enders (2003) *Applied Econometrics Time Series*, John Wiley & Sons, New York.

[40] R. Engelking (1977) *General Topology*, Polish Scientific Publ., Warszawa.

[41] R. F. Engle (1982) Autoregressive conditional heteroskedasticity with estimates of variance of United Kingdom inflation, *Econometrica* (50) 987–1008.

[42] H. Falk and H. Levy (1989) Market reaction to quarterly earnings' announcements: a stochastic dominance based test of market efficiency, *Management Science* (35) 425–446.

[43] P. C. Fishburn (1969) *Utility Theory for Decision Making*, John Wiley & Sons, New York.

[44] P. C. Fishburn (1970) *Utility Theory for Decision Making*, John Wiley & Sons, New York.

[45] P. C. Fishburn (1974) Convex stochastic dominance with continuous distribution functions, *Journal of Economic Theory* (7) 143–158.

[46] P. C. Fishburn (1980) Stochastic dominance and moments of distributions, *Mathematics of Operations Research* (5) 94–100.

[47] P. C. Fishburn (1988) *Nonlinear Preference and Utility Theory*, The Johns Hopkins University Press, Baltimore, MD.

[48] H. Föllmer and A. Schied (2004) *Stochastic Finance: An Introduction in Discrete Time*, de Gruyter Studies in Mathematics 27, Berlin.

[49] W. M. Fong, H. H. Lean, and W. K. Wong (2008) Stochastic dominance and behavior towards risk: the market for internet stocks, *J. Econ. Behav. and Organization*, 68(1) 194–208.

[50] W. M. Fong, W. K. Wong, and H. H. Lean (2005) International momentum strategies: A stochastic dominance approach, *J. Financial Markets* (8) 89–109.

[51] M. Frechet (1948) Les elements aleatoires de nature quelconque dans un espace distanciè, *Ann. Inst. Fourier* (IV), 215–310.

[52] W. A. Fuller (1987) *Measurement Error Models*, John Wiley & Sons, New York.

[53] D. Gasbarro, W. K. Wong, and J. K. Zumwalt (2007) Stochastic dominance analysis of iShares, *European J. Finance* (13) 89–101.

[54] J. D. Gibbons and S. Chakraborti (1992) *Nonparametric Statistical Inference*, Marcel Dekker.

[55] R. Gibbons (1992) *Game Theory for Applied Economists*, Princeton University Press.

[56] C. Gini (1921) Measurement of inequality of incomes, *The Economic Journal* (31) 124–126.

[57] P. Glasserman (2007) Calculating portfolio credit risk, In: *Handbooks in Operations Research and Management Science* (15) 437–470, Elsevier.

[58] L. R. Glosten, R. Jagannathan, D. E. Runkle, and E. David (1993) On the relation between the expected value and the volatility of the nominal excess returns on stocks, *Journal of Finance* (48)(5) 1779–1801.

[59] W. H. Greene (2003), *Econometric Analysis*, Prentice Hall.

[60] G. N. Gregoriou, K. Sedzro, and J. Zhu (2005) Hedge fund performance appraisal using data envelopment analysis, *European J. Oper. Research* (164) 555–571.

[61] J. M. Griffin, X. Ji, and J. S. Martin (2003) Momentum investing and business cycle risk: Evidence from pole to pole, *J. Finance* (58) 2515–2548.

[62] Z. Griliches and J. A. Hausman (1986) Errors in variables in panel data, *Journal of Econometrics* (31)(1) 93–118, Elsevier.

[63] D. Gujarati and D. Porter (2008) *Basic Econometrics*, McGraw-Hill.

[64] P. Gustafson (2004) *Measurement Error and Misclassification in Statistics and Epidemiology*, Chapman & Hall/CRC.

[65] J. Hadar and W. Russell (1969) Rules for ordering uncertain prospects, *Amer. Economic Review* (59) 25–34.

[66] J. S. Hammond (1974) Simplifying the choice between uncertain prospects where preference is nonlinear, *Management Science* 20(7) 1047–1072.

[67] G. Hanoch and H. Levy (1969) The efficiency analysis of choices involving risk, *Rev. Economic Studies* (36) 335–346.

[68] G. H. Hardy, J. E. Littlewood, and G. Polya (1934) *Inequalities*, Cambridge University Press.

[69] J. A. Hartigan (1987) Estimation of a convex density contour in two dimensions, *J. Amer. Statist. Assoc.* (82) 267–270.

[70] J. Heckman (1979) Sample selection bias as a specification error, *Econometrica* (47)153–161.

[71] P. J. Huber (1973) The use of Choquet capacities in statistics, *Bull. Inst. Intern. Stat.* (XLV)(4) 181–188.

[72] R. Jarrow (1986) The relationship between arbitrage and first-order stochastic dominance, *J. Finance* (41) 915–921.

[73] N. Jegadeesh and S. Titman (1993) Returns to buying winners and selling losers: Implications for stock market efficiency, *J. Finance* (48) 65–91.

[74] J. L. W. V. Jensen (1906) Sur les fonctions convexes et les inegalites entre les valeurs moyennes, *Acta. Math.* (30) 175–193.

[75] D. Kahneman and A. Tversky (1979) Prospect theory: An analysis of decision under risk, *Econometrica* (47) 263–292.

[76] H. M. Kat (2003) Ten things that investors should know about hedge funds, *J. Wealth Management* (5) 72–81.

[77] R. L. Keeney and H. Raiffa (1976) *Decisions with Multiple Objectives: Preferences and Value Tradeoffs*, John Wiley & Sons, New York.

[78] M. Kijima and Y. Muromachi (2008) An extension of the Wang transform derived from Bühlmann's economic premium principle for insurance risk, *Insurance: Mathematics and Economics* (42)(3) 887–896.

[79] A. N. Kolmogorov (1933) Sulla determinazione empirica di una legge di distributione, *Giorn. Inst. Ital. Attuari* (4) 83–91.

[80] V. Kreinovich, H. T. Nguyen, and S. Sriboonchitta (2009) A new justification of Wang transform operator in financial risk analysis, *International Journal of Intelligent Technologies and Applied Statistics* (2)(1) 45–57.

[81] D. M. Kreps (1988) *Notes on the Theory of Choice*, Westview.

[82] V. Krishna (2002) *Auction Theory*, Academic Press.

[83] J. L. Kuhle and J. R. Alvayay (2000) The efficiency of equity REIT prices, *J. Real Estate Portfolio Management* (6) 349–354.

[84] F. Lajeri and L.T. Nielsen (2000) Parametric characterizations of risk aversion and prudence, *Econ. Theory* (15) 469–476.

[85] G. A. Larsen and B. G. Resnick (1999) A performance comparison between cross-sectional stochastic dominance and traditional event study methodologies, *Review Quan. Finance and Accounting* (12) 103–112.

[86] H. H. Lean, W. K. Wong and X. B. Zhang (2008) The Size and Power of Some Stochastic Dominance Tests: A Monte Carlo Study for Correlated and Heteroskedastic Distributions, *Mathematics and Computers in Simulation* (79) 30-48.

[87] E. Lehmann (1955) Ordered families of distributions, *Ann. Math. Statist.* (26) 399–419.

[88] E. L. Lehmann (1998) *Theory of Point Estimation*, Second Edition, John Wiley & Sons, New York.

[89] E. L. Lehmann and J. P. Romano (2005) *Testing Statistical Hypotheses*, Springer-Verlag.

[90] M. Lesno and H. Levy (2002) Preferred by all and preferred by most decision makers: Almost stochastic dominance, *Management Science* (48) 1074–1085.

[91] P. L. Leung and W. K. Wong (2008) On testing the equality of the multiple Sharpe ratios with application on the evaluation of iShares, *J. of Risk* 10(3) 1–16.

[92] H. Levy and Z. Wiener (1998) Stochastic dominance and prospect dominance with subjective weighting functions, *J. Risk and Uncertainty* 16(2) 147–163.

[93] H. Levy and M. Levy (2004) Prospect theory and mean-variance analysis, *Review Finance Studies* (17) 1015–1041.

[94] H. Levy (2006) *Stochastic Dominance: Investment Decision Making under Uncertainty*, Springer-Verlag.

[95] M. Levy and H. Levy (2002) Prospect theory: Much ado about nothing?, *Management Science* 48(10) 1334–1349.

[96] C. K. Li and W. K. Wong (1999) Extension of stochastic dominance theory to random variables, *RAIRO Recherche Opérationelle* (33) 509–524.

[97] O. Linton, E. Maasoumi, and Y. J. Whang (2005) Consistent testing for stochastic dominance under general sampling schemes, *Review Econ. Studies* (72) 735–765.

[98] R. J. A. Little and D. B. Rubin (1987) *Statistical Analysis with Missing Data*, John Wiley & Sons, New York.

[99] M. O. Lorenz (1905) Methods of measuring concentration of wealth, *J. Amer. Statist. Assoc.* (9) 209–219.

[100] R. D. Luce (1959) *Individual Choice Behavior*, John Wiley & Sons, New York.

[101] R. D. Luce and P. Suppes (1965) Preference, utility, and subjective probability. In *Handbook on Mathematical Psychology* (D. Luce, R. Bush, and E. Galanter, Eds.), John Wiley & Sons, New York, 249–410.

[102] R. D. Luce and H. Raiffa (1989) *Games and Decisions: Introduction and Critical Survey*, Dover, New York.

[103] H. B. Mann and D. R. Whitney (1947) On a test of whether one of two random variables is stochastically larger than the other, *Ann. Mat. Statist.* (18) 50–60.

[104] H. Markowitz (1952) Portfolio selection, *Journal of Finance* (7)(1) 77–91.

[105] H. Markowitz (1970) *Portfolio selection: Efficient Diversification of Investments,* John Wiley, New York; (1991) Blackwell Publishers, Inc.

[106] A. W. Marshall and I. Olkin (1979) *Inequalities: Theory of Majoration and Its Applications,* Academic Press.

[107] A. W. Marshall and I. Olkin (2007) *Life Distributions,* Springer-Verlag.

[108] B. McCaig and A. Yatchew (2007) International welfare comparisons and nonparametric testing of multivariate stochastic dominance, *J. Applied Econometrics* (22) 951–969.

[109] P. McCullagh and J. A. Nelder (1989) *Generalized Linear Models,* Chapman & Hall.

[110] D. McFadden (1974) Conditional logit analysis of quantitative choice behavior. In *Frontiers in Econometrics* (P. Zarembka, Ed.), Academic Press, 105–142.

[111] D. McFadden (2001) Economic choices, *Amer. Econ. Review* (91) 351–378.

[112] J. Mcycr (1977) Second degree stochastic dominance with respect to a function, *International Economic Review* (18) 476–487.

[113] J. Meyer (1977) Further applications of stochastic dominance to mutual fund performance, *J. Finance and Quan. Analysis* (12) 235–242.

[114] J. Meyer (1987) Two-moment decision models and expected utility maximization, *Amer. Econ. Review* (77) 421–430.

[115] S. P. Mukherjee and A. Chatterjee (1992) Stochastic dominance of higher orders and its implications, *Communications in Statistics–Theory and Methods* (21) 1977–1986.

[116] A. Müller (1996) Orderings of risks: A comparative study via stop-loss transforms, *Insurance: Mathematics and Economics* (17) 215–222.

[117] A. Muller and D. Stoyan (2002) *Comparison Methods for Stochastic Models and Risks,* John Wiley & Sons, New York.

[118] D. B. Nelson (1991) Conditional heteroskedasticity in asset returns: A new approach, *Econometrica* (59) 347–370.

[119] R. B. Nelsen (1999) *An Introduction to Copulas*, Lecture Notes in Computer Science (139) Springer-Verlag.

[120] R. B. Nelsen (2006) *An Introduction to Copulas, Second Edition, Springer Series in Statistics*, Springer-Verlag.

[121] H. T. Nguyen (2006) *An Introduction to Random Sets*, Chapman & Hall/CRC Press/Taylor & Francis.

[122] H. T. Nguyen and V. Kreinovich (1997) *Applications of Continuous Mathematics to Computer Science*, Kluwer, Dordrecht.

[123] H. T. Nguyen, V. Kreinovich, and S. Sriboonchitta (2009) Stochastic volatility models and financial risk measures: Towards new justifications, *Proceedings of the Singapore Economic Review Conference, August 6-8, 2009*, to appear.

[124] H. T. Nguyen and T. Wang (2008) *A Graduate Course in Probability and Statistics, Volume I: Essentials of Probability for Statistics*, Tsinghua University Press, Beijing, China.

[125] T. Norberg (1992) On the existence of ordered couplings of random sets with applications, *Israel J. Math.* (77) 241–264.

[126] G. L. O'Brien (1984) Stochastic dominance and moments inequalities, *Mathematics of Operations Research* (9) 475–477.

[127] H. J. Paarsch and H. Hong (2006) *An Introduction to the Structural Econometrics of Auction Data*, The MIT Press.

[128] K. R. Parthasarathy (1967) *Probability Measures on Metric Spaces*, Academic Press.

[129] J. Pexider (1903) Notiz uber Funktionaltheoreme, *Monatsch. Math. Phys.* (14) 293–301.

[130] W. Polonik (1995) Density estimation under qualitative assumptions in higher dimensions, *J. Multivariate Analysis* (55) 61–81.

[131] T. Post and H. Levy (2005) Does risk loving drive asset prices?, *J. Financial Studies* 18(3) 925–953.

[132] B. L. S. Prasaka Rao (1983) *Nonparametric Functional Estimation*, Academic Press.

[133] J. W. Pratt (1964) Risk aversion in the small and in the large, *Econometrica* (320) 122–136.

[134] H. Raiffa (1970) *Decision Analysis*, Addison-Wesley.

[135] A. W. Roberts and A. E. Varberg (1973) *Convex Functions*, Academic Press.

[136] M. Rosenblatt (1956) Remarks on some nonparametric estimates of a density function, *Ann. Math. Statist.* (27) 832–837.

[137] S. A. Ross (1978) Mutual Fund Separation in Financial Theory–The Separating Distributions, *Journal of Economic Theory* (17) 254–286.

[138] D. Schmeidler (1986) Integral representation without additivity, *Proceedings Amer. Math. Soc.* (20) 223–235.

[139] F. Schmid (2005) A note on third degree stochastic dominance, *OR Spectrum* (27) 653–655.

[140] B. Schweizer and A. Sklar (1983) *Probabilistic Metric Spaces*, North Holland.

[141] R. Serfling (1980) *Approximation Theorems of Mathematical Statistics*, John Wiley & Sons, New York.

[142] M. Shaked and J. G. Shanthikumar (1994) *Stochastic Orders and Their Applications*, Academic Press.

[143] J. Shao (2003) *Mathematical Statistics, 2nd edition*, Springer-Verlag.

[144] W. F. Sharpe (1964) Capital asset prices: Theory of market equilibrium under conditions of risk, *J. Finance* (19) 425–442.

[145] A. N. Shiryaev (2008) *Essentials of Stochastic Finance*, World Scientific.

[146] G. R. Shorack and J. Wellner (1986) *Empirical Processes with Applications to Statistics*, John Wiley & Sons, New York.

[147] A. F. Shorrocks (1983) Ranking income distributions, *Economica* (50)(197) 3–17.

[148] A. Sklar (1959) Fonctions de répartition à $n$ dimensions et leurs marges, *Publications de l'Institut de Statistique de L'Université de Paris* (8) 229–231.

[149] N. V. Smirnov (1944) An approximation to the distribution laws of random quantiles determined by empirical data, *Uspehi. Mat. Nauk* (10) 179–206.

[150] S. Sriboonchitta and V. Kreinovich, Asymmetric heteroskedasticity models: A new justification, *International Journal of Intelligent Technologies and Applied Statistics* (2)(1) 1–12.

[151] D. Stoyan (1983) *Comparison Methods for Queues and Other Stochastic Models*, John Wiley & Sons, New York.

[152] H. E. Thompson and W. K. Wong (1991) On the unavoidability of "scientific" judgement in estimating the cost of capital, *Managerial and Decision Economics* (12) 27–42.

[153] H. E. Thompson and W. K. Wong (1996) Revisiting "dividend yield plus growth" and its applicability, *Engineering Economist* (41)(2) 123–147, 65–86.

[154] J. Tobin (1958) Liquidity preference as behavior towards risk, *Review of Economic Studies* (25), New York.

[155] J. Tobin (1958) Estimation of relationships for limited dependent variables, *Econometrica* (26) 24–36.

[156] K. Train (2003) *Discrete Choice Methods with Simulation*, Cambridge Univ. Press.

[157] A. Tversky (2003) *Preference, Belief, and Similarity: Selected Writings*, MIT Press, Cambridge, Massachusetts.

[158] A. Tversky and D. Kahneman (1992) Advances in prospect theory: cumulative representation of uncertainty, *J. Risk and Uncertainty* (5) 297–323.

[159] A. Van der Vaart and J. Wellner (1996) *Weak Convergence and Empirical Processes*, Springer-Verlag.

[160] J. von Neumann and O. Morgenstern (1944) *The Theory of Games and Economic Behavior*, Princeton University Press.

[161] S. S. Wang (1996) Premium calculations by transforming the layer premium density, *ASTIN Bulletin* (26) 71–92.

[162] S. S. Wang (2000) A class of distortion operators for pricing financial and insurance risks, *Actuarial Research Clearing House* (1) 19–44.

[163] S. S. Wang (2000) A class of distortion operators for pricing financial and insurance risks, *Journal of Risk and Insurance* (67) 15–36.

[164] S. S. Wang (2002) A universal framework for pricing financial and insurance risks, *Astin Bulletin* (32) 213–234.

[165] S. S. Wang (2003) Equilibrium pricing transforms: New results using Bühlmann's 1980 economic model, *Astin Bulletin* (33) 57–73.

[166] S. S. Wang and J. Dhaene (1998) Comonotonicity, correlation order and premium principles, *Insurance: Mathematics and Economics* (22) 235–242.

[167] G. A. Whitmore (1970) Third degree of stochastic dominance, *Amer. Economic Review* (60) 457–459.

[168] J. L. Wirch and M. R. Hardy (2001) Distortion risk measures: coherence and stochastic dominance, http://pascal.iseg.utl.pt/~cemapre/ime2002/main_page/papers/JuliaWirch.pdf, 1–13.

[169] W. K. Wong (2006) Stochastic dominance theory for location-scale family, *J. Applied Math. and Decision Sci.*, 1–10.

[170] W. K. Wong (2007) Stochastic dominance and mean-variance measures of profit and loss for business planning and investment, *European J. Oper. Research* (182) 829–843.

[171] W. K. Wong and R. Chan (2008) Markowitz and prospect stochastic dominance, *Ann. Finance* 4(1) 105–129.

[172] W. K. Wong, B. K. Chew, and D. Sikorski (2001) Can P/E ratio and bond yield be used to beat stock markets?, *Multinational Finance Journal* (5)(1) 59–86.

[173] W. K. Wong and C. K. Li (1999) A note on convex stochastic dominance theory, *Econ. Letters* (62) 293–300.

[174] W. K. Wong and C. Ma (2008) Preferences over Meyer's location-scale family, *Economic Theory* 37(1) 119–146.

[175] W. K. Wong, K. F. Phoon, and H. H. Lean (2008) Stochastic dominance analysis of Asian hedge funds, *Pacific-Basin Finance Journal* 16(3) 204–223.

[176] J.-M. Zakoian (1994) Threshold heteroskedastic models, *Journal of Economic Dynamics and Control* (18)(5) 931–955.

[177] W. T. Ziemba and R. G. Vickson, eds. (1975) *Stochastic Optimization Models in Finance*, Academic Press.

# Index